Analysis of Multivariate Social Science Data

Drawing on the authors' varied experiences researching and teaching in the field, *Analysis of Multivariate Social Science Data: Statistical Machine Learning Methods, Third Edition* enables a basic understanding of how to use key multivariate methods in the social sciences. With minimal mathematical and statistical knowledge required, this third edition expands its topics to include graphical modelling, models for longitudinal data, structural equation models for categorical variables, and latent class analysis for ordinal, nominal, and continuous variables. It also connects the topics to terminology and principles of machine learning, intended to help readers grasp the links between methods of multivariate analysis and advancements in the field of data science.

After describing methods for the summarisation of data in the first part of the book, the authors consider regression analysis. This chapter provides a link between the two halves of the book, signalling the move from descriptive to inferential methods. The remainder of the text deals with model-based methods that primarily make inferences about processes that generate data.

Relying heavily on numerical examples from a range of disciplines, the authors provide insight into the purpose and working of the methods as well as the interpretation of results from analyses. Many of the same examples are used throughout to illustrate connections between the methods. In most chapters, the authors present suggestions for further work that go beyond conventional practice, encouraging readers to explore new ground in social science research.

Features

- Contains new chapters on undirected graphical modelling and models for longitudinal data, as well as new material such as K-means, cross-validation, structural equation models for categorical variables, latent class analysis for categorical, nominal and continuous variables, and treatment of missing data.
- Connects topics with terminology and principles of machine learning.
- Presents numerous examples of real-world applications, including voting preferences, social attitudes, educational assessment, recidivism, and health.
- Covers methods that summarise, describe, and explore multivariate datasets, including longitudinal data.
- Establishes a unified approach to latent variable modelling by providing detailed coverage of methods such as item response theory, factor analysis for continuous and categorical data, and models for categorical latent variables.
- Covers models for hierarchical and longitudinal data and their connections to latent variable models.
- Offers a full version of the data sets in the text or the book's website, with software code for implementing the analyses on the website.

The book offers a balanced and accessible resource for students and researchers with limited mathematical and statistical training. It serves as a practical resource for courses in multivariate analysis and as a guide for applying these techniques in applied research.

Chapman & Hall/CRC

Statistics in the Social and Behavioral Sciences Series

Series Editors
Jeff Gill, Steven Heeringa, Wim J. van der Linden, and Tom Snijders

Recently Published Titles

Applied Regression Modeling: Bayesian and Frequentist Analysis of Categorical and Limited Response Variables with R and Stan
Jun Xu

The Psychometrics of Standard Setting: Connecting Policy and Test Scores
Mark Reckase

Crime Mapping and Spatial Data Analysis using R
Juanjo Medina and Reka Solymosi

Computational Aspects of Psychometric Methods: With R
Patricia Martinková and Adéla Hladká

Principles of Psychological Assessment
With Applied Examples in R
Isaac T. Petersen

Multilevel Modeling Using R, Third Edition
W. Holmes Finch, Jocelyn E. Bolin, and Ken Kelley

Polling, Prediction, and Testing, Second Edition
Ole J. Forsberg

Generalized Kernel Equating with Applications in R
Marie Wiberg, Jorge Gonzalez and Alina A. von Davier

Applied Survey Data Analysis, Third Edition
Brady T. West, Steven G. Heeringa, and Patricia A. Berglund

Visualization for Social Data Science
Roger Beecham

Introduction to Bayesian Data Analysis for Cognitive Science
Bruno Nicenboim, Daniel J. Schad and Shravan Vasishth

Understanding Structural Equation Models:
Models of Relationships Between Variables
Phillip K. Wood

Analysis of Multivariate Social Science Data:
Statistical Machine Learning Methods, Third Edition
Irini Moustaki, Fiona Steele, Yunxiao Chen, and David J. Bartholomew

For more information about this series, please visit: https://www.routledge.com/Chapman--HallCRC-Statistics-in-the-Social-and-Behavioral-Sciences/book-series/CHST-SOBESCI

Analysis of Multivariate Social Science Data

Statistical Machine Learning Methods

Third Edition

Irini Moustaki, Fiona Steele, Yunxiao Chen, and
David J. Bartholomew

CRC Press
Taylor & Francis Group
Boca Raton London New York

CRC Press is an imprint of the
Taylor & Francis Group, an **informa** business

A CHAPMAN & HALL BOOK

Third edition published 2026
by CRC Press
2385 NW Executive Center Drive, Suite 320, Boca Raton, FL 33431

and by CRC Press
4 Park Square, Milton Park, Abingdon, Oxon, OX14 4RN

CRC Press is an imprint of Taylor & Francis Group, LLC

© 2026 Irini Moustaki, Fiona Steele, Yunxiao Chen, and David J. Bartholomew

First edition published by CRC Press 2002
Second edition published by CRC Press 2008

Library of Congress Cataloging-in-Publication Data
Names: Moustaki, Irini author | Steele, Fiona (Professor of statistics)
author | Chen, Yunxiao author | Bartholomew, David J. author
Title: Analysis of multivariate social science data / Irini Moustaki, Fiona
Steele, Yunxiao Chen, and David J. Bartholomew.
Description: Third edition. | Boca Raton, FL : CRC Press, 2026. | Series:
Chapman and Hall/CRC statistics in the social and behavioral sciences |
Includes index. | Summary: "Enables a basic understanding of how to use
key multivariate methods in the social sciences. With minimal
mathematical and statistical knowledge required, this third edition
expands its topics to include graphical modelling, models for
longitudinal data, structural equation models for categorical variables,
and latent class analysis for ordinal, nominal and continuous variables.
It also connects the topics to terminology and principles of machine
learning, intended to help readers grasp the links between methods of
multivariate analysis and advancements in the field of data science"--
Provided by publisher.
Identifiers: LCCN 2025040179 (print) | LCCN 2025040180 (ebook) | ISBN
9781032774763 hardback | ISBN 9781032763729 paperback | ISBN
9781003483342 ebook
Subjects: LCSH: Social sciences--Statistical methods | Multivariate
analysis
Classification: LCC HA29 .A5824 2026 (print) | LCC HA29 (ebook)
LC record available at https://lccn.loc.gov/2025040179
LC ebook record available at https://lccn.loc.gov/2025040180

ISBN: 978-1-032-77476-3 (hbk)
ISBN: 978-1-032-76372-9 (pbk)
ISBN: 978-1-003-48334-2 (ebk)

DOI: 10.1201/9781003483342

Typeset in CMR10 font
by KnowledgeWorks Global Ltd.

Publisher's note: This book has been prepared from camera-ready copy provided by the authors.

Contents

4 Correspondence Analysis   92

5 Principal Components Analysis   128

6 Regression Analysis   158

Preface

Multivariate analysis is an important tool for the social researcher. This book aims to give students and social researchers with limited mathematical and statistical knowledge, a basic understanding of some of the main multivariate methods and the knowledge to implement them.

The third edition has a different title: *Analysis of Multivariate Social Science Data: Statistical Machine Learning Methods*. A major change from earlier editions is to connect topics, when appropriate, with terminology widely used in machine learning methods. The methods discussed in the book fall within the categories of supervised and unsupervised machine learning methods, and making this connection will be helpful to the reader to understand the links between methods of multivariate data analysis and recent developments in the data science field. As with the previous two editions of the book, interpretation of data analyses remains a key element.

The book originated from a course taught by David Bartholomew to MSc students in the social sciences at the London School of Economics for over 30 years. The content changed over the years, but the main objective remained the same. The other authors contributed to that course in its later years, in various capacities. In its new form, it can still be used as a course text, but we see an important role for it as a manual for social researchers.

The third edition introduces two new chapters and incorporates new methods and material into several existing chapters. The new chapters are: Chapter 12 on graphical models and Chapter 14 on longitudinal data analysis. Furthermore, the method of K−means has been added to Chapter 2 on cluster analysis, an introduction to prediction and cross-validation to Chapter 6 on regression analysis, latent class analysis for ordinal, nominal and continuous data to Chapter 10, confirmatory factor analysis and structural equation models for categorical data to Chapter 11, and a discussion of the connection between multilevel models and factor models to Chapter 13. Another new feature is discussions of approaches to handling missing values and their assumptions about the underlying nonresponse mechanism.

As with the second edition, the chapter on regression analysis comes at a pivotal point, providing both a link between the two halves of the book and a topic essential in its own right. Regression analysis also marks the point at which we move from descriptive methods (cluster analysis, multidimensional scaling, correspondence analysis and principal component analysis) to model-based methods, and where the emphasis shifts from interdependence to dependence as the key idea.

It is not easy to specify the exact prerequisites for a course of this kind. The best way is to list some of the topics which ought to have been covered in preparatory courses. These should certainly include: basic descriptive statistics, ideas of sampling and inference (including elementary hypothesis testing and interval estimation), the representation of categorical variables in two-way contingency tables, measures of association, tests for independence, correlation and, perhaps, basic analysis of variance and regression. This is by no means an exhaustive list, but the topics mentioned will serve as markers of the scope of the material students will need to have covered. Beyond this, one needs a general familiarity with simple mathematical formulae and their use and interpretation.

Because we make minimal mathematical demands, our presentation relies heavily on numerical examples and verbal and pictorial, rather than mathematical, exposition. It is designed to give insight into the purpose and working of the methods. In no sense is this a cookbook. Multivariate analysis is a sophisticated and delicate tool, which can best be learnt by working through detailed examples and attempting to interpret the results of the analysis under the guidance of an experienced practitioner. This is the pattern we have followed in our teaching, which has always been supported by practical work on a computer.

The book has several distinctive features. One, which we have already mentioned, is that it treats individual topics seriously and in depth in a non-mathematical way. A second is that we have emphasised the unity of the methods and presented them in such an order that practice gained in the early stages will contribute to the understanding of the more difficult methods that come later. There are many cross-linkages between different parts of the book. For this reason, and because some topics build cumulatively on previous topics, we recommend the book is read from beginning to end to gain the full benefit.

There are many examples, and the reader may notice that a majority are from applications in the UK or Europe. This is a consequence of our decision to concentrate on material on which we have worked or with which we have been associated. This is essential if the reader is to gain a sense of seeing the application from both the inside and the outside.

One special feature of the book is the detailed coverage we have given to latent variable methods and their connection to models with random effects employed in multilevel modelling and longitudinal data analysis. In the past, various branches of latent variable modelling were treated as watertight compartments. Thus, for example, Item Response Theory had evolved in the context of psychometrics and educational testing, with its own journals and books and very little contact with the field of factor analysis. This is regrettable because, conceptually, they are the same thing – the only difference being in the level of measurement of the variables. Fortunately, these different methods have since been brought together within a common framework, allowing for a more unified approach to latent variable modelling. Commercial software has also aligned more with the unification of the latent variable framework. More recently, there has been a substantial growth of interest in graphical modelling and its connection to latent variable models (network psychometrics), which has led to the inclusion of Chapter 12 in this edition. The second edition already included Chapter 13 on multilevel modelling, which offers a rich introduction to models for hierarchical data using random effects (essentially another name for latent variables) and we now make explicit the connection with factor analysis models. The new Chapter 14 on longitudinal data analysis utilises the multilevel and factor analysis framework to provide an extensive treatment for the analysis of repeated measurements using models with random effects/latent variables. Models such as the latent class growth analysis and growth mixture models are special cases of structural equation models. Chapters 7–14 are designed to provide a solid foundation in the full range of models currently available. We view them as a coherent group of closely related methods that are essential tools for use in a broad range of social scientific research.

This book provides only an introduction. Readers who become interested in a particular topic will need to pursue it further by consulting the specialist literature. To provide signposts to up-to-date and comprehensive treatments, we have included a short list of further reading at the end of each chapter. In addition to providing an authoritative and much more broadly based account than we can give here, such books will also offer a good point of entry to a much broader literature. There is also a list of references at the end of the book, which provides the sources of various research results we have drawn upon. These are included for the benefit of the reader who wishes to delve deeper; however, they are by no means a comprehensive listing of the relevant research literature.

The website linked to the book can be accessed from

https://github.com/AMSSD3ed/data-code

The website contains full versions of all the datasets discussed in the book, together with scripts for analysing the data in various commercial and open-access software. This has been updated extensively since the publication of the second edition.

David Bartholomew passed away in 2017 and was not involved in preparing the third edition of the book. He would have been eager to work on a third edition and add new materials to it. His tremendous knowledge and contributions to the book's subject, as well as his oversight of the entire book's production and his ability to keep us all on track, were greatly missed in preparing this edition. We would like to extend our gratitude to our colleague and former co-author, Jane Galbraith, who wrote the cluster analysis and regression analysis chapters in the first and second editions, respectively, and for providing insightful comments on all other chapters as well. In the third edition of the book, we invited one of our newest colleagues, Yunxiao Chen, who works on latent variable models, to join the book as a co-author. As with the previous two editions of the book, writing the book has been a joint operation. Each author has contributed chapters and sections to the book, and all authors provided feedback on the entire book.

In the Preface to the first and second editions, we thanked those who have assisted us over the years in various ways and, in particular (in alphabetical order): Nick Allum, Susannah Brown, Colin Chalmers, Elena Erosheva, Rex Galbraith, Olafur Gylfason, Anastasia Kakou, Martin Knott, Geoff Laslett, Colin Mills, Albert Satorra, Amani Siyam, Anders Skrondal, Panagiota Tzamourani and Melanie Wall. For the third edition, we would like to thank Camilo Cardenas-Hurtado for reviewing all the data examples and updating the code as necessary.

Irini Moustaki
Fiona Steele
Yunxiao Chen
David J. Bartholomew
January 2026

About the Authors

Irini Moustaki is a professor of statistics in the Department of Statistics at the London School of Economics and Political Science. She received her bachelor's in statistics and computer science from the Athens University of Economics and Business, and her MSc and PhD in statistics from the LSE. Her research interests are in latent variable models and structural equation models. Her methodological work includes treating missing data, longitudinal data, outlier detection, goodness-of-fit tests, and advanced estimation methods. Furthermore, she has made methodological and applied contributions to comparative cross-national studies and epidemiological studies on rare diseases. Prof. Moustaki received an honorary doctorate from the Faculty of Social Sciences, Uppsala University in 2014. She is a fellow of the British Academy. She was the executive editor of the journal *Psychometrika* from 2014 to 2018 and president of the Psychometric Society from 2021 to 2022.

Fiona Steele is a professor of statistics in the Department of Statistics at the London School of Economics and Political Science. She holds a PhD in social statistics from the University of Southampton. Her research interests are in the development of statistical methods that are motivated by social science problems. Her areas of expertise include longitudinal data analysis, multilevel and latent variable modelling, and dyadic data analysis. She has worked on a range of applications in demography, education, family psychology, and health. Prof. Steele has directed several research grants on methods for multilevel and longitudinal data analysis. She also led the development of, and contributed modules to, the popular online LEMMA course on multilevel modelling. Prof. Steele is a fellow of the British Academy and was awarded a CBE and the Royal Statistical Society Howard Medal for her contributions to social statistics.

Yunxiao Chen is an associate professor of statistics in the Department of Statistics at the London School of Economics and Political Science. He holds a PhD in statistics from Columbia University in the City of New York. His research focuses on the intersection of multivariate statistics and machine learning, where he develops models, computational algorithms, and statistical theories for learning from complex data and applies them to education, psychology, and other social science disciplines. Dr. Chen has received numerous awards, including the 2018 Brenda H. Lloyd Dissertation Award from the National Council on Measurement in Education and the 2022 Early

Career Award from the Psychometric Society. He was also a Spencer Foundation/NAEd Postdoctoral Fellow at the United States National Academy of Education from 2018 to 2020. His work has appeared in leading journals in statistics and machine learning, such as the *Journal of the American Statistical Association, Biometrika, Journal of the Royal Statistical Society, Series A (Statistics in Society)*, and the *Journal of Machine Learning Research*. Additionally, Dr. Chen serves as an associate editor for several prominent publications, including *Psychometrika, British Journal of Mathematical and Statistical Psychology, Journal of Educational and Behavioural Statistics*, and *Psychological Methods*. This book, *Analysis of Multivariate Social Science Data: Statistical Machine Learning Methods, Third Edition*, draws on his years of experience teaching and researching in the field.

David J. Bartholomew (deceased) was a professor of statistics in the Department of Statistics at the London School of Economics and Political Science from 1973 to 1996, when he became an emeritus professor. His research interests were in the areas of stochastic modelling, social measurement, factor analysis, and latent variable modelling. Prof. Bartholomew published some 25 books and more than 140 research papers. He served as pro-director of the LSE from 1988 to 1991. He served as co-editor of the *Journal of the Royal Statistical Society, Series B*, from 1966 to 1969 and president from 1993 to 1995. He was awarded the Guy medal in bronze in 1971. He was elected a fellow of the British Academy in 1987. Prof. Bartholomew passed away in 2017.

1

Setting the Scene

The analysis of multivariate data is part of most quantitative social research projects. Although elementary statistics tells us how to analyse data on single variables, much of the interest in social science lies in the interrelationships between many variables and predicting the value of one variable given the values of the others. We are often just as interested, for example, in whether people's views on global warming are related to their political views, as in their views on either topic in isolation, or in predicting their views on one topic given the other. When we come to consider many such variables simultaneously, the volume of data becomes large, and the pattern of possible interrelationships can be very complex indeed. This book gives methods for exploring such interrelationships and making predictions. These methods are based on machine learning algorithms and/or statistical models and, thus, are referred to as statistical machine learning methods.

1.1   Structure of the book

There is a progression from the relatively simple to the more difficult as we progress through the book. The major step occurs in Chapter 6, where we move from methods for summarisation to methods for generalisation. Prior to that point, the emphasis is on summarising the data we observe. From that point on, we are seeking to generalise beyond the particular sample we have to the population from which it was drawn. These two aspects are further discussed in the following two subsections. Teachers may find that, with judicious selection of material, the topics in Chapters 2–5 can each be covered in a single lecture, accompanied by supporting practical classes. The later chapters may require more teaching time, but by the time students have reached that stage, they will have a good deal of experience and should be capable of covering extra material by private study.

The methods covered in this book – cluster analysis, regression analysis, principal component analysis (PCA), multidimensional scaling (MDS), correspondence analysis, latent variable models, graphical models, multilevel models, and longitudinal analysis – are all powerful tools for making sense of the kinds of data we deal with in social sciences. These methods form the statistical foundation upon which many modern machine learning algorithms are built. Unsupervised learning techniques like clustering, PCA, and MDS align with tasks such as dimensionality reduction, feature extraction, and classification in high-dimensional data. At the same time, regression models remain essential for supervised learning tasks and predictive modelling, providing interpretable estimates of covariate effects and forming the basis for more flexible approaches like regularised regression.

Latent variable models and graphical models enable the understanding of complex dependencies. Multilevel and longitudinal models offer frameworks for handling hierarchical data and temporal dynamics, which are increasingly relevant in real-world machine learning applications. By understanding these classical multivariate techniques, readers gain a deep statistical intuition that enhances their ability to apply, interpret, innovate, and bridge the statistics and machine learning landscape.

Summarisation

All of the methods that we shall describe have as one of their objectives the summarisation of a set of multivariate data. Summarisation is concerned with condensing a large mass of data into some simpler form which is more readily understandable. This is often done by learning a low-dimensional data representation using computational algorithms or statistical models and presenting it using graphs and tables. In all our waking lives, we are continually receiving large amounts of information through the senses. The only way we can make sense of that vast amount of data is to extract the salient features. For the most part, this is done unconsciously, and we are hardly aware of the complexity of what is taking place in our brains. Sometimes we carry out this summarisation by selecting elements which seem significant or representative in some way. In other cases, we may do it by grouping similar things and treating them as single entities. In multivariate analysis, we are trying to make some of those processes explicit as they relate to quantitative information. The apparatus for doing this mathematically may seem unfamiliar and abstruse, but the ideas are often simple enough. For example, the notion of projecting an image from a high-dimensional space into two dimensions may seem entirely foreign until we remember that this is what happens in photography. A three-dimensional object in the real world is presented as a two-dimensional picture. Something is lost in the process, but the art of photography lies in choosing the angle of the shot to reveal what is judged to be the essence of the object. There are multivariate techniques which are designed to do essentially the same kind of thing. Many of the methods which we shall use have close parallels in everyday experience, even though we may meet them in an unfamiliar guise. In explaining the methods, we shall make frequent use of such familiar analogies.

However, it may be simpler to begin with the similarities with univariate statistics, where the ideas of summarisation are much more familiar. The concepts of an average and standard deviation are well known to anyone who has ever taken a statistics course. These numbers are capable of telling us two important things about very large collections of numbers. An average gives us a good indication of where the numbers are located along the scale of measurement, but it tells us nothing about how dispersed they are. For that, we need a measure of dispersion such as the standard deviation. Both numbers are summary measures, each revealing an important feature of the data. The two together give an economical summarisation of the raw data. When a second variable is introduced, the strength of the linear relationship between the two variables may be measured by the correlation coefficient. In all of these examples, we are extracting salient features of the data in a form which is easier to grasp than would be the case if we merely looked at the large mass of raw data.

Summarisation does not have to be in the form of numbers. Pictures or diagrams can serve equally well and, sometimes, even better. Pie charts are very familiar to readers of company reports, where they are often used as a visual way of presenting a set of percentages. Such diagrams can be made to convey more information by the judicious use of shading or colour. The aim here is to find a way of summarising a large amount of data in a form which our eyes can grasp immediately.

The term "descriptive" is often used, as at the beginning of this section, for that part of methods which is concerned with summarising data. We shall use both terms from time to time, but "summarisation" more adequately conveys the idea that a substantial reduction of the volume of data is involved in the analysis. Similarly, people speak about "exploratory" (data) analysis, referring to the fact that they are looking for the main message that the data has to convey without any prior conceptions about what they expect to find. Exploratory analysis is typically performed at an early stage of the data analysis process, before any formal modelling or hypothesis testing, while summarisation can happen throughout the process. Summarisation is also closely related to the concept of "representation" learning, a terminology commonly used in machine learning, which concerns extracting features automatically from data for subsequent learning tasks, such as regression, classification, and cluster analysis.

The first part of the book, comprising Chapters 2–5, and the first part of Chapter 6, covers methods which aim to summarise, describe, explore, and represent multivariate datasets.

Generalisation

Generalisation goes beyond summarisation by aiming to discover something about the process which has generated the data, and hence to discover something about a wider class of data, of which the present set is only a sample. Generalisation involves two closely related but different tasks – inference and prediction, where inference refers to learning about the mechanism of the data generation process, while prediction concerns predicting previously unseen data coming from it. We frequently perform these tasks in our daily lives, sometimes unconsciously in our brains. For example, a property developer may want to understand how characteristics of the local area (e.g. crime rate and school ratings) and of a house (e.g. number of bedrooms and bathrooms) affect house prices. In this case, they may be interested in knowing the added value of a house if a bedroom were converted to an additional bathroom. This is an inference task, and the prices and characteristics may inform the property developer of previously sold houses. On the other hand, a house buyer may be interested in predicting the value of a house given its characteristics without having to understand how this value depends on each of these characteristics. Of course, the knowledge that we gain from inference can often help us make predictions. For example, imagine the buyer is interested in purchasing a house with similar characteristics to a recently sold house, except for having an extra bathroom. If the property developer informs the buyer of the expected added value of the additional bathroom, they can then add this value to the price of the sold house to provide a reasonable estimate of the house's value.

Much statistical investigation is based on samples, and the interest is not then in the sample itself but in the wider population from which it has been drawn. To establish the link between the population and the sample, we need to know how the sample was selected. In experimental work, the data may be obtained in a controlled experiment with treatments allocated to experimental units using a randomisation device. In a sample survey, the sample members may be selected from a larger population using, for example, random numbers. These methods establish a probabilistic link between sample and population, and so the theory of statistical inference and prediction depends on the theory of probability. To make an inference from the sample mean to the population mean, for example, we need to know about sampling distributions and the t-distribution in particular. The set of assumptions on which all this depends is often described collectively as the "model", and the methods are described as model-based methods. For the statistical inference results to be sensible, we need the assumed model to approximate the data generation process sufficiently well. This is often checked by a model goodness-of-fit test. The performance of prediction methods also depends on the process that has generated the data. However, suppose we are only interested in making predictions. In that case, it matters less whether the model assumptions hold as long as we can accurately predict the values of previously unseen observations from the population. Consequently, model assumptions are of less concern when the model is used purely for prediction purposes. Predictions based on a complex model that involves more parameters do not always outperform those based on a simple model, even though the complex model typically better approximates the data generation process. This is because the complex model is harder to estimate given a sample, and a poorly estimated model leads to poor prediction performance.

In many circumstances, however, there is no formal sampling or allocation process. We simply observe what nature or society puts before us. What we observe is undoubtedly subject to uncertainty, but it is now less clear how to describe that uncertainty in precise probabilistic terms. This is the case, for example, where a number of schools are required to take part in a testing programme. Random selection may be impossible because of the need to obtain cooperation, and so we have a self-selected sample of those willing and able to take part. We still want to be able to generalise from this particular group, but lack the rigorous basis of sampling theory to do it. We may then treat the sample as if it were truly random to get some "feel" for the uncertainties involved. This is a defensible procedure as long as we recognise the tentative basis for our generalisations.

The second part of the book, starting with Chapter 6 and continuing to the end, deals with model-based methods. These methods can be used for both inference and prediction tasks. We will put more emphasis on the inference task, given its heavier use in social science research. We also note that this book focuses on methods that are based on statistical models, although many methods are algorithm-based and do not involve a statistical model, such as random forest and boosting. These algorithm-based statistical learning methods will not be covered in the current book.

The division we have made between summarisation and generalisation is not as hard and fast as we have made it appear. The assumptions on which the model-based methods are based are not always well-founded. Although we shall use the whole paraphernalia of inference, the samples we shall use are not always random, and the populations from which they are supposedly drawn are not always well defined. We shall often, therefore, find ourselves using things like goodness-of-fit tests in descriptive mode as measures of agreement between observed and expected frequencies rather than as formal tests. Conversely, our choice of methods in descriptive analysis and the conclusions we shall draw from them sometimes go beyond what can be strictly justified. Nevertheless, the perspective that this dichotomy gives should help readers to follow the progression of chapters.

1.2   Our limited use of mathematics

Mathematics is an enormously powerful tool which enables a quantitative argument to be conducted with precision and rigour. It is therefore indispensable in developing the theory of multivariate analysis. However, very little mathematical skill is needed to grasp the main ideas behind the methods or to use them intelligently. There is very little use of formal mathematics in this book, though sometimes mathematical ideas are expressed in words – less precisely but, we hope, more understandably. There are no mathematical derivations or proofs. There are, however, a number of formulae and equations and a certain amount of mathematical terminology. For those who are ill at ease with any mathematics, we believe that the effort to become, at least, familiar with the language will be amply repaid. Later, we shall summarise the essentials that are necessary to obtain full benefit from the text.

Most of the main textbooks on multivariate analysis are heavily dependent on mathematics, and it is worth pausing to consider why we are able to dispense with so much of it. Modern multivariate analysis developed around the middle of the twentieth century as a natural outgrowth of univariate (one variable) and bivariate (two variables) statistics. It was natural and fairly easy to generalise the *t*-test, simple linear regression, and analysis of variance to deal with many variables rather than one or two. This involved moving from the normal distribution, which was the backbone of elementary statistical inference, to the multivariate normal distribution. The special mathematical properties of this distribution lent themselves to many elegant manipulations which drew upon much advanced mathematics. At the same time, it was challenging to implement any of these methods because of the minimal computing facilities available. This tended to encourage mathematical as opposed to applied developments of the subject. With the rather notable exception of factor analysis, which originated outside statistics, very little use could be made of multivariate analysis by social scientists. Perhaps this was just as well because the assumption of multivariate normality on which most of the early theory was founded is very rarely applicable in social science. In fact, continuous variables are the exception rather than the rule in the social field.

The position has been totally changed with the coming of powerful desktop computers and high-performance computing systems. Whereas once computing struggled to keep up with theory, the reverse is true now, with computing often leading the way. Computers not only made it possible to implement existing theory but also stimulated the development of so-called computer-intensive methods, such as those used in the early chapters of this book, which involve very little formal theory. This transformation is the main reason it is now possible to give a sound, if somewhat incomplete, account of applied multivariate methods without overburdening students with mathematics.

Matrices and vectors

Although we shall use no mathematical arguments or proofs, it will be useful to be familiar with some mathematical notation and terminology. It will be convenient to speak of *matrices* and *vectors* as a shorthand for arrays of numbers. A matrix is simply a rectangular array of numbers. For example, the following array is a 3×4 matrix, so-called because it has 3 rows and 4 columns:

$$\begin{pmatrix} 21 & 33 & 17 & 9 \\ 13 & 41 & 12 & 37 \\ 11 & 25 & 6 & 19 \end{pmatrix}.$$

A vector is a matrix with only one row (called a row vector) or with only one column (called a column vector). Vectors and matrices are often enclosed in brackets to emphasise that they are being treated as single entities. In practice, of course, the rows, columns, and elements all have substantive meanings but, for the moment, these can be left on one side.

Very often we shall wish to speak not of individual matrices, but of families of matrices. One of the most important matrices we shall meet is the *data matrix*, and there are things we shall wish to say about all such matrices. Imagine a social survey in which data are collected by questionnaire. Suppose there are ten questions addressed to 100 respondents. The results can be set out as a matrix in which the rows represent respondents and the columns the answers to questions. The first row will contain the answers given by the first individual, in order from left to right. The second row, the results for the second individual, and so on. The result is a 100×10 data matrix. Such data matrices will be the starting point for many of the analyses carried out in this book. For that reason, it is convenient to have a notation which is sufficiently flexible to accommodate matrices of all sizes. We do this by using symbols instead of numbers to indicate the various components of the matrix. We might use y, for example, to represent any element of the matrix. We identify which element we are talking about by using a pair of subscripts to identify the row and column from which it comes. Thus y_{24} is the element in row 2 and column 4; in the matrix given above, y_{24} has the value 37. We need a notation capable of representing matrices of any size. In the case of a data matrix, we shall use n for the number of rows and p for the number of columns. We cannot write a row of p numbers when p is unspecified, but we can write the first few elements and the last as in the data matrix below; the columns are dealt with similarly.

$$\begin{pmatrix} y_{11} & y_{12} & \cdots & y_{1p} \\ y_{21} & y_{22} & \cdots & y_{2p} \\ \vdots & & & \vdots \\ y_{n1} & y_{n2} & \cdots & y_{np} \end{pmatrix}$$

We can refer to a matrix by a symbol chosen to be suggestive of the contents. To distinguish matrices and vectors from other quantities, they will be printed in bold type. Thus, the above matrix might be denoted by $\mathbf{Y}$. To save writing out the full matrix, we can simply give the typical element in the ith row and jth column. For example, the above matrix could then be written

$\{y_{ij}\}$. This is only sensible if the dimensions of the matrix are clear from the context.

Three other important matrices we shall meet are *distance, correlation,* and *covariance* matrices. These are square symmetrical matrices in which the element in row i and column j is equal to the element in row j and column i. The upper right triangle is thus a reflection of the lower left triangle, and so it is not essential to write out both parts.

Matrices and vectors can be multiplied just like ordinary numbers, but only if they are "conformable". The only case we need to consider arises when we multiply a matrix by a vector to obtain a new vector. Thus $\mathbf{Ay}$ means

$$
\begin{pmatrix}
\alpha_{11} & \alpha_{12} & \cdots & \alpha_{1p} \\
\alpha_{21} & \alpha_{22} & \cdots & \alpha_{2p} \\
\cdots & \cdots & \cdots & \cdots \\
\alpha_{n1} & \alpha_{n2} & \cdots & \alpha_{np}
\end{pmatrix}
\begin{pmatrix}
y_1 \\
y_2 \\
\cdots \\
y_p
\end{pmatrix}
=
\begin{pmatrix}
\alpha_{11}y_1 + \alpha_{12}y_2 + \cdots + \alpha_{1p}y_p \\
\alpha_{21}y_1 + \alpha_{22}y_2 + \cdots + \alpha_{2p}y_p \\
\cdots \\
\alpha_{n1}y_1 + \alpha_{n2}y_2 + \cdots + \alpha_{np}y_p
\end{pmatrix}.
$$

The matrix and vector on the left-hand side are *conformable* because $\mathbf{A}$ has a number of columns equal to the number of elements of $\mathbf{y}$.

Formulae and equations

The most formidable expressions which the reader will meet are some of the equations. A typical example might appear as follows:

$$
g_i(\mathbf{f}) = \alpha_{i0} + \alpha_{i1}f_1 + \cdots + \alpha_{iq}f_q \qquad (i = 1, \ldots, p).
$$

This equation conveys a great amount of information in a very compact form. It would be much harder to express its meaning in words, and it is therefore worth spending time to unravel its meaning.

An equation, or formula, like this can be thought of as a recipe for calculating the quantity on the left-hand side of the "=" sign from the elements which appear on the right. The quantity on the left, called g, is referred to as a *function* because its value depends on a number of other quantities (which appear on the right). The fact that it has a subscript means that the formula tells us how to calculate a set of functions, not just one. Looking to the extreme right, we have the range of i, which in this case runs from 1 to p. So it is really a formula for calculating p different functions. The expression $(\mathbf{f})$ after g_i identifies the variables on which the value of g depends, namely the elements of the column vector $\mathbf{f}$, $f_1, f_2, \ldots, f_q$. The right-hand side spells out the form of this dependence. It involves the q fs already mentioned, together with a collection of constants, the αs. (Because they are constants, they are not mentioned on the left-hand side.) These constants have a pair of subscripts which indicate that they are elements of a matrix. This matrix has p rows (because i runs from 1 to p) and $q + 1$ columns (because the second subscript runs from 0 to q). Given numerical values for the αs and the fs, it is then a straightforward matter to calculate the gs by taking the αs, in turn, multiplying by the appropriate f and adding the results. A single line of text conveys all of that! It should be noticed that, in addition to the use of symbols, the pattern of the layout helps to convey the meaning of the equation. A practised reader will immediately recognise the *form* of the equation and be able to take in its main meaning without paying too much attention to the detail.

The equation could be made even more compact by the use of the "sigma" notation, which is used to indicate summation (Σ is the Greek "S" for "sum"). Instead of having to write every term in a sum, we then only have to write one of them. Using this notation, we could write

$$\alpha_{i0} + \alpha_{i1}f_1 + \cdots + \alpha_{iq}f_q = \alpha_{i0} + \sum_{j=1}^{q} \alpha_{ij}f_j.$$

The summation sign (Σ) indicates that we add up all the terms following the sign, obtained by letting j take each of the values in turn between the "limits of summation" indicated by the numbers 1 and q. In this particular case, the gain is very modest, and we have preferred to write out the expression in full, but there are occasions where the "sigma" notation offers greater advantages. A further step towards simplifying the expression is to use the matrix notation used at the end of the last subsection. We may then write the whole set of formulae as

$$\mathbf{g(f)} = \mathbf{Af},$$

where $\mathbf{g(f)}$ is a column vector.

Many equations that we shall meet are *linear* like the one we have used as an illustration in this section. Others, like the following, from Chapter 9 (eq. (9.6)),

$$\text{Corr}(y_i^*, f_j) = \frac{\alpha_{ij}^*}{\sqrt{\sum_{j=1}^{q} \alpha_{ij}^{*2} + \sigma_i^2}},$$

look more complicated but if broken down into their basic elements, involve similar operations.

Much of the apparent complexity of mathematical expressions arises from a statistician's propensity to add embellishments like $*$, $\hat{\ }$, or $\sim$ to symbols already encumbered with subscripts and superscripts. The formula immediately above is a good example. This practice is actually intended to make things simpler! Having learnt, for example, that α represents an element of a particular matrix, the addition of a pair of subscripts will tell us which element, a circumflex accent ($\hat{\ }$) will add the information that it is an estimate we are dealing with, the asterisk that the original α has been standardised, and so on. All of this information is conveyed while preserving the family relationship with the original α.

Notation

Readers familiar with the first two editions will find little change in the notation with which they are familiar. The only significant change in the original chapters has been to replace x by y (for observed variables) in Chapters 2, 5, and 7–10.

In the second part of the book, we shall occasionally need to say that some variable has a normal distribution and that it has a particular mean and variance. The conventional shorthand for this is $y \sim N(m, v)$, which is read as y is normally distributed with mean m and variance v. In latent variable modelling, we often wish to speak of the conditional distribution of one variable, y say, *given* the value of another variable f. We do this by writing that $y \mid f$ has such and such a distribution.

Before leaving the topic of mathematics, it is important to emphasise that "non-mathematical" does not mean the same as "easy" in this context. The fundamental ideas remain the same whether we express them mathematically or in words.

1.3   Variables

The numbers we observe and record in a data matrix may have very different meanings, and this fundamentally affects the inferences we can draw from them. They may be measurements of continuous variables such as length or time and, if so, it is meaningful to calculate summary measures such as means, standard deviations, and correlations. On the other hand, they may be codes indicating, for example, whether someone answered "yes" or "no" to a question. In the latter case, it is immaterial whether we assign "1" and "0" to the two possibilities or "25" and "7" because the only purpose of the codes is to distinguish one response from the other. Nothing we do subsequently with the codes should depend on this arbitrary feature. Issues of this kind are often discussed in the social science literature under the heading "levels of measurement".

Level of measurement

The main distinction required in this book is between *continuous* and *categorical* variables. Continuous variables are those which can be recorded on some scale, like response times, lengths or examination scores, where the interval between two values on the scale has a quantitative interpretation. This makes it legitimate to calculate such things as correlations between pairs of continuous variables and to assume that they have, at least approximately, continuous probability distributions. It is common in statistics to distinguish between continuous and discrete variables. The former can take any value in an interval, the latter only particular values (usually the positive integers). This distinction has no real significance for our purposes, so we use the term "continuous" to cover both. Note that we do not cover count data in the book.

Categorical variables arise when individuals are allocated to categories. Country of birth is a categorical variable, as is the highest level of educational qualification. We can use numbers to code the categories into which individuals fall, but those numbers are nothing more than codes. It would make little sense, for example, to calculate correlations from them. It is not unusual to find this advice ignored and sometimes in ways which are not immediately obvious. For example, if respondents are asked to say whether they "strongly agree", "agree", "disagree" or "strongly disagree" with some proposition, it is not unusual for equally spaced scores such as 1, 2, 3, and 4 to be assigned to these categories, thus appearing to turn them into continuous variables. In such cases, the codes are being made to carry more meaning than their arbitrary assignment justifies.

Categorical variables may be *ordered* or unordered. The highest level of educational qualification is ordered, but the country of birth is not. (Countries can, of course, be ranked by population, gross domestic product and so forth, but if that were the case, "country" would be merely being used as a crude proxy for a continuous variable.) The expressions of agreement or disagreement mentioned above provide another example of an ordered variable. Although the numbers 1, 2, 3, and 4 assigned to categories are not true variate values, the "error" in treating them as continuous variables may not be as severe in practice as we have implied. To do this is rather like changing the distribution of the variable, and the results of the analysis may not depend critically on the exact form of the distribution.

The traditional classification of levels of measurement is as follows: nominal, ordinal, interval and ratio. A nominal variable is the same as an unordered categorical variable. A special case of both nominal and ordinal variables is a binary variable that can only have two categories. Binary variables are commonly encountered in social science research, for example, sex (male/female), result of an exam (pass/fail), and voting in a two-party political system (party A/party B).

Interval and ratio level variables are both special cases of what we have called continuous variables, and it will rarely be necessary to distinguish between them. Ratio level variables have a natural zero point and are necessarily non-negative. Amounts of money, length and weight are familiar examples. The units in which they are measured are arbitrary – money may be measured in dinars or dollars, for example. Conversion factors are available to convert from dinars to dollars, or feet to meters for that matter. They are termed ratio level variables because ratios are independent of the units of measurement. Thus, for example, a camera which costs three times as much as another when priced in dollars will still be three times as costly if the ratio is expressed in any other currency.

Interval level variables also have an arbitrary scale, but, in addition, have an arbitrary origin. Many scales constructed for educational or social purposes are of this kind. Measures of clinical depression or social need are usually constructed by summing scores assigned based on psychological tests or demographic characteristics. There is usually no natural origin or unit of measurement for such scales, and so we may choose them to suit our convenience. The usual example given for an interval level measure is temperature, where the Celsius and Fahrenheit scales, for example, have different units and origins. (Strictly speaking, temperature is a ratio level variable because there is an absolute zero, but it is so far from the normal temperatures of meteorology that the fact can be ignored for practical purposes.)

The arbitrariness of the scale and location of many social variables is closely linked to the process of *standardisation*. If there is no natural origin or scale, then the results of our analysis should not depend on what values we choose for these quantities. Otherwise, we are adding something to the data which is irrelevant. We are therefore able to choose whatever is most convenient, knowing that it will not affect the answer. For many purposes, the best choice is to choose the origin at the mean and to use the standard deviation as the unit of measurement. This is achieved for each column of the data matrix by subtracting the column mean from each element and dividing the result by the standard deviation. (Formulae for means and standard deviations are given in Chapter 5, Section 5.3). For present purposes, the important point to remember is that correlation coefficients depend on neither the scale nor the origin of the variables and, therefore, the fact that one or the other of these is often arbitrary has no effect on the subsequent analysis.

In what follows, where we shall introduce measures of association for categorical variables, and in Chapter 7 and subsequently, where we shall be dealing with factor analysis and related methods, we shall meet an entirely different classification of variables. This is concerned with whether or not we can observe the variable. Variables are called *manifest* or *observable* if they can be directly observed and *latent* or *hidden* if they cannot. Sometimes a variable can only be partially observed as, for example, when we can only observe whether or not it exceeds some threshold. Such a variable may be described as *incompletely observed*. The term *underlying variable* is also used but as this is also used as a synonym for *latent variable*, there is a risk of confusion.

1.4   The geometry of multivariate analysis

At many points, we shall wish to express the patterns in multivariate data on two-dimensional diagrams. In that sense, therefore, there is a strong geometrical element in our approach. But geometrical ideas go much deeper than this, and that fact is reflected in the language of the subject. Since we shall wish to use that language, it is helpful to know something about the geometry of multivariate analysis, even if we are going to make no direct use of n-dimensional geometry.

At the most elementary level, we learn how the ideas of simple correlation can be expressed geometrically. The scatter diagram is a two-dimensional representation of pairs of numbers. An example will be found in Figure 2.1. Such a diagram is a representation of an $n \times 2$ data matrix. The "rows" of the matrix are treated as coordinates and become the "points" of the scatter plot. The form of the relationship between the two variables can be "seen" by looking at the diagram in a way which is more immediate than what we learn from inspecting the two columns of the data matrix. It is natural, therefore, to speak of the rows of the table as points and the relationship between the variables as linear, curvilinear or whatever it turns out to be. Similarly, we speak of the two-dimensional "space" in which the points are located. The same terminology can be used for a $n \times 3$ data matrix, though it is not quite so easy to "see" what is going on in a three-dimensional space. However, when we move on to four or more dimensions, our ability to visualise the points fails. Nevertheless, it is still convenient to continue using geometrical terminology. So we continue to call the rows of the data matrix points and refer to distances between points just as if they were distances we could visualise. In some chapters, especially 5 and 7, we shall carry this use of geometrical terminology further. There, we shall need to think of rotation of the axes and projections from a higher to a lower dimension. If we wished to develop the theory of these manipulations, it would be perfectly possible to do so algebraically without any reference to any geometrical notions. All that we shall need here is some intuitive insight into what the theory is trying to achieve and, for this purpose, elementary geometrical terminology enables us to visualise what is going on in simple problems with two or three dimensions and to take on trust that something analogous is happening in higher dimensions.

1.5   Use of examples

The examples are a distinctive feature of this book, and they play a central role in our exposition. They serve several distinct purposes.

i) Simple examples are used to explain and illustrate the methods. These are a partial substitute for the theoretical treatment, which would be given in a more mathematical text. Each step in the argument is executed on an example so that the reader can learn by seeing in detail what is being done. These examples are often unrealistically small to prevent the essentials of the argument from becoming lost in a mass of computation. To that extent, therefore, they can give the impression that the usefulness of the technique is limited since it seems to do little more than reveal what was obvious at the outset.

ii) Once the method has been explained, we shall usually work through one or more substantial examples chosen to illustrate the main stages in the analysis. At this stage, the emphasis changes from how the analysis is done to how it is to be interpreted. This is not easy to do in a textbook because interpretation depends on a thorough knowledge of the context of a study and the purpose for which it is being carried out. Most of the "real" examples we shall use were parts of larger research programmes and something is inevitably lost by removing them from their context. We have attempted to minimise these drawbacks by choosing examples which do not require too much detailed knowledge of the background, and which are small enough to deal with adequately without making them trivial.

iii) We also give more substantial examples, often based on data given on the website. These examples are intended to extend the reader's experience of the use of the technique by seeing them at work in a variety of situations. It frequently happens that the same dataset can be analysed by several techniques, each one illuminating a different aspect of the problem. It is possible, and desirable, for the reader to go beyond the particular analyses which we have chosen to present. The availability of the datasets either in the book or on the website will enable the reader to explore the many options that the various software packages offer. Our suggestions for further work are intended to take the reader beyond the conventional idea of "exercises", which often appear at the end of chapters, into more of a research mode where new ground can be explored. The first step should be to try to reproduce the results we have given in the text. This will enable you to become familiar with the software you are using and, in particular, to find your way around the output. The layout of results has not been linked to any specific software package because these constantly change. The next step should be to try variations on our analyses, some of which are suggested in the text, but others can easily be invented. For example, trying other techniques, omitting one or more variables or, where possible, drawing sub-samples will provide additional practice and insight.

We particularly wish to counter the idea, common among students, that there is one and only one "correct" technique for any given problem. The various methods presented in this book are better seen as ways of revealing different aspects of the data which help in answering substantive research questions.

A word is in order about our choice of examples. We have tried to avoid overuse of what might be called "textbook" examples. That is, examples which have been selected or tailored to show the techniques to their best advantage. In real research, such examples are rare, and the student brought up on them is likely to think that there is something wrong if the technique does not "work" properly. We think that a method can be a useful research tool even if it leaves much unexplained, or if the model does not fit very well. There are occasions, especially under (i) above, where we need simple examples to illustrate particular points of a method unencumbered by complications. In such cases, we have not hesitated to construct artificial examples or borrow well-worn examples from the literature. But when we come to the "Additional examples", we have used only real datasets, some old and some new. These require minimal background knowledge to appreciate the motive for carrying out the analysis. We hope that you will find them interesting in their own right and find that your analyses reveal things which were not evident at the outset.

1.6 Data inspection, transformations and missing data

This is an introductory text, and we have not thought it desirable to burden the reader with many of the practical issues on which good applied research depends. Only when one has a clear idea of where one is going is it possible to know the important questions which arise under this head.

It is rarely possible to analyse a dataset in the form in which it is collected. There may be obvious copying errors or omissions. Recoding of some categorical variables may have to be done. A few summary calculations of means, proportions and variances and the plotting of histograms will help one to get a "feel" for the data and to identify potential problems. If one is proposing to use a technique such as regression analysis or factor analysis, which involves assumptions about the form of the distribution of the observed variables, it is desirable to check that they hold approximately. If not, it may be possible to transform the raw data, perhaps by taking logarithms or square roots, to make them more nearly satisfied. All of this we shall take as read, though you should look carefully at any data before starting an analysis.

Undoubtedly, one of the most common problems at this stage concerns missing data. In practice, some of the elements in the data matrix will be missing for one reason or another. In social research, especially, there are many reasons why this happens. Individuals may refuse to answer some questions or their recorded answers may be obviously "wrong". Candidates in an examination may not have time to answer all the questions. Data may be lost or forgotten. It may, therefore, not be possible to apply the basic operations which we describe without some preliminary adjustment of the data matrix.

For some summarisation methods, it is possible to modify the technique to make use of whatever information is available. For example, those methods like cluster analysis and multidimensional scaling, which start from a set of distances between objects, can still function if some of those distances are missing or estimated from incomplete information.

The most serious consequence of missing data is that the results may be biased when generalising from a sample to a population. People who refuse to answer particular questions, for example, may be the very people who hold extreme views on the matter at issue. Dealing with potential biases of this kind is not a straightforward matter, especially if there is a large amount of missing data. The problems are more acute in those techniques which are model-based, and so depend on assumptions about the nature of the sample. The reader who encounters this situation must recognise that the problem needs special treatment and cannot be ignored. We provide guidance on the treatment of missing data without going into much detail in some of the chapters.

A final word

Although this chapter contains introductory material, the relevance of some of it may not be immediately apparent until the reader has some first-hand experience of the methods. We recommend referring back to it as necessary and then, perhaps, reading it again at the end as if it were the last chapter.

1.7   Reading

Readers who wish to refresh their knowledge of elementary statistics might usefully consult the following:

> Agresti, A. and Finlay, B. (2008). *Statistical Methods for the Social Sciences* (3rd ed.). New Jersey: Prentice Hall.

This contains examples to work through, drawn from the social sciences. It is also listed at the end of Chapter 6 because of its coverage of regression analysis.

Although there are many books on multivariate data analysis which cover the topics of this book, most are relatively strong on the mathematics and weak on the practical side.

Mathematical treatments of some of the topics covered here are given, for example, in:

Bartholomew, D. J., Knott, M. and Moustaki, I. (2011). *Latent Variable Models and Factor Analysis: A Unified Approach* (3rd ed.). John Wiley & Sons.

Krzanowski, W. J. and Marriott, F. H. C. (1995a). *Multivariate Analysis, Part 1: Distributions, Ordination and Inference*. London: Arnold.

Krzanowski, W. J. and Marriott, F. H. C. (1995b). *Multivariate Analysis, Part 2: Classification, Covariance Structures, and Repeated Measurements*. London: Arnold.

At an intermediate mathematical level, with more emphasis on applications we would still recommend:

Everitt, B. S. and Dunn, G. (2001). *Applied Multivariate Data Analysis* (2nd ed.). London: Arnold.

2

Cluster Analysis

2.1 Classification in social sciences

Classification is one of the most basic operations in scientific inquiry. It is particularly important in social science, where comprehensive theory is often lacking, and the first step in the enquiry is usually to detect some sort of pattern in the data. Methods of classification have long been used in biology, where the grouping of individuals according to species and genus has been the foundation of much subsequent work. Although some early work in cluster analysis was done by biologists seeking to classify plants, much of the stimulus for items developing the subject has come from problems in the social sciences, broadly interpreted. The following highly selective list illustrates why we might be interested in finding clusters and what practical purposes they might serve.

i) *Marketing.* Direct mailing is likely to be more effective if it is directed to people with similar characteristics who are likely to respond in the same way. Market segmentation, as it is called, aims to divide the target population into clusters (segments) so that each can be targeted in the manner most likely to achieve a positive response.

ii) *Archaeology.* Artefacts made at about the same time or by the same group of people are likely to be more similar than those originating from different times or peoples. By forming clusters of similar objects, it may be possible to reconstruct something of the history of a region.

iii) *Education.* Schools vary in their performance, and when seeking the reasons for that variation, it may be useful to cluster schools so that one can ask what those which appear to be broadly similar have in common.

In this chapter, we shall describe and illustrate a number of methods of cluster analysis which are both easy to apply and widely applied.

The problem which cluster analysis aims to solve is to group individuals in such a way that those allocated to a particular group are, in some sense, close together. It is straightforward to do this if objects are characterised by a single measurable quantity such as income. All that we have to do is group together those individuals who have similar income. It is true that we shall have to decide what we mean by "similar", but that will be governed by the use to which we intend to put the classification. The problem is more difficult if judgements of similarity are subjective or based upon a large number of characteristics of the objects. For example, when judging the similarity of two schools, there will typically be a whole set of possibly relevant characteristics like size, location, ethnic mix, and so on. The question then is how we summarise these diverse bits of information so that we can make defensible judgements of similarity. It is this feature which makes cluster analysis a multivariate technique.

To see what is involved in basing distance judgements on more than one variable, consider the case where we have two variables, each measured on a continuous scale. If the "objects" were people, we might imagine that we have records giving their ages and incomes and that we want to group them based on those two variables. Suppose we plot individuals as points in the plane. Then their position might appear as in Figure 2.1.

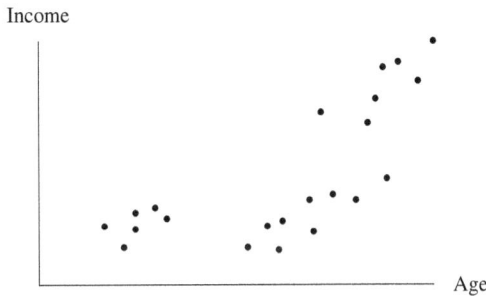

FIGURE 2.1: Fictitious data illustrating income and age for several people

If there were any clustering present, we would immediately recognise it in the figure. There are, in this case, three such clusters which we identify by using the ability of the eye to detect clustering patterns. With three variables, we could imagine points plotted in three dimensions, but beyond that, our ability to visualise fails, and we need some other way of recognising clusters. Notice, incidentally, that if we had information on just one of these variables — income, say — we would see only two clusters, as the two age groups with low income would be hard to separate. This illustrates how multivariate data analysis can reveal more than the analysis of each variable separately.

In this example, the distance between a pair of individuals is simply defined as their distance apart on the figure. However, the information which we have may not be in the form of measurements on a continuous scale. It may, for example, merely indicate whether or not individuals possess a particular attribute. In such cases, the geometrical representation is not available and some preliminary work has to be done in deciding how we are to measure their distance apart.

Although this example is rudimentary, it serves to identify the two basic steps in any cluster analysis:

i) The measurement of the distance apart of all pairs of objects

ii) The development of a routine, or algorithm, for forming clusters based on those distances.

The distances under (i) may be determined either subjectively or by devising a measure of distance based on the observation of a collection of variables. In the former case, it is the human brain which processes the multivariate information available on each object; in the latter, the distance is constructed according to some rational principle.

Before proceeding, we need to clarify the meaning of the term *distance* which we have used to describe how far apart objects are. Sometimes it is more natural to think in terms of closeness or *proximity*. Alternatively, the terms *similarity* and *dissimilarity* are used. The latter have the advantage that they suggest a looser and more subjective assessment of distance, which is more appropriate for some of the applications we shall meet. Proximity and similarity stand in inverse relationship to distance and dissimilarity, and so measures of one can easily be converted into measures of the other. We shall use the various terms interchangeably, but take "distance" as the primary term because it is also central to multidimensional scaling, which we meet in Chapter 3.

The first stage of cluster analysis is the construction of distances between pairs of objects. We shall defer the discussion of how this is done until Section 2.4 by which point we shall have a clearer idea of how they are to be used. For the present, we note that the clustering process itself begins with a *distance matrix* which is an array in which the distance between object i and object j appears in the ith row and the jth column. For example, if we have four objects, we have a 4×4 distance matrix:

$$\begin{pmatrix} - & \delta_{12} & \delta_{13} & \delta_{14} \\ \delta_{21} & - & \delta_{23} & \delta_{24} \\ \delta_{31} & \delta_{32} & - & \delta_{34} \\ \delta_{41} & \delta_{42} & \delta_{43} & - \end{pmatrix},$$

where δ_{ij} is the distance between object i and object j. Usually, the distance matrix will be symmetrical, that is $\delta_{21} = \delta_{12}$, $\delta_{31} = \delta_{13}$ and so on. This is because assessments of distance do not usually depend on the order in which we take the two objects. For this reason, it is only necessary to write down half of the δs — either those in the upper triangle or those in the lower triangle of the matrix. The diagonal may be left blank because these elements play no role in the clustering process. The δ_{ij}s are sometimes referred to as the observed distances or, simply, as the *observations*.

Methods of cluster analysis may be broadly classified as *hierarchical* and *non-hierarchical*. In a hierarchical method, the clustering process yields a hierarchy in which subsets of clusters at one level are aggregated to form the clusters at the next, higher, level. Hierarchical methods can themselves be divided into *agglomerative* and *divisive* methods. In an agglomerative method, we start by treating each object as a one-member cluster, and then proceed in a series of steps to amalgamate clusters. In such a method, once a pair of individuals has been put together in a cluster, they can never be subsequently separated. This is because any new cluster is formed from clusters already created at previous stages of the process. In a divisive method, we start at the other end, treating the whole set of individuals as a single cluster and then proceeding by splitting up existing clusters. Once a pair of individuals has been separated in such a process, they can never come together again. This makes it possible, as we shall see below, to represent the stages in the process by a tree diagram with the branch points indicating where clusters come together or are separated. We shall briefly indicate how both agglomerative and divisive methods might work, but shall concentrate mainly on agglomerative methods.

In non-hierarchical methods, clusters are formed by adjusting the membership of those clusters existing at any stage in the process by moving individuals in or out. A popular method of this kind is the K-means clustering method, to be introduced in Section 2.7. This method classifies all the objects into K clusters for some pre-specified K by minimising the total within-cluster dissimilarity with respect to the clusters. To solve this optimisation problem, the K-means method uses an iterative algorithm that adaptively adjusts for the clusters at each iteration to reduce the total within-cluster dissimilarity. Besides K-means clustering, there exist many other non-hierarchical methods, such as K-medians clustering and spectral clustering. We shall give a brief introduction to these methods in Section 2.7.

In this chapter, we shall treat cluster analysis as a purely descriptive method but the term also covers a wide range of methods, some of which are model-based. We shall meet an example in Chapter 10 on models with latent variables where individuals are required to be allocated to categories specified within the framework of statistical models.

2.2   Hierarchical methods for cluster analysis

We shall use a very simple example consisting of only five individuals to illustrate two of the most commonly used agglomerative hierarchical methods, namely the *nearest neighbour* and *farthest neighbour* methods. These are sometimes referred to as *single linkage* and *complete linkage* methods, respectively. Later, we will briefly mention some other methods.

In an example with only five individuals, there is very little that the formal analysis can add to what we can see by inspecting the matrix of distances. We must not, therefore, expect to learn very much about interpretation from this particular example. Its purpose is solely to define the steps which have to be followed. Imagine that the data concern five customers of a supermarket: Adam, Brian, Carmen, Donna, and Eve, and that information has been obtained from a survey of their tastes and preferences. Suppose also, this information has been analysed and summarised in the table of distances, Table 2.1. The entries represent how far apart individuals are in regard to their potential buying patterns. The smaller the number, the closer the pair. The row and column headings indicate the customer, abbreviated by his or her initial. In a larger scale study with many more customers, the supermarket's objective might be to target its advertising to groups of like-minded customers.

TABLE 2.1: Distance table for illustrating clustering methods

	A	B	C	D	E
A	–				
B	3	–			
C	8	7	–		
D	11	9	6	–	
E	10	9	7	5	–

The table has been set out in lower triangular form as the distances are assumed to be symmetrical, and so there is no need to repeat the information in the upper half of the table.

The nearest neighbour (or single linkage) method

This is an agglomerative method in which each customer is initially treated as a separate cluster. First we look for the closest pair of individuals, which involves scanning all the numbers in the table to find the smallest entry. This is the 3 in row B and column A. At the first stage, therefore, Adam and Brian are brought together in a cluster. At the next stage, we construct a new distance table appropriate for the four clusters existing at the end of the first stage. To do this, we must specify how distance is to be measured between groups that contain more than one individual. In the nearest neighbour method, the distance between two clusters is defined as the distance between their nearest members. Thus, for example, the distance between the cluster (Adam, Brian) and the individual Carmen is the smaller of the distances from A to C and B to C. That is the smaller of 8 and 7. So in constructing the new distance table, the entry for the row labelled C and the column labelled (A,B) is 7 as shown in Table 2.2.

At the second stage, we simply repeat the procedure for Table 2.2 that we applied in Table 2.1; namely, we look for the smallest value. This is the value 5, which is the distance between Donna and Eve. At the second stage,

TABLE 2.2: First stage distance table for nearest neighbour clustering

	(A,B)	C	D	E
(A,B)	–			
C	7	–		
D	9	6	–	
E	9	7	5	–

therefore, we combine D and E into a single cluster. We now have three clusters and require a new distance table giving the distances between these three clusters. This is given in Table 2.3. The only novelty at this stage is finding the distance between the clusters (A,B) and (D,E), which both have more than one member. Adam and Donna are separated by a distance of 11, Adam and Eve are 10 units apart, but Brian is 9 units away from both Donna and Eve. So 9 is the smallest distance and thus goes into the third row, and first column of Table 2.3.

TABLE 2.3: Second stage distance table for nearest neighbour clustering

	(A,B)	C	(D,E)
(A,B)	–		
C	7	–	
(D,E)	9	6	–

The smallest entry in Table 2.3 is 6, indicating that at the next stage we should amalgamate Carmen with the group (Donna, Eve). We then reach a position where there are only two clusters and the only further step possible is to amalgamate them into a single cluster. The foregoing analysis can be conveniently brought together in an *agglomeration table* as follows (Table 2.4).

TABLE 2.4: Agglomeration table for nearest neighbour clustering

Stage	Number of clusters	Clusters	Distance level
Initial	5	(A) (B) (C) (D) (E)	0
1	4	(A,B) (C) (D) (E)	3
2	3	(A,B) (C) (D,E)	5
3	2	(A,B) (C,D,E)	6
4	1	(A,B,C,D,E)	7

What we have achieved by this operation is not quite what we set out to produce. Instead of having arrived at a single set of clusters, we have a hierarchical sequence ranging from the set of five individual clusters with which we started, to the single cluster at which the process ends. We therefore need some way of judging whether any particular set in this sequence has particular claims on our attention. We return to this question after describing a second method.

Farthest neighbour (or complete linkage) method

This is the same as the nearest neighbour method except that the distance between two groups is now defined as the distance between their most remote members. We illustrate the method using the same supermarket example.

The first two stages proceed exactly as in the nearest neighbour method because, up to that point, we were only dealing with clusters containing a single member. The difference arises at the third stage, and for this, we need the table of distances between the three groups then existing. This is given in Table 2.5. When judging the distance between the cluster (A,B) and individual C, we now choose the *larger* of the distance between Adam and Carmen and the distance between Brian and Carmen in Table 2.1, namely 8 rather than 7. Similarly, the distance between the cluster (D,E) and individual C is now 7. Finally, the distance between the clusters (A,B) and (D,E) is 11, which is the largest of the four distances in the left-hand bottom corner of Table 2.1.

TABLE 2.5: Second stage distance table for farthest neighbour clustering

	(A,B)	C	(D,E)
(A,B)	–		
C	8	–	
(D,E)	11	7	–

The smallest distance in Table 2.5 is 7. This indicates that at the next stage, we should amalgamate Carmen with the cluster (Donna, Eve). The resulting agglomeration table is as follows (Table 2.6).

TABLE 2.6: Agglomeration table for farthest neighbour clustering

Stage	Number of clusters	Clusters	Distance level
Initial	5	(A) (B) (C) (D) (E)	0
1	4	(A,B) (C) (D) (E)	3
2	3	(A,B) (C) (D,E)	5
3	2	(A,B) (C,D,E)	7
4	1	(A,B,C,D,E)	11

It so happens that the sets of clusters produced by the farthest neighbour method coincide with those from the nearest neighbour method, although the distance levels differ at which the clusters merge. Generally, the nearest and farthest neighbour methods give different results, sometimes very different, especially when there are many objects or individuals to be clustered. When the two methods (and other methods) give similar results, we may be reassured that the clusters found reflect some true structure in the data.

It is clear that both of the methods considered so far would be very tedious to apply manually to distance tables of a more realistic size. However, each step consists of the very simple operations of selecting the smallest of a set of numbers and constructing the distance table for the next step. This makes it ideal for a computer which can perform very large numbers of such operations accurately and speedily.

Furthermore, an important point to notice about both methods is that they only depend on the ordinal properties of the distances. If we were to change all of the distances without changing their order, for example, by squaring them, it would make no difference to the clustering. This is a great advantage in many social science applications where the distances are often determined subjectively, and it would not be justifiable to treat them as continuous.

Other agglomerative hierarchical methods

The foregoing methods made rather extreme assumptions about how to measure the distance between two groups. Rather than choosing the nearest or farthest neighbours, it might seem more natural to take distances measured to somewhere near the centre of the cluster. This is what the centroid method does. However, to use the centroid method, we must make stronger assumptions about the distances and the variables on which they are based.

If the distances are Euclidean (i.e. straight line distances between the points), we can measure the distances between two clusters by the distance between their averages (centroids). In doing this, we are treating the distances as continuous rather than ordinal. Sometimes the distances may not have been constructed from the co-ordinates of p continuous variables, but have been arrived at directly and subjectively. Is there then any justification for treating them as if they were "real" distances? To answer this question, we need to remind ourselves that cluster analysis is a descriptive method. Its success or otherwise is to be judged by whether or not it produces "meaningful" clusters rather than the means used for their construction. Obviously, we would prefer a method which reliably produces clustering which can be interpreted sensibly, but this does not rule out *ad hoc* methods of whatever kind. In fact, we shall recommend that several different methods should be used, since well-defined clusters are likely to show up using any reasonable method.

Of the many other agglomerative hierarchical methods available in the various software packages, we mention only one, namely Ward's method. This is of interest partly because it embodies a new general idea about the approach to clustering and partly because, in practice, it often seems to yield the clearest picture of any clustering which is present. At each stage in the clustering process, Ward's method considers all pairs of clusters and asks how much "information" would be lost if that pair were to be amalgamated. The pair chosen is then the one which involves the least loss of information. Information in this case is measured by a sum of squares. If a set of numbers is replaced by their mean, the information loss is defined to be their sum of squares about the mean. The method is designed for the case where the elements of the data matrix are continuous variables since only then is it sensible to compute the sums of squares involved. However, the standard software will produce an answer from the distance table alone by treating them as if they had been computed from continuous variables. The remark we made about the centroid method applies equally here — the usefulness of the method is to be judged by what it produces rather than by the assumptions made along the way.

Hierarchical divisive methods

A hierarchical divisive method starts with the complete set of objects and, at each stage, divides existing clusters. The division of an existing cluster into two is itself a clustering task. A particularly simple case arises when the data are binary, in which case one can divide the objects into two clusters on the basis of a single variable — according to whether or not they possess that particular attribute. The question is then which variable to choose for making the division. This can be done by determining which variable is the best proxy for all the variables. One solution is to base this choice on the strength of each variable's association with all the others. A slightly more complicated case arises when the data are continuous. In this case, one may choose a sensible non-hierarchical clustering method to divide an existing cluster. For example, one may apply the K-means clustering algorithm with $K = 2$ to the objects in the existing cluster to divide it into two clusters; see Section 2.7 for an introduction to K-means clustering. One can keep dividing the existing clusters until all the clusters are sufficiently small. Of course, the result of hierarchical divisive clustering will depend on the method being used to divide the existing clusters.

2.3   Graphical presentation of results

We now return to the question of how to select a set of clusters from the set of possibilities offered by the foregoing methods. There are two types of diagrams which are frequently used for this purpose, the *dendrogram*, or tree diagram, and the *icicle plot*.

The dendrogram

This has a tree structure with respect to a scale of distances, with individuals, or groups of individuals, located on branches emanating from stems higher up the scale. It can be drawn with the distance scale either horizontal or vertical. Drawn vertically, it resembles an artistic "mobile". It has the property that any two branches hanging from the same stem can be interchanged without affecting the structure (as if the two branches of the mobile were rotated through 180 degrees). However, we will use the horizontal version because it allows easier annotation and labelling of individuals or objects and is consequently often easier to interpret. Again, any two branches meeting at the same stem may be interchanged.

The dendrogram for the nearest neighbour method applied to the data in Table 2.1 is shown in Figure 2.2. There is a horizontal axis showing distance, corresponding to what we called the distance level in Table 2.4. The linking of Adam and Brian takes place at distance 3, which was the distance at which the first amalgamation was made. Donna and Eve come together at distance 5, which was where the second amalgamation was made. Carmen joins Donna and Eve at distance level 6, and finally, all five people join at level 7.

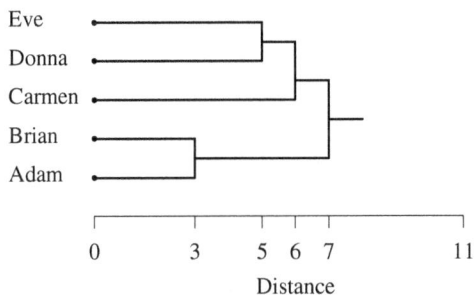

FIGURE 2.2: Dendrogram for nearest neighbour (single linkage) clustering

Figure 2.3 shows the corresponding dendrogram using the farthest neighbour method, drawn with the same distance scale as in Figure 2.2. The two diagrams are very similar in this case and differ only in the distance levels at which the various clusters join. In general, though, individuals or groups may merge in a different order for the two methods.

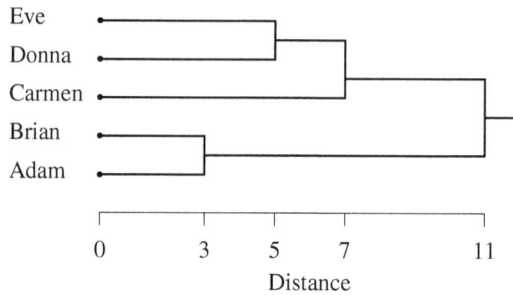

FIGURE 2.3: Dendrogram for farthest neighbour (complete linkage) clustering

The clustering in this example is too ill-defined to give a clear idea how such a dendrogram might help us to choose a set of clusters. We therefore use an artificial example designed to make the point more clearly. The distance matrix in Table 2.7 relates to seven individuals who fall into two distinct clusters. Individuals are labelled numerically here, as is often done in computer programs.

TABLE 2.7: Distance table showing a clear pattern of clustering

	1	2	3	4	5	6	7
1	–						
2	1.5	–					
3	2	2.5	–				
4	10	9	11	–			
5	9	13	12	3.5	–		
6	14	10	11	4	4.75	–	
7	13	12	11	4.5	4.25	3.75	–

It is obvious from inspection of the table, without further analysis, that there are two clear clusters. Individuals 1, 2, and 3 are all close to one another as are 4, 5, 6, and 7. The members of the first group are more tightly packed than members of the second group but the between-group distances are much greater than the within-group distances.

When the results are expressed in the form of a dendrogram for nearest neighbour clustering, we obtain Figure 2.4. The two clusters are clearly identifiable as two groups of branches "hanging" from a single root at the right-hand end. It is rare in practice to find anything as clear cut as this but Figure 2.4 gives us a clue about what we should be looking for in general. A tendency for a group of branches to come together at around the same point and then not to be involved in further amalgamations for some distance indicates a possible cluster.

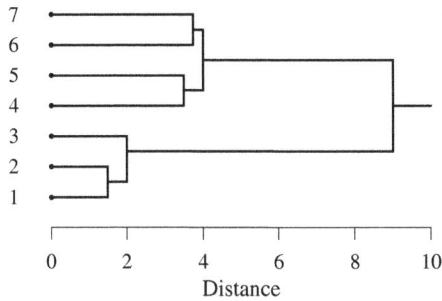

FIGURE 2.4: Dendrogram for the data in Table 2.7

These matters will become clearer when we come to real examples when the clusterings which appear can be related to common characteristics that the members share.

The icicle plot

Whereas the dendrogram can be thought of as a plot of individual against distance, the icicle plot has individual plotted against the stage of merging. Again, these diagrams may be plotted horizontally or vertically, but the name derives from the vertical form which (with some imagination) looks like hanging icicles. We illustrate the vertical icicle plot using the data in Table 2.8. These are extracted from the example on English dialects, which we shall use in Section 2.5. For our present purpose, we merely need to know that the figures in Table 2.8 are *similarities* rather than distances. This means that when looking for the closest pair, we look for the largest number rather than the smallest. Here the individuals are villages labelled V13 to V19.

TABLE 2.8: Similarities for a subset from the dialect data

	V13	V14	V15	V16	V17	V18	V19
V13	—						
V14	63	—					
V15	64	63	—				
V16	54	62	68	—			
V17	72	57	61	59	—		
V18	59	51	56	47	54	—	
V19	38	46	51	42	44	53	—

Number of   Similarity
clusters   level

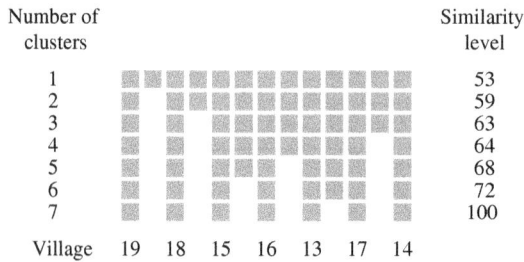

| Village | 19 | 18 | 15 | 16 | 13 | 17 | 14 |

with similarity levels:

1 — 53
2 — 59
3 — 63
4 — 64
5 — 68
6 — 72
7 — 100

FIGURE 2.5: Icicle plot for the data in Table 2.8

Carrying out nearest neighbour cluster analysis, we obtain the vertical icicle plot given in Figure 2.5. The numbers involved here are too small to demonstrate adequately the usefulness of the icicle plot, and we simply use this example to show how the figure is constructed. The plot enables us to read off the composition of the clusters at each stage of the process. In the diagram, the stage in the clustering process is designated by the number of clusters present. If we look in the first row, when there is only one cluster, we see an unbroken row of shaded boxes extending across the whole table. This is the way the plot shows that all members form a single cluster. In the second row, when there are two clusters, we see a continuous row of shaded boxes extending from village 18 to village 14 with a separate single shaded box under village 19. This means that at this stage of the process, there is one cluster containing the six right-hand members and a single cluster consisting of the one member V19. Continuing to move down the table to the row with five clusters, we see that there are two 2-member clusters: (villages 15 and 16) and (villages 13 and 17) along with three single-member clusters (village 19), (village 18) and (village 14). The other rows are interpreted similarly. The final row of the table indicates the position when there are seven clusters each consisting of a single village.

To see how a well-designed icicle plot can display clusters, refer to Figure 2.8 where all 25 villages are shown grouped into three main clusters, along with two or three stragglers. By adding the distance or similarity level on the right-hand side of the icicle plot, we incorporate all the information given in an agglomeration table.

2.4   Derivation of the distance matrix

The first step in cluster analysis is to obtain or construct the distance matrix $\{\delta_{ij}\}$. Broadly speaking, there are two approaches: one that aims to estimate the δs directly and the other that computes distances or similarities from variables measured on the objects, that is from the data matrix.

In the former case, distance data may be collected directly in an experiment where subjects are asked to give subjective assessments of the similarity (or dissimilarity) between pairs of objects. The colour data, which we shall meet in the next chapter, were obtained by asking subjects to judge differences in colour on a five-point scale. In similar experiments, subjects have been asked to compare pieces of music or the taste of wines. In such experiments, subjects are not usually told what criteria to use to make their similarity judgements.

The other possible starting point is a *data matrix* (see Chapter 1). For each of n individuals, we have values of p variables, giving the $n \times p$ data matrix:

$$
\begin{pmatrix}
y_{11} & y_{12} & \cdots & y_{1p} \\
y_{21} & y_{22} & \cdots & y_{2p} \\
\vdots & & & \vdots \\
y_{n1} & y_{n2} & \cdots & y_{np}
\end{pmatrix},
$$

where y_{ik} is the value of variable k for individual i.

For each pair of individuals, we now ask for some measure of how far apart their respective rows are. The problem is to measure the distance, in some appropriate sense based on the elements of the rows, between any two rows of the table. Note that when distances are constructed from a data matrix, we are implicitly imposing the criteria by which objects are assessed as similar or dissimilar, rather than discovering the criteria through the analysis itself.

Given a data matrix, there are many ways in which the distance (or similarity) between two objects i and j may be defined. We shall review a few of them, beginning with the case when all of the ys are continuous.

Distance and similarity measures between objects based on continuous variables

The most commonly used distance measure is Euclidean distance. The Euclidean distance between objects i and j is

$$
\delta_{ij} = \sqrt{\sum_{k=1}^{p}(y_{ik} - y_{jk})^2}.
$$

For $p = 2$, the Euclidean distance corresponds to the "straight line" distance between the two points (y_{i1}, y_{i2}) and (y_{j1}, y_{j2}).

Often, the variables (columns in the data matrix) will be standardised prior to calculating distances. If desired, a weight w_k could be assigned to variable k if it was believed that more importance should be attached to some variables than to others giving:

$$\delta_{ij} = \sqrt{\sum_{k=1}^{p} w_k (y_{ik} - y_{jk})^2}.$$

Another common measure is the so-called "city block" measure, which is the sum of the absolute differences between the pairs of points. Compared to Euclidean distance, this gives less relative weight to large differences. A large difference for a single variable can easily have a dominating effect on the Euclidean distance.

Sometimes, the most relevant information for the substantive issues under investigation may be in the comparison of the "shapes" or the profiles of the objects rather than of their "sizes". Similarity need not be judged only by the distance apart of two rows of the data matrix. For example, consider the following examination scores (out of 100) for two students in Mathematics, English, French, and Physics:

Student	Maths	English	French	Physics	Total
1	21	34	17	42	114
2	62	75	58	85	280

Clearly, the second student has performed much better than the first in all subjects, so the Euclidean distance between the two students is relatively large. However, although the students differ considerably in their *level* of performance, their profiles have a similar "shape". In fact, the rank orders of their performances in the four subjects are identical, and in this respect they are similar. One way to bring out this feature would be to express each score as a percentage of the total for that student. Thus, the row profile would not show how well a student had done, but only the relative contribution of each examination subject. This would only be a sensible thing to do for variables measured on the same ratio scale (here the examinations were all marked on a scale from 0 to 100).

A more general way of eliminating irrelevant differences in size is to scale the scores for each student by subtracting his/her row mean and dividing by his/her row standard deviation. If we then proceed to construct a Euclidean distance between the scaled scores, we would have a measure of how different the shapes of the original profiles were. Interestingly, as simple algebra will show, such a measure is equivalent to using the "correlation" between the two rows as a measure of similarity. (In standard statistics, the Pearson product-moment correlation coefficient is calculated from several individuals or objects to measure the similarity between two variables. Here we refer to the same calculation done on several variables to measure similarity between two individuals or objects.) Put another way, we are arguing that similarity in profile of two objects could be revealed by plotting the variables (using the original scores for the two objects as co-ordinates) and seeing how close they lie to a straight line. For this example, they are very close and the calculated "correlation" is 0.999 compared with a maximum possible value of 1.

This discussion illustrates the important point that the measure we use to define δ_{ij} will depend on what we wish to regard as similar: in this case, students with similar scores on each subject, or students with similarly shaped profiles.

Distance and similarity measures between objects based on categorical variables

If the ys are binary, a measure of similarity could be based on the proportion of variables on which two objects (or individuals) match. Suppose there are p binary variables, each taking the value 0 or 1 for each of n objects. Suppose there are eight variables $y_1, \ldots, y_8$, which for objects 1 and 2, say, take the following values:

Object	y_1	y_2	y_3	y_4	y_5	y_6	y_7	y_8
1	0	1	1	1	0	1	0	0
2	0	1	1	1	0	0	1	1

There are four possible combinations of values for a single variable: $(1,1)$, $(1,0)$, $(0,1)$ and $(0,0)$, where the first value in the pair is the value of y_k for object 1 and the second value is the value of y_k for object 2. Adding up over all variables gives the following frequency table of the number of times each combination occurs:

Combination of values	Frequency for objects 1 and 2	Frequency for two arbitrary objects
$(1,1)$	3	a
$(1,0)$	1	b
$(0,1)$	2	c
$(0,0)$	2	d
Total	8	p

The final column gives the standard notation for a pair of objects and p binary variables: a is the number of variables for which both objects take the value 1, and so on.

We could use the ratio as a measure of similarity:

$$r = \frac{a+d}{p},$$

where $p = a+b+c+d$ is the number of variables. This measure varies between 0 (no matches at all) and 1 (matches on all variables). It is not difficult to show that the Euclidean distance between the two objects is given by

$$\delta = \sqrt{p(1-r)},$$

which is a monotonic decreasing function of r. For example, for objects 1 and 2 above, the proportion of matches is $r = 5/8$ and the Euclidean distance is $\sqrt{3} = \sqrt{8-5} = \sqrt{8}\sqrt{1-5/8}$.

However, some care should be taken because 1 and 0 may have different meanings for different variables. Sometimes, we may only wish to consider $(1,1)$ as a match. For example, suppose we have a binary variable for nationality where $y = 1$ if a person is a UK citizen, and $y = 0$ otherwise. In this case, if we are really interested in nationality (rather than just whether UK or not), we cannot tell whether two people with $y = 0$ match on nationality. If we do not wish to consider $(0,0)$ as a match, a suitable measure of similarity is a/p. Another measure is Jaccard's coefficient in which $(0,0)$ responses are excluded from both the numerator and the denominator giving:

$$\frac{a}{a+b+c}.$$

For categorical data, where the ys have more than two categories, a full set of dummy variables can be constructed for each y. For example, suppose a variable y represents one of four colours red, blue, green or yellow. Then four binary variables y_1, y_2, y_3 and y_4 can be constructed with $y_1 = 1$ when y is red and $y_1 = 0$ otherwise; $y_2 = 1$ when y is blue and $y_2 = 0$ otherwise; and so on. Jaccard's coefficient applied to the full set of binary variables might then be a suitable measure of similarity.

2.5   Example on English dialects

This example is based on a study carried out at the University of Leeds on English dialects. The particular dataset used here is taken from Morgan (1981) and is based on results for 25 villages in the East Midlands, shown on the map in Figure 2.6. The study was concerned with the similarities in dialects among the various villages. A set of 60 items was chosen and members of each village were asked the name which they gave to each item. The data matrix in this case can be thought of as a two-way array with one column for each item and one row for each village. The entry in a typical cell of the table would be the name used for the item heading that column by the villagers represented by the row. The entries in this case are words rather than numbers. If the inhabitants of two villages gave exactly the same names to all items then there would be no difference between villages. The fewer the number of names the villages have in common, the greater the difference between their dialects. The percentage of the items described by the same dialect word in any pair of villagers was used as a measure of similarity.

The complete table of similarities is given in Table 2.9. The 100s on the diagonal represent the maximum similarity (corresponding to agreement on

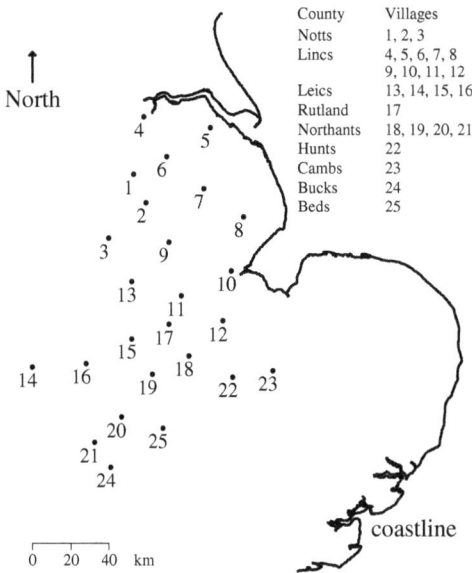

County	Villages
Notts	1, 2, 3
Lincs	4, 5, 6, 7, 8
	9, 10, 11, 12
Leics	13, 14, 15, 16
Rutland	17
Northants	18, 19, 20, 21
Hunts	22
Cambs	23
Bucks	24
Beds	25

FIGURE 2.6: Map of locations of 25 East Midland villages

every item) but are not used in the clustering process. We shall analyse the same set of data in Chapter 3 using multidimensional scaling which, in retrospect, will turn out to be more informative. Nevertheless, as a first step we might want to ask whether we can identify groups, or clusters, of villages which share a common vocabulary.

We begin by applying nearest neighbour cluster analysis to the similarity matrix. Table 2.10 shows the agglomeration table. A different format is used for the agglomeration table compared to Table 2.4 because of the larger number of villages. Instead of listing all clusters at each stage, columns 3 and 4 of Table 2.10 just show which two clusters are merged. A label prefaced by an "m" denotes the cluster formed at the stage indicated, while a label without an "m" denotes a single village. Thus at stage 1, villages 2 and 7 merge; at stage 2, villages 21 and 24 merge; and at stage 3, villages 13 and 17 merge. At stage 4, the clusters merged are denoted 25 and m 2, which means that village 25 merges with the cluster formed at stage 2 (i.e. villages 21 and 24) to form a cluster consisting of villages 21, 24, and 25. One may continue down this table identifying which villages merged at each stage and at what similarity level. The dendrogram (Figure 2.7) shows the information graphically, where it is easier to see patterns of interest.

The style of dendrogram drawn in Figure 2.7 highlights which villages merged at an early stage (small distance, large similarity level) and which did not merge until a later stage (large distance, small similarity). For example, villages 1, 2, 5, 6, 7, and 8 all merged at a fairly early stage, while 3 and 4

TABLE 2.9: Similarity matrix for the English dialect data

	V1	V2	V3	V4	V5	V6	V7	V8	V9	V10	V11	V12	V13	V14	V15	V16	V17	V18	V19	V20	V21	V22	V23	V24	V25
V1	100																								
V2	71	100																							
V3	58	57	100																						
V4	49	45	48	100																					
V5	63	63	47	59	100																				
V6	64	66	50	53	71	100																			
V7	71	75	52	53	71	68	100																		
V8	52	56	36	34	60	58	69	100																	
V9	46	50	57	33	42	43	43	44	100																
V10	61	49	52	40	58	61	56	61	52	100															
V11	57	60	56	35	53	48	55	48	63	59	100														
V12	39	46	45	30	42	40	47	44	50	53	60	100													
V13	42	50	53	28	41	36	43	39	48	47	58	48	100												
V14	32	34	47	20	27	29	31	23	44	39	43	39	63	100											
V15	32	39	50	19	25	25	36	37	43	41	48	49	64	63	100										
V16	23	27	42	14	20	22	24	28	36	38	41	35	54	62	68	100									
V17	41	47	56	25	38	42	46	38	54	54	42	48	72	57	61	59	100								
V18	39	42	48	24	37	34	36	42	49	49	44	56	59	51	56	47	54	100							
V19	32	36	43	22	22	24	34	29	34	40	53	47	38	46	51	42	44	53	100						
V20	27	36	38	19	20	20	25	25	25	25	40	40	45	49	54	49	42	44	63	100					
V21	28	37	37	20	25	25	31	33	29	41	37	37	46	48	49	47	43	44	58	59	100				
V22	26	26	30	20	21	28	28	28	33	39	37	46	34	33	40	33	38	40	58	54	47	100			
V23	30	33	32	16	25	26	33	32	37	37	37	46	47	49	56	39	46	58	42	44	42	50	100		
V24	36	49	45	26	29	31	41	47	32	52	52	57	57	49	56	49	54	53	63	68	73	51	51	100	
V25	31	44	40	23	29	32	32	31	33	43	43	45	45	47	53	43	46	53	60	61	62	55	54	72	100

TABLE 2.10: Nearest neighbour agglomeration table, English dialect data

Stage	Number of clusters	Clusters merged		Distance level	Similarity level
1	24	2	7	25	75
2	23	21	24	27	73
3	22	13	17	28	72
4	21	25	m 2	28	72
5	20	1	m 1	29	71
6	19	5	6	29	71
7	18	m 6	m 5	29	71
8	17	8	m 7	31	69
9	16	15	16	32	68
10	15	20	m 4	32	68
11	14	m 3	m 9	36	64
12	13	9	11	37	63
13	12	14	m 11	37	63
14	11	19	m 10	37	63
15	10	10	m 8	39	61
16	9	m 12	m 15	40	60
17	8	12	m 16	40	60
18	7	18	m 17	40	60
19	6	4	m 18	41	59
20	5	m 13	m 19	41	59
21	4	3	m 20	42	58
22	3	23	m 21	42	58
23	2	22	m 14	42	58
24	1	m 22	m 23	43	57

joined them only at a much later stage. Villages 19 to 25 all merged with each other before merging with other villages, though not all at an early stage. Other patterns are also evident from Figure 2.7, which seems to be quite informative.

For completeness, Figure 2.8 shows an icicle plot for this analysis. This shows three clear clusters that together include all but three villages, numbers 23, 3, and 22, which only merge at similarity level 58. The similarity level at each stage of merging is listed on the right-hand side. Because the original data are discrete, there are several ties in the similarities and consequently, several mergers take place at the same level.

Note that when there are ties, different computer packages may produce different sets of clusters even when using the same clustering method. This is because an arbitrary rule is needed to decide which clusters to merge first. Different rules can result in different sets of clusters being formed (see Morgan and Ray, 1995).

We may also apply farthest neighbour cluster analysis to these data. This produces the dendrogram in Figure 2.9. Although the dendrograms for the two clustering methods differ in the levels at which clusters merge there is a good measure of agreement between them.

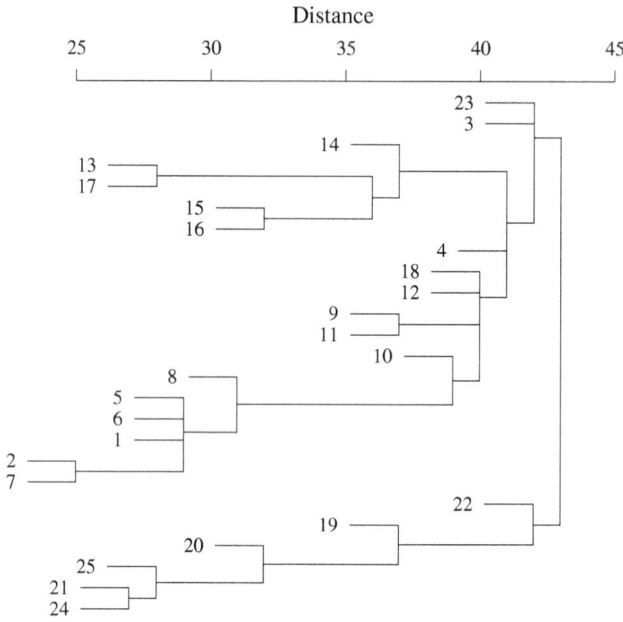

FIGURE 2.7: Dendrogram for nearest neighbour (single linkage) cluster analysis for the English dialect data (distance = 100 − similarity)

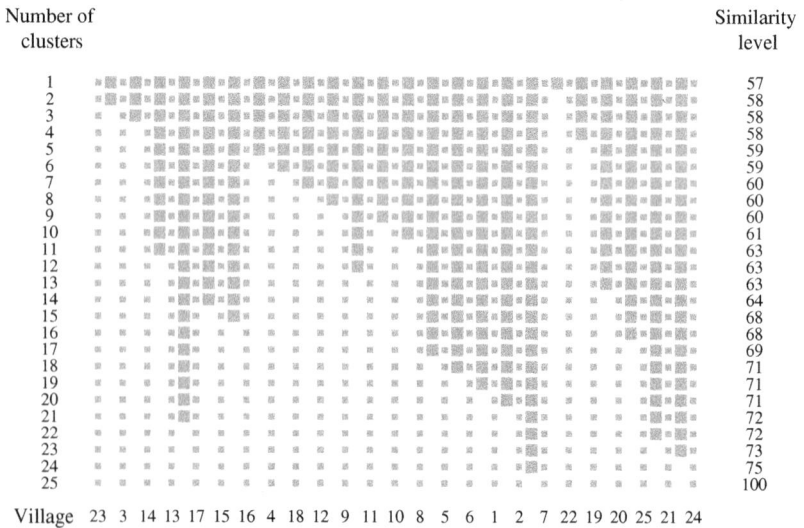

FIGURE 2.8: Icicle plot for nearest neighbour clustering, English dialect data

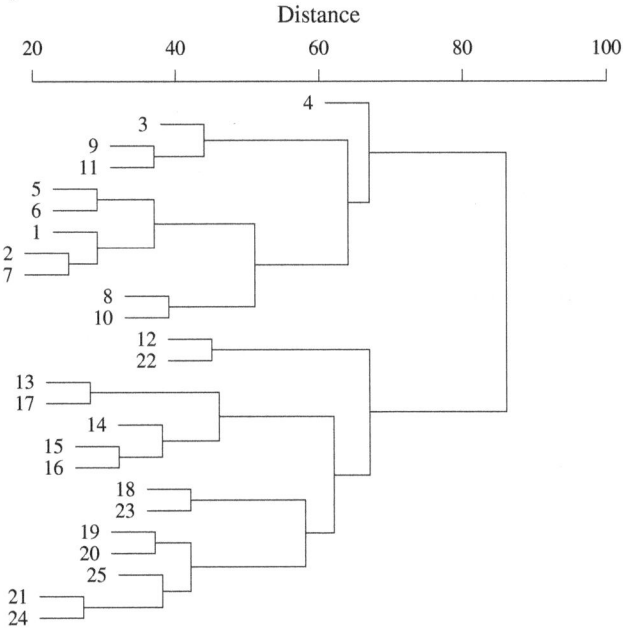

FIGURE 2.9: Dendrogram for farthest neighbour (complete linkage) cluster analysis for the English dialect data (distance = 100 − similarity)

Our interpretations should be read alongside Figure 2.6, which is a map of the East Midlands on which the villages are identified by numbers. Roughly speaking, the identifier codes increase as we move from north to south. Both methods suggest three fairly well defined clusters. Villages 1, 2, 5, 6, 7, and 8 constitute a first cluster of contiguous villages in Lincolnshire and part of Nottinghamshire. Villages 13, 14, 15, 16, and 17 are all in Leicestershire and Rutland. A group consisting of villages 19, 20, 21, 24, and 25 includes parts of Northamptonshire, Buckinghamshire, and Bedfordshire. There is then a less well defined grouping centred on villages 9 and 11, in south Lincolnshire, but the two methods each append different nearby villages in their neighbourhood. There are several villages which do not fit readily into this scheme or are classified differently by the two methods, in particular, villages 3, 4, 12, 22, and 23.

In order to investigate the position further, we have applied five other methods. These are known as the centroid, median, Baverage, Waverage, and Ward's method, the first and last of which have already been mentioned. The median method is similar to the centroid except that it uses the median instead of the arithmetic mean for the cluster centres. The Baverage method uses the between-cluster average and Waverage uses the within-cluster average. It must be remembered that some of these methods assume that the distances (or similarities in this case) have been constructed from a data matrix of continuous variables which is not the case here. In order to apply the methods, the software needs to convert the similarities to distances and then determine what data matrix would have given rise to them. There is no need to go into the details of this since we are using the methods in a purely empirical fashion to suggest possible clusterings. The results are summarised, along with those for nearest and farthest neighbour methods, in Table 2.11.

TABLE 2.11: Comparison of the clusters arrived at when various methods are applied to the English dialect data (the numbers refer to the villages)

| Method | Cluster | | | | |
	I	II	III	IV	V
Nearest	1,2,5,6,7 8	9,11 10,18,12	13,14,15,16,17	3,4,(23)	19,20,21,24,25 22
Farthest	1,2,5,6,7 8,(10)	9,11 (4)	13,14,15,16,17	12,22	19,20,21,24,25 18,23
Centroid	1,2,5,6,7 4,8,10	9,11 18,(23)	13,14,15,16,17		19,20,21,24,25 22
Median	1,2,5,6,7 3,4	9,11 8,10,12	13,14,15,16,17 (18,23)		19,20,21,24,25 22
Baverage	1,2,5,6,7 8,(10)	9,11 3,(4)	13,14,15,16,17 (18,23)		19,20,21,24,25 (12)(22)
Waverage	1,2,5,6,7 4,8,10	9,11 3,12	13,14,15,16,17 18		19,20,21,24,25 22,23
Ward's	1,2,5,6,7 (4),5,8,10	9,11,13,14,15,16,17 3,12,18			19,20,21,24,25 22,23

The members of the clusters are set out in the columns of the table. Most methods yield four clusters (labelled I, II, III, and V) but with the nearest and farthest neighbour methods we identified small residual clusters which did not seem to fit anywhere else. These are listed as cluster IV. In the case of Ward's method, the geographically northerly and southerly clusters reappear but it fails to distinguish clearly between those in the middle. This middle group has been listed so as to span the columns allocated to clusters II and III. In order to emphasise the common elements in the clusters yielded by the various methods, those that are common are placed on the first line of each cell with the other members on a second line. The cluster membership of villages in brackets is less clear but they have been allocated somewhat arbitrarily to what seemed to be the nearest cluster. Broadly speaking, these further analyses confirm the conclusions we drew from farthest and nearest neighbour methods, but the variations are interesting and should be studied carefully. One particularly interesting case is the only village in Cambridgeshire, village number 23. This does not fit easily into the general scheme of things. Sometimes it is linked with the villages to the west in Northamptonshire and Bedfordshire and sometimes it goes with the more northerly group in Leicestershire. This raises interesting questions about the affinities between various dialects which might be worth pursuing.

In spite of the ambiguities at the edges of the clusters, it is clear that cluster analysis confirms that there are regional groupings in the use of dialect words.

2.6 Comparisons

Faced with such a variety of hierarchical clustering methods, it is natural to ask whether there are any guiding principles that may be adopted in choosing a method. The problem is that the effectiveness of any method will depend on the "shape" and "location" of any underlying clusters and on the presence or absence of intervening points. These are things we are not likely to know. However, some insight can be gained from looking at the position in two dimensions. Figure 2.10 is designed to illustrate circumstances where the nearest and farthest neighbour methods will differ.

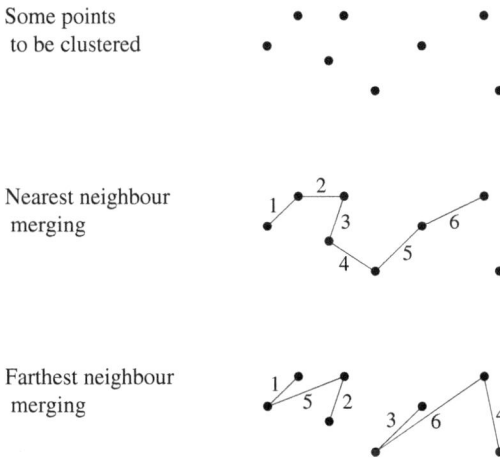

FIGURE 2.10: An example showing how points (top panel) are clustered differently using nearest neighbour (middle panel) and farthest neighbour (bottom panel) methods

The top panel shows eight points in a plane to be combined into two clusters. The middle panel gives the nearest neighbour clustering. The numbers on the lines joining the points indicate the order in which the points were linked into the cluster, using single linkage. The two clusters consist of a chain snaking from left to right and a single point forming its own cluster at the lower right. The bottom panel gives the farthest neighbour clustering. Again the numbers show the order in which points or clusters were linked together, this time using complete linkage. Farthest neighbour gives two clusters of equal size, one to the left and one to the right.

Generally, nearest and farthest neighbour give similar results when the clusters are compact and well separated, as in Figure 2.11, but will give different results when the objects are spread out or when there are intermediate objects strung out between the (farthest neighbour) clusters. In Figure 2.11, there are three compact clusters with no intermediate points. So, at the next stage of clustering, the same two clusters would merge using either method.

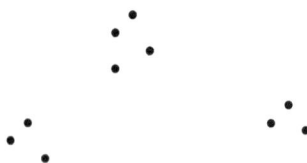

FIGURE 2.11: An arrangement of points that give the same result using either nearest neighbour or farthest neighbour clustering

There has been some theoretical work on the question of choice of method. One can start by laying down criteria that any "reasonable" clustering should meet. This approach was taken by Jardine and Sibson (1971) who arrived at the conclusion that the nearest neighbour method was the only one that satisfied all their criteria. However, the middle panel of Figure 2.10 illustrates that the nearest neighbour method may give results that are not very useful. Points close to the existing cluster are successively added to form a set in which the more distant members are so far apart as to raise the question of whether they can be properly regarded as belonging to the same cluster. This phenomenon is known as "chaining", and it can easily lead to non-interpretable clusters.

2.7 Non-hierarchical methods for cluster analysis

K-means clustering method

The K-means method is probably the most widely used non-hierarchical clustering method. It classifies all the objects into K clusters for some prespecified number of clusters K. It aims to find the clusters such that the total within-cluster dissimilarity is minimised. Consider an $n \times p$ data matrix as in Section 2.4, and let $\mathbf{y}_i = (y_{i1}, \cdots, y_{ip})$ be the data vector of object i. The dissimilarity within cluster k is calculated as

$$\text{Dis}_k = \sum_{i \text{ in cluster } k} \sum_{j=1}^{p} (y_{ij} - \bar{y}_{kj})^2,$$

where $\bar{y}_{kj}$ is the average of variable j for objects in the kth cluster. We refer to the point $\bar{\mathbf{y}}_k = (\bar{y}_{k1}, ..., \bar{y}_{kp})$ as the centre or the centroid of the kth cluster. Note that the dissimilarity within cluster k is defined as the sum of the squared Euclidean distances between all the data points in the kth cluster and the centroid of the cluster. The total within-cluster dissimilarity is defined as the sum of Dis$_k$ across clusters, i.e. $\sum_{k=1}^{K}$ Dis$_k$. Consider the fictitious example in Figure 2.1, in which age is recorded in years, and income is recorded in thousands of pounds. In Panel (a) of Figure 2.12, we randomly assign each individual into three clusters, where individuals in the same cluster are represented using the same plotting symbol. The total within-cluster dissimilarity for these clusters is 28,069. Panel (b) of Figure 2.12 shows the clusters discovered by the K-means method. The observations in the same cluster in Panel (b) are closer to each other than those in Panel (a). This visual impression is confirmed by the total within-cluster dissimilarity for these clusters in Panel (b), which is 4,213, about 85% smaller than that of Panel (a).

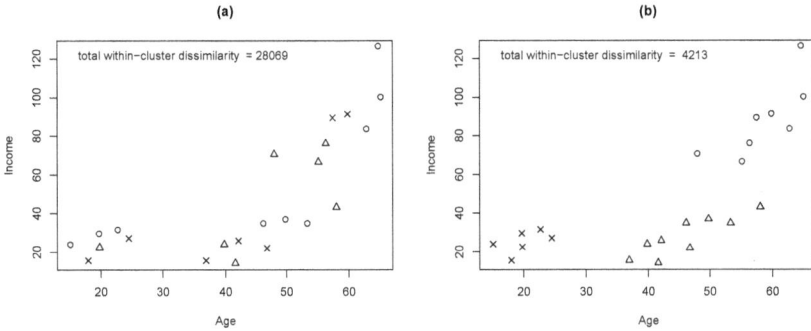

FIGURE 2.12: Clustering of fictitious data of age and income as in Figure 2.1. In each plot, points in the same cluster are represented using the same plotting symbol. Panel (a): The individuals are randomly assigned into three clusters. Panel (b): The individuals are assigned into three clusters by the K-means method.

The K-means method uses a computationally efficient iterative algorithm to minimise the total within-cluster dissimilarity. It often converges quickly. However, its solution is not guaranteed to be the global solution. In other words, there may exist another clustering of the objects for which the within-cluster dissimilarity is even smaller.

This iterative algorithm starts with an initial cluster assignment and then performs the following two steps at each iteration:

i) Calculate the centroids for the current clusters, $\bar{\mathbf{y}}_k$, $k = 1, ..., K$.

ii) Calculate the Euclidean distance between each individual point and each centroid, based on which each object is assigned to the cluster whose centroid is closest in Euclidean distance. For example, if object i's distance to $\bar{\mathbf{y}}_1$ is smaller than its distances to $\bar{\mathbf{y}}_2$, ..., $\bar{\mathbf{y}}_K$, then we assign it to cluster 1.

The total within-cluster dissimilarity is reduced at each iteration, and the algorithm stops when it remains the same in two consecutive steps. Panels (a)–(f) in Figure 2.13 show the first three iterations of the K-means algorithm applied to the previous example. The initial cluster assignment is given in Panel (a) of Figure 2.12. Two panels in the same row correspond to the two steps in the same iteration, where the left panel shows the centroids of the current clusters calculated in step i), and the right panel shows the new clusters obtained in step ii). We notice that the total within-cluster dissimilarity reduces quickly from 28,069 to 4,631 in these iterations. The algorithm has not yet converged after the third iteration, and more iterations will follow.

We note that the total within-cluster dissimilarity that K-means clustering aims to minimise relies on the squared Euclidean distance. This choice of dissimilarity measure is key to ensure that the total within-cluster dissimilarity decreases at each iteration of the K-means algorithm. One may wonder if we can replace the squared Euclidean distance with a different dissimilarity measure, noting that the hierarchical agglomeration methods introduced earlier are very flexible in the choice of dissimilarity measure. The answer is yes. Later in this chapter, we shall briefly introduce some other non-hierarchical clustering methods that involve minimising alternative total within-cluster dissimilarities defined by different dissimilarity measures.

We now comment on some practical considerations when using the K-means clustering algorithm. As mentioned previously, the K-means algorithm does not guarantee finding the global solution to the optimisation problem it aims to solve. In fact, the solution that the K-means algorithm finds depends on the initial cluster assignment. To increase the chance of finding a global solution, it is recommended to obtain multiple results by using multiple random cluster assignments and then choose the solution that yields the smallest total within-cluster dissimilarity.

In addition, K-means clustering is not scale-invariant in the sense that the resulting clusters may be different when changing the scales of some of the variables. This is because the squared Euclidean distance depends on the scales of the variables. The same phenomenon has also been discussed in Section 2.4 for hierarchical clustering methods. For example, imagine that we record the variable Income in pounds instead of thousands of pounds. In that case, the variable Income will have a much larger variance than the variable Age (in years). Consequently, income will be dominant when calculating the squared Euclidean distance, and the resulting clusters will be mainly determined by income. In practice, we recommend standardising the data before performing K-means clustering to avoid some variables being dominant in the squared Euclidean distance. For example, one may standardise each variable such that all its values range between zero and one. Alternatively, one may standardise each variable so that it has a sample mean of zero and a sample standard deviation of one. See Section 2.4 for a related discussion about the construction of distances and their standardisation.

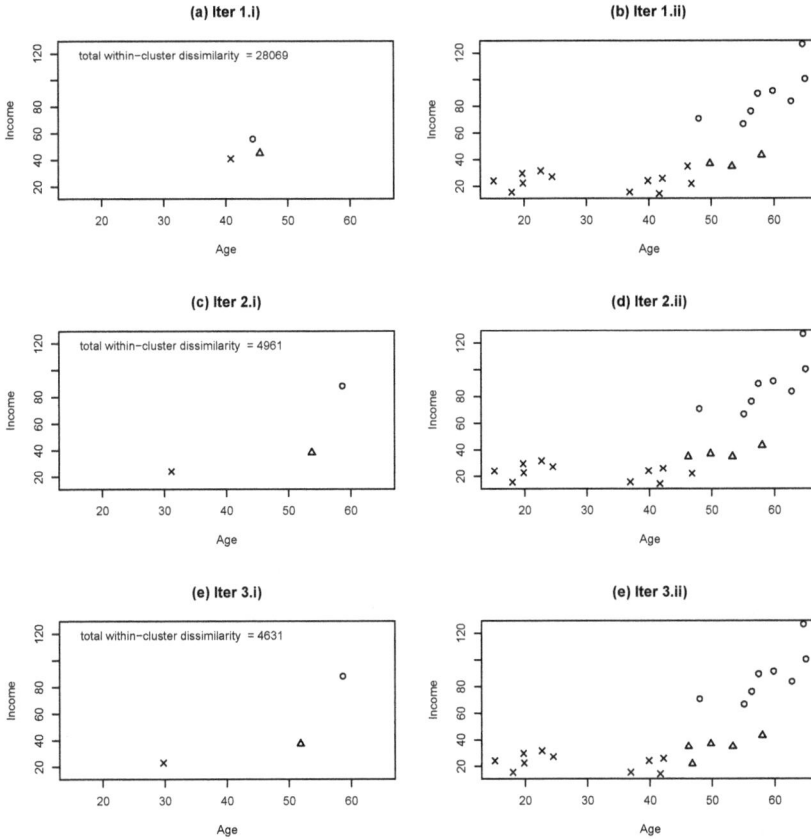

FIGURE 2.13: The first three iterations of the K-means algorithm when applied to the fictitious example of age and income. Panels (a), (c), and (e) show the calculated centroids in the first three iterations. Panels (b), (d) and (f) show the cluster assignment based on the calculated centroids

Finally, we discuss the choice of K, which typically has no underlying population value. As a method for exploratory data analysis, the purpose of applying K-means clustering is to better understand the structure of the data. Therefore, it is suggested to run the K-means algorithm with multiple choices of K and examine their results. In this case, it is useful to draw a scree plot of the total within-cluster dissimilarity as a function of K. The scree plot for the age and income example is given in Figure 2.14. The total within-cluster dissimilarity decreases with K due to the way the optimisation problem for K-means clustering is defined. The curve obtained by connecting the points in the scree plot typically has an elbow shape, and the value of K at which the elbow appears is called the elbow point (which requires a subjective judgement). Arguably, the elbow point in Figure 2.14 is $K = 6$. Beyond the elbow point, increasing the number of clusters does not lead to a significant reduction in the total within-cluster dissimilarity, and the resulting solutions also tend to be unstable (i.e. the resulting clusters can vary substantially when perturbing the data slightly, for example, adding a small number of objects into the data). Therefore, it is suggested to focus on the cluster structure for the values of K that are before the elbow point, especially those with meaningful substantive interpretations.

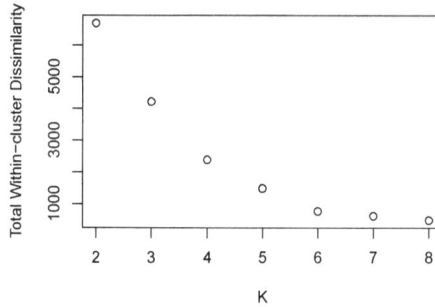

FIGURE 2.14: Scree plot of K-means clustering for the fictitious example of age and income. The X-axis shows the number of clusters, and the Y-axis shows the total within-cluster dissimilarity of the corresponding K-means solution.

Other methods

There are a great many other methods available for cluster analysis. Here we merely list some of the possibilities in order to illustrate the rich diversity of methods on offer.

i) *Optimisation-based methods.* Like the K-means method, an optimisation-based method defines a measure of total within-cluster dissimilarity and finds the clusters by minimising this measure. For example, the K-medians method defines a total within-cluster dissimilarity measure similar to that of the K-means method but replaces the squared Euclidean distance with the (unsquared) Euclidean distance and uses the medians of the clusters as their centroids. As the median tends to be more robust against outliers than the mean, the K-medians method tends to be more robust and allows for better interpretability of its cluster centroids.

Another popular method of this type is the K-metroids method. Unlike the K-means and K-medians methods, the K-metroids method uses the actual data points as the cluster centroids and uses a sum of pairwise dissimilarities to measure the total within-cluster dissimilarity. This construction has some advantages. First, the data points identified as cluster centroids serve as the exemplars of the clusters, yielding greater interpretability. Second, any dissimilarity measure can be used to define the total within-cluster dissimilarity, making the method very flexible. Finally, similar to the K-medians method, the use of actual data points as cluster centroids also makes it more robust against outliers than the K-means method.

We note that finding the global solution for the optimisation problems in these methods is computationally infeasible when the number of objects is large. Similar to the K-means method, these methods are based on iterative algorithms that make locally optimal choices at each iteration to minimise the total within-cluster dissimilarity measure. These algorithms often converge quickly but do not guarantee a global solution.

ii) *Clustering after representation learning.* We are often faced with clustering tasks involving complex data, such as photos, text, and videos. The dimensionality of such data is very high, for which the standard optimisation-based methods may not perform well. There are also situations when we need to do clustering without access to the raw data but instead have only a similarity/dissimilarity matrix of the objects. In these cases, we can first use representation learning methods to transform the raw data or similarity/dissimilarity matrix into continuous data. A good representation learning method finds an informative low-dimensional representation without losing much information about the cluster structure in the original data. Examples of relevant representation learning methods include principal component analysis (PCA) in Chapter 5 and multidimensional scaling (MDS) in Chapter 3. PCA is suitable for reducing the dimension of high-dimensional data, while MDS can learn a low-dimensional representation from a dissimilarity matrix. Once a low-dimensional representation has been learned, we can then use a standard optimisation-based method to cluster the objects.

A popular clustering method in this category is spectral clustering, which has many applications, including social network analysis, document classification, and video segmentation. This method takes a nonnegative similarity matrix as the input and uses an eigendecomposition-based method for representation learning. It typically, but not always, uses the K-means method for clustering.

iii) *Model-based methods.* Here we start with the hypothesis that the objects have been randomly sampled from a population made up of several sub-populations. The aim is to classify the objects according to their sub-population of origin. Models involving discrete latent variables, including latent class models and Gaussian mixture models (which we shall meet in Chapter 10) can be regarded as cluster analysis models and we shall return to this point in that chapter.

2.8   Clustering variables

There are two types of clustering which can be carried out on a data matrix. One, which has been the subject of this chapter up to this point, is to cluster objects, or individuals. Another type of clustering is clustering variables where we consider whether variables can be grouped because they are distributed similarly across individuals. This duality arises with all analyses that start from a data matrix. If we wished to carry out a cluster analysis on variables we would need measures of similarity between columns of the data matrix instead of between the rows.

In this case, similarity between variables (columns) is typically measured by their correlation. For continuous variables, the product moment correlation will serve very well; for binary or polytomous variables, the tetrachoric and polychoric coefficients respectively, achieve the same end. Alternative measures, such as the Euclidean distance between columns that have not been standardised, might sometimes seem more appropriate for a given problem. This is analogous to carrying out a principal component analysis of the covariance matrix, using unstandardised variables.

2.9   Additional examples and further work

In this section, we present briefly three further applications of cluster analysis. The first has 21 attributes observed for 24 objects — ancient carvings of archers at Persepolis. The second has responses of 379 people on four binary variables (from the 1986 British Social Attitudes Survey), and the third is an example of clustering educational variables using their correlations. Our analyses are not intended to be complete, and you should explore other methods.

Persian archers

There are twenty-four archers carved in bas-relief going up the south stairs on the west side of the east face of the Apadana at Persepolis in Southern Iran. All of the archers look similar, but they differ in minor details, such as the way the beard curls and the way the head-dress is decorated. Figure 2.15 is a picture of the ninth archer (from the top of the stairs) and identifies 21 features or attributes that may differ between archers. Each attribute has only a small number of variants, usually two or three. Details are given in Roaf (1983).

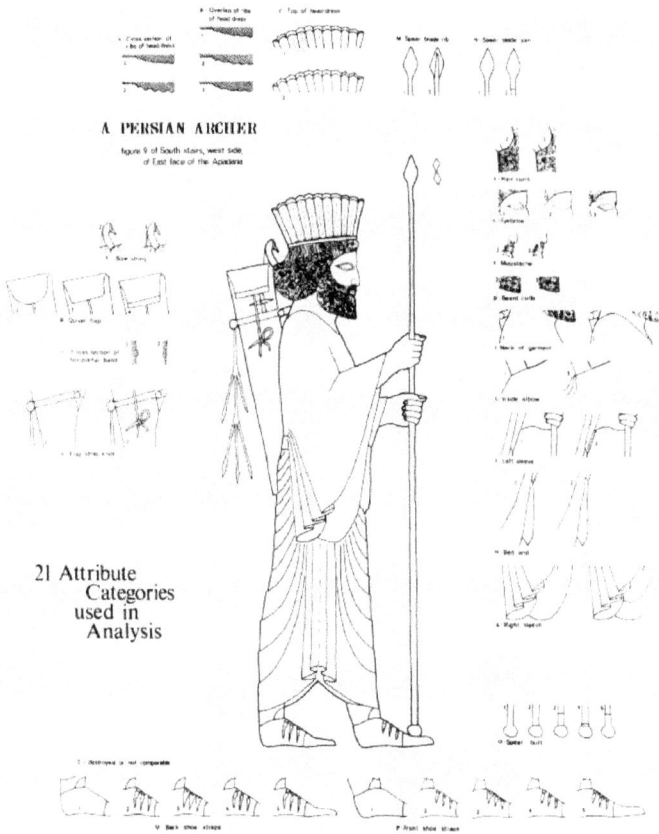

FIGURE 2.15: Archer number 9 at Persepolis, surrounded by variants of 21 attributes (reproduced from Roaf, 1978 with permission)

Table 2.12 gives the data matrix showing which archer has which variant of each attribute. The attributes are labelled A to U and variants of an attribute

TABLE 2.12: Data matrix showing for each of 24 archers which variant of each of 21 attributes A to U he possesses (a zero indicates that the attribute for that archer is missing)

Archer	A	B	C	D	E	F	G	H	I	J	K	L	M	N	O	P	Q	R	S	T	U
1	2	2	1	2	3	2	1	1	2	1	1	3	0	0	2	3	4	2	2	2	0
2	2	2	1	2	2	2	1	1	2	1	1	3	1	1	2	0	4	2	2	1	2
3	2	2	1	2	2	2	1	1	2	1	2	3	2	2	2	3	0	3	2	1	3
4	2	2	1	2	3	1	1	1	2	1	2	3	2	2	2	3	4	2	2	1	3
5	2	3	1	3	2	2	1	1	2	1	2	0	2	2	4	3	4	3	2	1	3
6	2	3	1	2	2	2	1	1	2	1	2	3	2	2	4	3	4	3	2	1	3
7	2	3	1	3	3	2	1	1	2	2	2	3	2	2	3	3	4	3	2	1	3
8	2	3	1	3	3	2	2	1	2	2	1	2	2	2	3	3	4	3	2	1	3
9	3	1	1	3	2	2	1	2	2	1	1	2	1	1	3	2	2	4	3	2	2
10	3	1	1	3	2	2	1	2	2	2	1	2	2	2	3	2	2	2	2	2	2
11	3	1	1	3	2	2	1	2	2	1	2	2	2	2	0	2	4	3	2	2	2
12	3	1	2	2	2	2	1	2	2	1	2	2	1	1	1	5	2	2	3	1	3
13	3	1	1	3	2	2	1	2	2	2	2	1	1	0	0	3	2	2	2	2	3
14	3	1	1	3	2	2	1	2	2	2	2	2	2	2	5	3	2	2	2	2	3
15	3	1	1	3	2	2	1	2	2	1	2	2	2	2	3	2	2	4	3	2	2
16	3	1	1	3	2	2	1	2	1	1	2	2	1	1	3	2	4	4	3	2	2
17	3	1	2	3	2	3	1	2	2	2	2	1	2	2	3	4	2	2	2	2	2
18	3	1	1	3	2	2	1	2	2	1	2	2	2	2	6	1	1	2	2	2	2
19	2	1	1	3	2	2	1	2	2	1	2	2	2	2	3	1	4	2	2	2	2
20	2	1	2	2	2	2	3	1	2	1	2	2	2	2	1	3	4	2	2	1	3
21	2	1	2	2	2	2	3	1	2	1	2	2	2	2	0	3	4	2	2	2	3
22	2	1	2	2	2	2	3	1	1	1	2	2	2	1	0	3	4	2	2	2	3
23	2	1	2	2	2	2	3	1	2	1	2	2	2	1	1	3	4	2	2	2	3
24	2	1	1	2	2	2	3	1	2	1	2	2	2	1	2	3	4	2	2	2	3

are labelled by integers between 1 and 6. A zero indicates the attribute being missing from the bas-relief of that archer, due to damage over time.

Archaeologists and art historians are interested in whether the archers were carved by a single sculptor or by several, and in whether these data suggest any groupings that might bear on this question. Roaf (1978, 1983) reported several cluster analyses identifying clusters of archers which might have been carved by the same sculptor or team of sculptors. We use these data to illustrate farthest neighbour cluster analysis, and we invite you to try other methods.

Table 2.13 gives a table of similarities for all pairs of archers. The similarity of each pair of archers is calculated from Table 2.12 to be the proportion of attributes for which each archer has the same variant, multiplied by 21. For a pair of archers with no attributes missing, this is just the count of the number of attributes for which they share the same variant. For two archers with m missing attributes from one or both of them, the similarity is the corresponding count but scaled up by a factor of $21/(21 - m)$. For example, the similarity between archers 1 and 2 is $15 \times 21/17 = 18.5$ because attributes M, N, P, and U are missing for one or other archer and there are 15 matches among the remaining 17 attributes. The rationale for this scaling up is that archers with several features missing would otherwise have misleadingly low similarities with the other archers.

TABLE 2.13: Table of similarities for 24 Persian archers

Archer	1	2	3	4	5	6	7	8	9	10	11	12	13	14	15	16	17	18	19	20	21	22	23	24
1	21																							
2	18.5	21																						
3	16.1	15.5	21																					
4	17.5	14.7	17.8	21																				
5	12.4	12.2	17.7	14.0	21																			
6	14.0	13.6	18.9	16.0	20.0	21																		
7	12.8	10.5	15.8	15.0	17.8	17.0	21																	
8	11.7	9.4	12.6	12.0	15.8	14.0	18.0	21																
9	8.2	10.5	6.3	4.0	7.4	6.0	6.0	7.0	21															
10	9.3	9.4	8.4	7.0	9.4	8.0	10.0	11.0	16.0	21														
11	7.4	7.7	9.9	7.0	11.1	9.0	8.0	7.0	17.0	16.0	21													
12	7.0	10.5	9.4	8.0	8.4	9.0	6.0	5.0	13.0	10.0	12.6	21												
13	9.2	10.5	9.3	7.0	10.5	8.0	9.0	8.0	13.0	14.0	13.3	13.0	21											
14	9.3	8.4	11.6	10.0	11.6	10.0	11.0	10.0	11.0	16.0	14.7	12.0	18.8	21										
15	7.0	7.4	9.4	7.0	10.5	9.0	9.0	8.0	18.0	17.0	21.0	12.0	13.3	14.0	21									
16	7.0	9.4	6.3	5.0	8.4	7.0	7.0	6.0	18.0	13.0	16.8	12.0	14.4	11.0	17.0	21								
17	9.3	9.4	8.4	9.0	9.4	8.0	9.0	8.0	11.0	14.0	12.6	11.0	15.5	15.0	12.0	12.0	21							
18	9.3	9.4	10.5	9.0	11.6	10.0	9.0	8.0	13.0	16.0	16.8	11.0	15.5	16.0	16.0	13.0	14.0	21						
19	11.7	11.6	11.6	11.0	13.6	12.0	12.0	11.0	13.0	16.0	15.8	10.0	14.4	15.0	16.0	14.0	14.0	18.0	21					
20	11.7	11.6	14.7	14.0	14.7	15.0	12.0	12.0	6.0	9.0	9.4	13.0	9.9	11.0	9.0	7.0	11.0	11.0	13.0	21				
21	13.6	11.1	14.4	13.0	14.4	14.0	11.0	11.0	7.0	10.0	10.5	11.0	11.1	12.0	10.0	8.0	12.0	12.0	14.0	19.0	21			
22	12.4	11.1	12.2	11.0	12.2	12.0	9.0	9.0	7.0	8.0	8.4	11.0	11.1	10.0	8.0	10.0	10.0	10.0	12.0	17.0	18.9	21		
23	12.8	11.6	12.6	12.0	12.6	13.0	10.0	10.0	8.0	9.0	9.4	13.0	12.2	11.0	9.0	9.0	11.0	11.0	13.0	19.0	20.0	20.0	21	
24	15.2	13.6	14.7	14.0	13.6	14.0	11.0	11.0	9.0	10.0	10.5	11.0	13.3	13.0	10.0	10.0	11.0	12.0	14.0	17.0	18.9	18.9	19.0	21

Figure 2.16 gives the dendrogram using farthest neighbour clustering. This shows two major clusters. The first consists of archers 9 to 19 in the middle section of the staircase and the second consists of archers 1 to 8 at the top and archers 20 to 24 at the bottom.

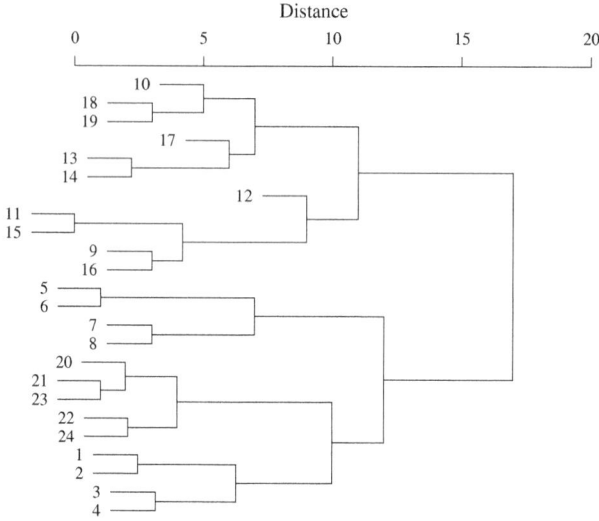

FIGURE 2.16: Dendrogram for farthest neighbour (complete linkage) cluster analysis for the Persian archers data (distance $= 21 -$ similarity)

The major clusters are divided into smaller groups of (mainly) contiguous archers — in particular the bottom five archers (20 to 24) form a tight cluster. Within this overall pattern, there are two relatively isolated archers, 12 and 17. By examining the 12th and 17th rows and columns of the similarity matrix (Table 2.13), you can confirm that all of their similarities are small. You could try other methods of cluster analysis on these data and compare the results. You might also look at Section 3.7 where the same data are analysed using multidimensional scaling.

An alternative visual presentation of the archers and their similarities is shown in Figure 2.17. This diagram, reproduced from Roaf (1978) (Figure 5), shows the positions on the staircase and links individuals with 15 or more attribute variants in common. Here three groups of archers are clearly identified, the bottom five being very similar. This simple analysis reinforces the main conclusions from the dendrogram and shows that a customised analysis and presentation can sometimes provide a better summary of the similarity structure than standard cluster analysis.

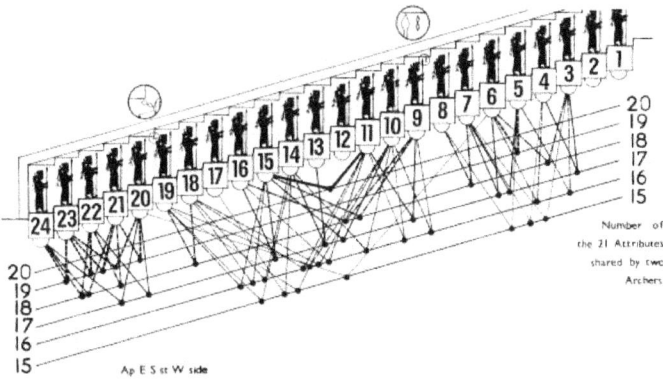

FIGURE 2.17: A diagram of archers on the staircase linking those with 15 or more attribute variants in common (reproduced from Roaf 1978 with permission)

Attitude to abortion

In this example, we consider some data extracted from the 1986 British Social Attitudes Survey. Binary responses to four items are given for 379 respondents. The items are questions relating to the circumstances under which a respondent would consider that an abortion should be allowed under law. The four circumstances are (briefly): the woman decides on her own that she does not wish to have the child; the couple agree that they do not want to have the child; the woman is not married and does not wish to marry the man; and the couple cannot afford to have any more children. Table 2.14 shows the numbers of people responding positively (the law should allow an abortion) and negatively (an abortion should not be allowed) for each item, together with a short item name. Positive responses are coded as 1 and negative responses as 0.

TABLE 2.14: Attitude to abortion marginal frequencies for four binary variables

		Response		
Variable	Item	1	0	Total
y_1	WomanDecide	166	213	379
y_2	CoupleDecide	225	154	379
y_3	NotMarried	241	138	379
y_4	CannotAfford	234	145	379

Table 2.15 sets out the 16 possible response patterns and their frequencies. The patterns are arranged by the number of 1s they contain. The data may be regarded as 379 observations on four binary variables y_1, y_2, y_3, and y_4. Thus, for example, 103 respondents thought an abortion should not be allowed in any of the four situations; 13 thought an abortion should be allowed if the couple could not afford have the child; and so on to the last row, where 141 respondents thought that an abortion should be allowed in all four situations.

TABLE 2.15: Response patterns, frequencies and cluster allocations using farthest neighbour (complete linkage) clustering and K-means clustering for the attitude to abortion data

Pattern	y_1	y_2	y_3	y_4	Total	Frequency	Cluster (farthest)	Cluster (K-means)
1	0	0	0	0	0	103	1	1
2	0	0	0	1	1	13	1	1
3	0	0	1	0	1	10	2	1
4	0	1	0	0	1	9	1	1
5	1	0	0	0	1	1	1	1
6	0	0	1	1	2	21	2	2
7	0	1	0	1	2	6	1	2
8	1	0	0	1	2	0	1	1
9	0	1	1	0	2	7	2	2
10	1	0	1	0	2	0	2	2
11	1	1	0	0	2	3	1	1
12	0	1	1	1	3	44	2	2
13	1	0	1	1	3	6	2	2
14	1	1	0	1	3	3	1	2
15	1	1	1	0	3	12	2	2
16	1	1	1	1	4	141	2	2
					Total	379		

We apply hierarchical clustering to the data. Usually in a cluster analysis, we begin with n objects or individuals. But here the 379 people are already grouped according to which of the 16 response patterns they have. So now we will try to cluster the response patterns (and thereby implicitly to cluster individuals). For example, we may find evidence that people fall into two groups (essentially pro- or anti-abortion), or that attitudes are more diverse.

As there are only four binary variables, the similarity measure $r = (a+d)/p$ has just four distinct values $(0, \frac{1}{4}, \frac{1}{2}$ and $\frac{3}{4})$ corresponding to whether two patterns match on 0, 1, 2 or 3 responses. This could only give very limited possibilities for clustering response patterns. Instead, we use the similarities shown in Table 2.16, and which we explain below.

The similarities in Table 2.16 are obtained by weighting each response for each variable to give the following measure of similarity between patterns i and j:

$$\sum_{k=1}^{4} w_{k1} y_{ik} y_{jk} + \sum_{k=1}^{4} w_{k0}(1 - y_{ik})(1 - y_{jk}),$$

TABLE 2.16: Similarities between response patterns for the attitude to abortion data

Pattern		1	2	3	4	5	6	7	8	9	10	11	12	13	14	15	16
0000	1	9.60															
0001	2	6.99	8.61														
0010	3	6.85	4.24	8.43													
0100	4	7.14	4.53	4.39	8.82												
1000	5	7.82	5.21	5.07	5.36	10.10											
0011	6	4.24	5.86	5.81	1.78	2.46	7.43										
0101	7	4.53	6.15	1.78	6.21	2.75	3.40	7.83									
1001	8	5.21	6.83	2.46	2.75	7.49	4.08	4.37	9.11								
0110	9	4.39	1.78	5.97	6.08	2.61	3.35	3.46	0.00	7.65							
1010	10	5.07	2.46	6.65	2.61	7.36	4.03	0.00	4.74	4.19	8.93						
1100	11	5.36	2.75	2.61	7.04	7.64	0.00	4.43	5.03	4.30	4.90	9.33					
0111	12	1.78	3.40	3.35	3.46	0.00	4.97	5.08	1.62	5.04	1.57	1.68	6.66				
1011	13	2.46	4.08	4.03	0.00	4.74	5.65	1.62	6.36	1.57	6.32	2.28	3.19	7.94			
1101	14	2.75	4.37	0.00	4.43	5.03	1.62	6.05	6.65	1.68	2.28	6.71	3.30	3.90	8.33		
1110	15	2.61	0.00	4.19	4.30	4.90	1.57	1.68	2.28	5.87	6.47	6.58	3.26	3.86	3.97	8.15	
1111	16	0.00	1.62	1.57	1.68	2.28	3.19	3.30	3.90	3.26	3.86	3.97	4.88	5.48	5.59	5.54	7.16

where y_{ik} and y_{jk} are the responses to item k for patterns i and j, respectively, and where w_{k1} and w_{k0} are weights. We use $w_{k1} = n/n_k$ and $w_{k0} = n/(n-n_k)$ where n_k is the number of positive responses for item k and $n - n_k$ is the number of negative responses. The product $y_{ik}y_{jk}$ will equal 1 if y_k equals 1 in both patterns, and will be zero otherwise. Likewise $(1 - y_{ik})(1 - y_{jk})$ will equal 1 if y_k equals 0 in both patterns. So using weights $w_{k1} = n/n_k$ and $w_{k0} = n/(n - n_k)$ has the effect of giving more weight to agreement on rarer responses.

For example, from Table 2.15, response patterns 6 and 7 both agree on $y_1 = 0$ and $y_4 = 1$. So $(1 - y_{61})(1 - y_{71}) = 1$ and $y_{64}y_{74} = 1$ are the only non-zero terms in the above formula. From Table 2.14, the relative frequency of 0 for item 1 is $213/379$ and the relative frequency of 1 for item 4 is $234/379$. Hence, the similarity between patterns 6 and 7 is

$$\frac{379}{213} \times 1 \ + \ \frac{379}{234} \times 1 \ = \ 3.40.$$

Likewise for patterns 7 and 8, which agree on $y_3 = 0$ and $y_4 = 1$, the similarity is

$$\frac{379}{138} \times 1 \ + \ \frac{379}{234} \times 1 \ = \ 4.37.$$

Thus, patterns 7 and 8 are judged to be more similar than patterns 7 and 6, because $y_3 = 0$ is a rarer response than $y_1 = 0$.

You could consider the appropriateness of this measure of similarity which gives a larger weight to less frequent responses than to more frequent responses. Another feature of this measure is that the diagonal elements of the similarity matrix are not all equal, but, of course, they do not affect the clustering process.

Figure 2.18 shows the dendrogram for nearest neighbour (single linkage) cluster analysis using the similarity matrix in Table 2.16. This is an extreme example of *chaining*. The first two patterns that merge, numbers 1 and 5, have pattern 11 as their nearest neighbour so 11 then merges. This process continues, with each pattern successively merging into the same group. Only a single link is required to add a new object (response pattern) to the existing cluster, resulting in a single chain.

Figure 2.19 shows the dendrogram for farthest neighbour (complete linkage) clustering. This looks more successful. Here a new pattern can only be joined to a cluster if it is close to *all* members of that cluster, that is, complete linkage to every member is required.

In Figure 2.19, two clusters each of eight patterns can be identified. Cluster 1 (in the top half of the dendrogram) consists of response pattern 1 (0000) and other patterns with several 0s, and cluster 2 (in the bottom half) has pattern 16 (1111) and other patterns with several 1s. This allocation of patterns to clusters is listed in the second last column of Table 2.15. Perhaps respondents in cluster 1 have a generally unfavourable attitude to abortion while those in cluster 2 have a more favourable attitude. But there are two

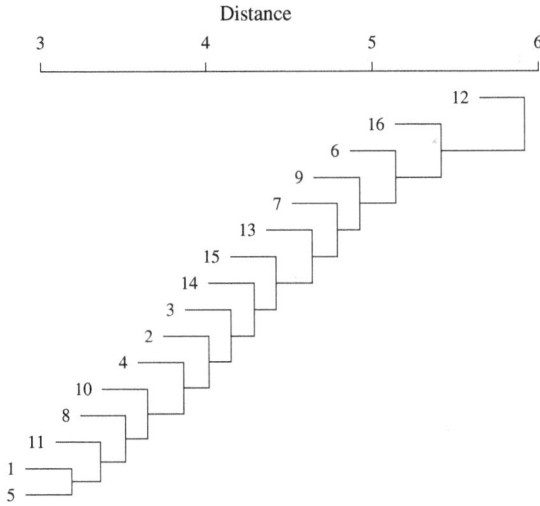

FIGURE 2.18: Dendrogram for nearest neighbour (single linkage) cluster analysis for the attitude to abortion data (distance $= 11 -$ similarity)

exceptions. Pattern 3 (0010) has been allocated to cluster 2 and pattern 14 (1101) to cluster 1. Closer inspection reveals that all patterns in cluster 1 have $y_3 = 0$ and all in cluster 2 have $y_3 = 1$. The division into these two clusters could therefore have been achieved on the basis of a single item relating to

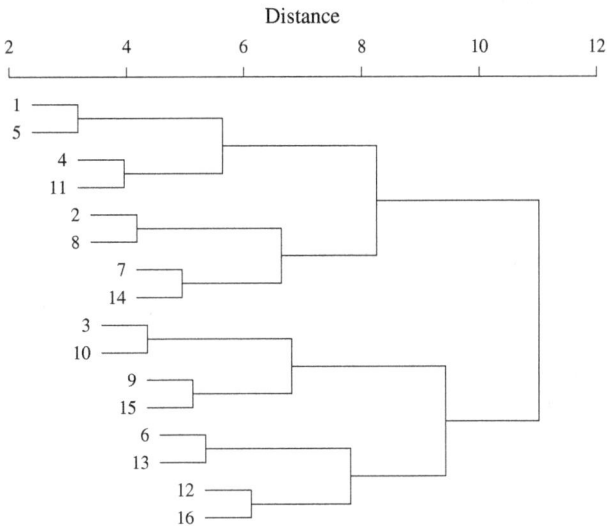

FIGURE 2.19: Dendrogram for farthest neighbour (complete linkage) cluster analysis for the attitude to abortion data (distance $= 11 -$ similarity)

whether or not the respondent thought abortion should be allowed when the woman is not married. Does this make sense? To what extent is it the result of our choice of similarity measure?

We further analyse the data with the K-means method. When performing this analysis, we treat the binary data as continuous. Specifically, we compare the two-cluster structure obtained by K-means clustering with that obtained by the farthest neighbour hierarchical clustering. For K-means clustering, the two cluster centroids are (0.03, 0.09, 0.07, 0.09) and (0.68, 0.89, 0.96, 0.92), and the resulting cluster allocations are given in the last column of Table 2.15. This result is arguably more sensible than that from the farthest neighbour hierarchical clustering. More specifically, the clusters are no longer determined by a single item. Response patterns with zero or one positive response now all belong to cluster 1, while those with three or four positive responses now belong to cluster 2. For response patterns with two positive responses, only patterns 8 (1001) and 11 (1100), which have low frequencies (only three participants had these patterns), are classified into cluster 1, and the rest are all classified into cluster 2. Thus, most of the respondents with two positive responses belong to cluster 2. Although the above application to binary data seems successful, we note that, in general, K-means clustering should be applied with caution when data are categorical, as the Euclidean distance may not be a sensible dissimilarity measure (see discussions in Section 2.4).

The result from K-means clustering is quite similar to the result from latent class analysis of the same data, as given in Table 10.7. In this result, all patterns with three or more 0s are allocated to class 1 and all with two or more 1s are allocated to class 2. This result could also be obtained using factor analysis for binary data, by allocating those with fitted factor scores below -0.35 to class 1 and those with fitted scores above -0.35 to class 2 (see Table 8.6).

Clustering educational variables

Rather than clustering individuals or objects, this last example aims to examine how five measurements made on secondary school girls in 1964 relate to four measurements (three the same and one new) made on the same girls in 1968 (see Peaker 1971). The dataset, which we will refer to as "educational circumstances", comes from a national survey of primary school children in 1964 and a follow-up survey in 1968. About one quarter of the children could not be traced which could introduce bias as the missing children may differ from those who were followed up. We use data for 398 girls in their final year of primary school in 1964 and in their fourth year of secondary school in 1968. The nine variables we analyse are composite measures described in Table 4 of Peaker (1971). The variables for 1964 are (briefly): parental circumstances (y_1), details of class teacher (y_2), school-parent interaction (y_3), girl's attitude (y_4) and test score (y_5), and in 1968: type of school (y_6), parental circumstances (y_7), school-parent interaction (y_8), and test score (y_9). The correlations given in Table 2.17 are taken from Table 7 of Peaker (1971) with one correction to an obvious typographical error.

TABLE 2.17: Correlation matrix, educational circumstances

	y_1	y_2	y_3	y_4	y_5	y_6	y_7	y_8	y_9
y_1	1								
y_2	0.177	1							
y_3	0.305	0.155	1						
y_4	0.193	0.124	0.243	1					
y_5	0.501	0.134	0.556	0.317	1				
y_6	0.423	0.124	0.308	0.308	0.572	1			
y_7	0.770	0.184	0.351	0.193	0.436	0.388	1		
y_8	0.206	−0.050	0.149	0.128	0.252	0.382	0.206	1	
y_9	0.499	0.127	0.413	0.339	0.758	0.613	0.459	0.315	1

It is interesting to see whether a cluster analysis on the correlation matrix will combine measurements made at the same time, or measurements of the same characteristic made at different times, or a mixture of these. A further question is which other variables might be associated with success in the two tests. Before proceeding with this analysis, we make two cautionary remarks. First, a cluster analysis is not an ideal way of investigating how one set of variables depends on another, and secondly, the use of product moment correlation as a measure of association between variables some of which are ordinal could be improved upon (see Section 9.5). However, the use of simple descriptive methods may sometimes reveal interesting aspects that a more focused analysis might miss.

The dendrogram in Figure 2.20 shows the result of a nearest neighbour cluster analysis and gives the short names of the variables. Two pairs of variables, parental circumstances in 1964 and in 1968, and total test scores 1964 and 1968, are each closely linked but the two measurements for school-parent interaction are linked only at the sixth out of eight steps. There is some chaining of variables, but to determine whether this is due to the use of single linkage or whether it would still appear with other methods we leave as an exercise for you. Overall, from the nearest neighbour analysis, we might conclude that the teacher's characteristics, the girl's attitude in 1964, and school-parent interaction in 1968 are only weakly associated with the test scores, whereas the other four variables have stronger associations with the test scores. You can confirm these conclusions by examining the correlation matrix.

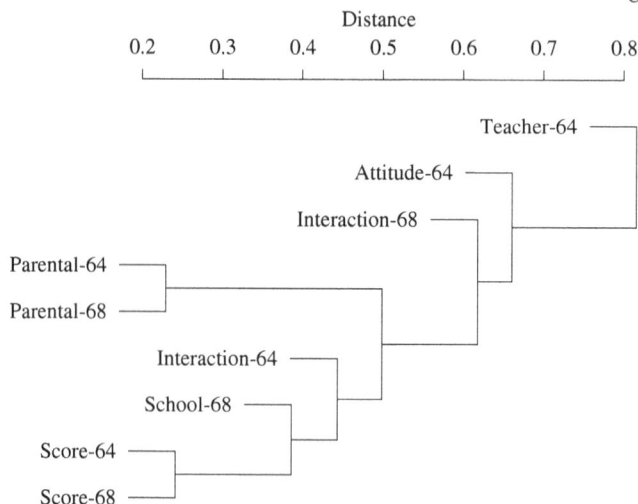

FIGURE 2.20: Dendrogram for nearest neighbour (single linkage) cluster analysis for the correlation matrix in Table 2.17 (distance = 1 − correlation)

Finally

The analyses reported in this chapter have mainly used either hierarchical agglomerative clustering methods or the K-means method. You are recommended to try out other methods on these datasets and on some of the other datasets given in later chapters. You should also compare the results of the analyses in this chapter with analyses using some of the same datasets in Section 3.7 (archers and English dialect data), Section 5.9 (educational circumstances), Chapter 8 (attitude to abortion data), and especially the latent class analysis in Section 10.4 (attitude to abortion data).

When conducting cluster analysis in practice, it is advisable to use several methods and compare the results. Where different methods give similar clusters, as in Table 2.11, the analyst can feel confident that they are reflecting some aspect of the data structure. Where they give very different clusters, the analyst may wish to investigate why.

2.10   Further reading

Everitt, B. S. and Landau, S. and Leese, M. (2001). *Cluster Analysis* (4th ed.). London: Arnold.

Everitt, B. S. and Rabe-Hesketh, S. (1997). *The Analysis of Proximity Data*. London: Arnold.

Gordon, A. D. (1999). *Classification* (2nd ed.). London: Chapman and Hall/CRC.

3

Multidimensional Scaling

3.1 Introduction

Multidimensional scaling is one of several multivariate techniques that aim to reveal the structure of a dataset by representing the data points using a small number of dimensions, typically one or two. The basic idea can be motivated by a geographical example. Suppose we are given the distances between pairs of cities and are asked to reconstruct the two-dimensional map from which those distances were derived. We could attempt to do this by a process of trial and error by moving points about on a sheet of paper until we got the distances right. A procedure that does this automatically is called multidimensional scaling (MDS). The "multi" part of the name refers to the fact that we are not restricted to constructing maps in one or two dimensions.

The simple example above differs in two important ways from the typical MDS problem. In the first place, there is no ambiguity about what we mean by the "distance" between two cities (measured in miles or kilometres in a straight line), whereas in the typical MDS problem, there is often a degree of arbitrariness in the definition of distance which, in some cases, may be based on subjective assessments rather than precise measurement. Secondly, we know that the cities can be located on a two-dimensional map (provided that the curvature of the earth and other topographical features can be ignored), whereas in the typical MDS problem, we would have little idea about how many dimensions would be necessary in order to reproduce, even approximately, the given distances between objects of interest. Indeed one of the prime objects of the analysis will be to discover whether such a representation is possible in a small number of dimensions. Unless this can be done, preferably in one or two dimensions, we shall not be able to take advantage of the eye's ability to spot patterns in the plots. Even if it turns out that more than two dimensions are necessary, the main way we can view the points is by projecting them onto two-dimensional space.

The input data for MDS is in the form of a distance matrix representing the distances between pairs of objects. We have already discussed the construction of such matrices in Chapter 2, and there is nothing to add here. However, whereas the choice between distance and proximity was largely a matter of indifference in several popular clustering methods (e.g. agglomerative hierarchical clustering methods), distance is the prime concept in MDS. Thus although we may start with a proximity or similarity matrix, it may need to be converted to a distance matrix in the course of the analysis; the output will be expressed in terms of distance.

As we have said, MDS is used to determine whether the distance matrix may be represented by a map or configuration in a small number of dimensions such that distances on the map reproduce, approximately, the original distance matrix $\{\delta_{ij}\}$. For example, we would aim to have the two objects that are closest together according to the distance matrix closest together on the map, and so on. As we have posed the problem, the distances on the map would be in the same metric (scale of measurement) as the original δ_{ij}s. This is often known as *classical* multidimensional scaling. However, it is often the case, particularly in social science research, that the values of the δ_{ij}s may be interpreted only in an ordinal sense as if, for example, the distances come from subjective similarity ratings. In such cases, it may be more reasonable only to attempt to produce a map on which the distances have the right rank order. This is called *ordinal* or *non-metrical* multidimensional scaling. In this chapter, we shall be mainly concerned with ordinal MDS. In the second example in Section 3.2 below, students were asked to rate the degree of similarity between pairs of countries on a nine-point scale. Similarity, here, is a subjective thing for which there is no natural underlying "space" reflected in the similarities. Part of the interest in the analysis is to try to uncover which attributes of the countries appear to carry weight in the students' judgement of similarity.

Returning to classical scaling, suppose that we have four cities labelled A, B, C, and D and that the distances (in hundreds of miles) between the pairs of cities are as given by the following matrix:

$$
\begin{array}{c}
A \\ B \\ C \\ D
\end{array}
\left(
\begin{array}{cccc}
- & & & \\
2 & - & & \\
1 & 3 & - & \\
5 & 3 & 6 & -
\end{array}
\right).
$$

Using multidimensional scaling (or by inspection), it is possible to represent this distance matrix exactly in one dimension. A possible solution is given in Figure 3.1.

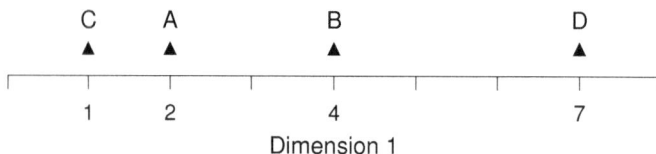

FIGURE 3.1: A one-dimensional configuration of four cities using classical MDS

We shall denote the distance between objects i and j in the above configuration by d_{ij} and in this case, these distances are precisely equal to the δ_{ij}s.

In classical MDS, we seek a configuration such that the d_{ij}s, the inter-point distances in the configuration, will be approximately equal to the corresponding δ_{ij}s, as given in the distance matrix; whereas in ordinal MDS, the object is only to find a configuration such that the d_{ij}s are in the same rank order as the corresponding δ_{ij}s.

There are at least two reasons why we hope to find a good low-dimensional representation. First, a low-dimensional representation allows us to visualise the data, which is the main purpose of MDS as mentioned previously. Second, if the input distance matrix is a noisy measurement of a (approximately) low-dimensional configuration, a low-dimensional MDS solution tends to reduce the measurement error more than a higher-dimensional solution and thus may better approximate the true distance matrix.

An MDS solution may be used as input data in subsequent analysis. For example, the K-means method cannot directly cluster objects based on a non-Euclidean dissimilarity matrix. However, we could first perform MDS on the dissimilarity matrix and then apply K-means clustering to the MDS solution. For another example, we could use the MDS solution as features of the objects and include them as regressor variables in a regression model.

Measures of similarity between variables

We have already remarked in Section 2.8 that one can reverse the roles of variables and objects. Instead of clustering objects, which was our main concern, we could have clustered variables. This duality arises with all analyses that start from a data matrix. If we wished to carry out an MDS analysis on variables, we would need measures of similarity between columns of the data matrix instead of between the rows.

3.2   Examples

Reproducing a two-dimensional map from air distances between pairs of cities

MDS was carried out to determine whether a two-dimensional map could be produced from a matrix of pairwise distances between ten cities in Europe and Asia. The dissimilarity or distance matrix is shown in Table 3.1.

The solution from a classical MDS in two dimensions is shown in Figure 3.2. The MDS has mapped points in two-dimensional space such that the "straight line" (Euclidean) distances between the points d_{ij} match the observed distances δ_{ij}. The d_{ij}s are very close to the (rescaled) δ_{ij}s. They are not precisely equal because the δ_{ij}s are not "straight line" distances but distances across the surface of a sphere.

TABLE 3.1: Distances between ten cities in air miles

	London	Berlin	Oslo	Moscow	Paris	Rome	Beijing	Istanbul	Gibraltar	Reykjavik
London	–									
Berlin	570	–								
Oslo	710	520	–							
Moscow	1550	1000	1020	–						
Paris	210	540	830	1540	–					
Rome	890	730	1240	1470	680	–				
Beijing	5050	4570	4360	3600	5100	5050	–			
Istanbul	1550	1080	1520	1090	1040	850	4380	–		
Gibraltar	1090	1450	1790	2410	960	1030	6010	1870	–	
Reykjavik	1170	1480	1080	2060	1380	2040	4900	2560	2050	–

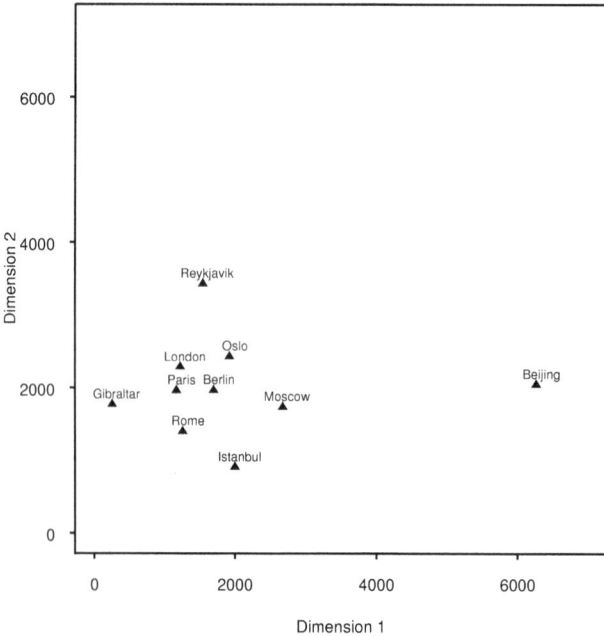

FIGURE 3.2: Two-dimensional plot of 10 cities from a classical MDS

Figure 3.2 is recognisable as a map of Europe and Asia. However, in general a configuration may need to be rotated and/or reflected in order to clarify the interpretation. Three important points about interpreting MDS solutions are:

   i) The configuration can be reflected without changing the inter-point distances.

   ii) The inter-point distances are not affected if we change the origin by adding or subtracting a constant from the row or column coordinates.

   iii) The set of points can be rotated without affecting the inter-point distances. This comes to the same thing as rotating the axes.

We must therefore be prepared to look for the most meaningful set of axes when interpreting an MDS solution. This idea will become clearer when we come to the next example. To summarise, the interpretation we put upon any MDS solution must be invariant under reflection, translation, and rotation.

An attempt to determine the dimensions underlying similarity judgements for pairs of 12 countries

In 1968, a group of 18 students was asked to rate the degree of similarity between each pair of 12 countries on a scale from 1 ("very different") to 9 ("very similar"). The study is described in Kruskal and Wish (1978), but our analysis is slightly different. The mean similarity ratings were calculated across students to obtain the similarity matrix in Table 3.2. The resulting solution in two dimensions is shown in Figure 3.3.

TABLE 3.2: Subjective similarities between pairs of 12 countries

	Brazil	Congo	Cuba	Egypt	France	India	Israel	Japan	China	Russia	USA	Yugo-slavia
Brazil	–											
Congo	4.83	–										
Cuba	5.28	4.56	–									
Egypt	3.44	5.00	5.17	–								
France	4.72	4.00	4.11	4.78	–							
India	4.50	4.83	4.00	5.83	3.44	–						
Israel	3.83	3.33	3.61	4.67	4.00	4.11	–					
Japan	3.50	3.39	2.94	3.83	4.22	4.50	4.83	–				
China	2.39	4.00	5.50	4.39	3.67	4.11	3.00	4.17	–			
Russia	3.06	3.39	5.44	4.39	5.06	4.50	4.17	4.61	5.72	–		
USA	5.39	2.39	3.17	3.33	5.94	4.28	5.94	6.06	2.56	5.00	–	
Yugo-slavia	3.17	3.50	5.11	4.28	4.72	4.00	4.44	4.28	5.06	6.67	3.56	–

We have to consider whether we can identify what is varying as we move along the two axes. Thus, for example, what do those countries on the right of the diagram have more of than those on the left, or those at the top than those at the bottom? Nothing very obvious seems to emerge from such comparisons, but we must remember that the orientation is arbitrary and maybe the message will be clearer if we consider other rotations. The dotted axes shown on Figure 3.3 correspond to a rotation that does seem to have an interpretation in terms of meaningful variables. Kruskal and Wish (1978) note that variation in the direction of the axis that runs from bottom left to top right corresponds to a tendency to be pro-Western or pro-Communist. Those at the top right are the more pro-Communist and those at the bottom left are the more pro-Western. Variation in the direction at right angles separates the developed (top left) from the developing (bottom right) countries. It thus appears that when making their judgements in 1968, the students were taking account, consciously or unconsciously, of two types of difference, and the analysis has helped us to identify what those two dimensions were.

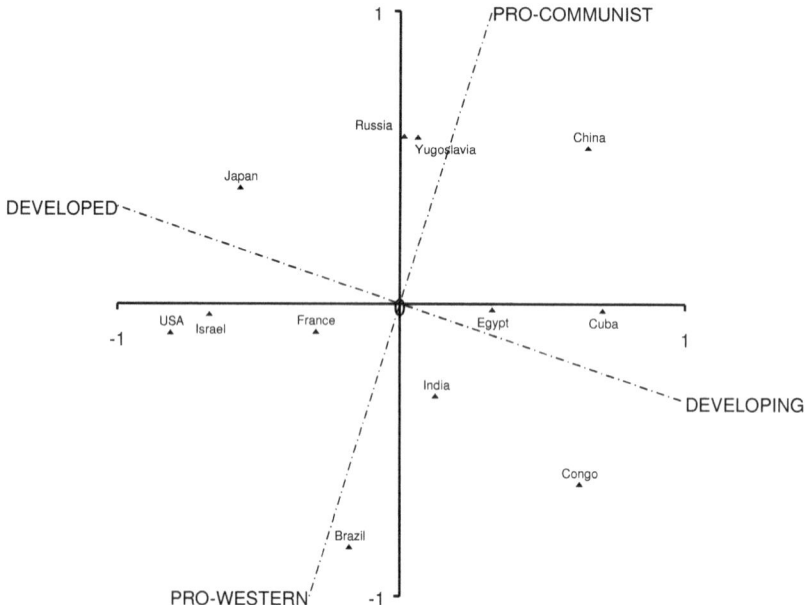

FIGURE 3.3: Two-dimensional plot of countries from ordinal MDS

It is worth adding two cautionary remarks about this example. The similarities were obtained by averaging the assessments of the 18 students. Implicitly, therefore, we are assuming that all are using the same two dimensions and that they are giving them the same relative weight. This may not be the case and it would be useful to have a method of discovering whether this was true. Such methods, known as Individual Scaling or Three-Way Scaling, are available but are outside the scope of this book (see, for example, Borg and Groenen 2005 or Kruskal and Wish 1978).

The second remark is that identifying interpretable axes for a plot is not always the best way to discern interesting patterns. It may be that we can identify clusters of points which have practical significance, as in the acoustic confusion example in Section 3.7, or, as in the colour data example in Section 3.6, the clue may be in the "horseshoe" shape of the two-dimensional plot.

3.3 Classical, ordinal, and metrical multidimensional scaling

We now pose the problem of multidimensional scaling in more formal terms so that we can outline the algorithms used to arrive at a solution.

Classical scaling

In *classical* MDS, the aim is to find a configuration in a low number of dimensions such that the distances between the points in the configuration, d_{ij}, are close in value to the observed distances δ_{ij}. The method treats the distances as Euclidean distances. We saw in Chapter 2 how to go from a data matrix to a Euclidean distance matrix; here we have to go in the reverse direction and recover the data matrix from the distances. We cannot recover everything because information about location and orientation is lost in the process of calculating distances, but we can determine the configuration. This problem can be tackled algebraically, and it turns out that the solution gives us a series of approximations starting with one dimension, then two, and so on. It also happens, however, that the mathematics involved is equivalent to that for another problem for which the solution is already known. This establishes an interesting link with principal components analysis that we shall discuss in Chapter 5. We shall return to this link in that chapter, but we can prepare the ground by expressing the classical MDS problem in a slightly different way. If we start with an $n \times p$ data matrix, we first construct a distance table and then might seek to find a two- or three-dimensional map on which the inter-point distances are as close as possible to the original distances. Another way of putting this is to say that we are looking for a new data matrix, with two or three columns, which is close to the original matrix in the sense that it gives rise to (nearly) the same distance matrix.

Having found a solution, we may wish to have a measure of how good the fit is. This would be particularly useful for helping us to judge how many dimensions are necessary to get a good enough fit. An obvious way to do this is to look at the sum of squares $\sum_{i<j}(d_{ij} - \delta_{ij})^2$. (This is mathematically appropriate since the fits obtained are best in a least squares sense.) However, the simple sum of squares depends on the scale in which the distances are measured. It is, therefore, preferable to normalise the sum of squares and, in order to reduce it to the same units as the distances, to take the square root. Our goodness-of-fit measure is then

$$\sqrt{\frac{\sum_{i<j}(d_{ij} - \delta_{ij})^2}{\sum_{i<j} d_{ij}^2}}. \tag{3.1}$$

This measure is called the *stress* or, sometimes, the normalised stress. There are other ways of calculating a normalised stress measure. For example, an alternative measure of stress may be obtained by replacing d_{ij} with δ_{ij} in the denominator of eq. (3.1). Values of stress that are close to zero would indicate that the MDS solution is a good fit to the original δ_{ij}s.

Ordinal (non-metrical) scaling

Very often it is not the actual value of δ_{ij} that is important or meaningful, but its value in relation to the distances between other pairs of objects. This is particularly true when the δ_{ij}s are the result of an experiment where subjects are asked to give their subjective assessments of the distance between objects. In such cases, the δ_{ij}s can be interpreted only in an ordinal sense. In *ordinal* MDS, the aim is to find a configuration such that the d_{ij}s are in the same rank order as the original δ_{ij}s. So, for example, if the distance between objects 1 and 3 ranks fifth among the δ_{ij}s then they should also rank fifth in the MDS configuration. The emphasis in this chapter, as noted in Section 3.1, is on ordinal MDS.

In ordinal MDS, we construct fitted distances, often called *disparities*, $\hat{d}_{ij}$, from the d_{ij}s such that the $\hat{d}_{ij}$s are in the same rank order as the δ_{ij}s (for dissimilarities) or reverse rank order (for similarities). We can think of the $\hat{d}_{ij}$s as "smoothed" versions of the d_{ij}s. This smoothing process is carried out using a method called least-squares monotonic regression ("monotonic" means that the regression curve is either non-decreasing or non-increasing). Using this method, the d_{ij}s are regressed on the δ_{ij}s. In a plot of d_{ij} versus δ_{ij}, we would like to see a monotonic curve (one where the lines joining adjacent points are flat/increasing if δ_{ij} are dissimilarities or flat/decreasing if δ_{ij} are similarities). If the d_{ij}s and the δ_{ij}s have the same rank order, then the plot will show such a monotonic curve and the d_{ij}s will not require any smoothing. Usually, however, there will be some departures from monotonicity and some smoothing will be necessary. The aim of monotonic regression is to fit a monotonic curve to the points (d_{ij}, δ_{ij}), while making the sum of squared vertical deviations as small as possible (as in least-squares linear regression). The point on the monotonic curve, $\hat{d}_{ij}$, is the fitted or predicted value of d_{ij} from the monotonic regression. In judging how good the fit is, we are now interested in how close the distances, d_{ij}, are to the disparities, $\hat{d}_{ij}$, rather than the observed distances, δ_{ij}. This is because we are only aiming to reproduce the rank order of the observed distances and not the distances themselves. Hence, our measure of fit is obtained by cleverly replacing δ_{ij} by $\hat{d}_{ij}$ in the formula for the stress ($\hat{d}_{ij}$ and δ_{ij} having the same rank order). Thus in ordinal MDS, the stress is calculated as

$$\sqrt{\frac{\sum_{i<j}(d_{ij} - \hat{d}_{ij})^2}{\sum_{i<j} d_{ij}^2}}.$$

This is also known as Kruskal's stress, type I (which we shall refer to simply as stress). The optimum configuration is determined by minimising this measure of stress or some variant of it.

The points (δ_{ij}, d_{ij}) are shown by a cross in Figure 3.4. Note that while the first and second points (counting from left to right) follow a monotonic pattern, the third does not. To achieve monotonicity, the values of d_{ij} for the second and third points are replaced by their mean. Similarly, the values of d_{ij} for the fourth and fifth points are replaced by their mean. This leads to the monotonic regression curve consisting of the series of solid lines shown in the plot. The vertical dotted lines represent the distances $d_{ij} - \hat{d}_{ij}$.

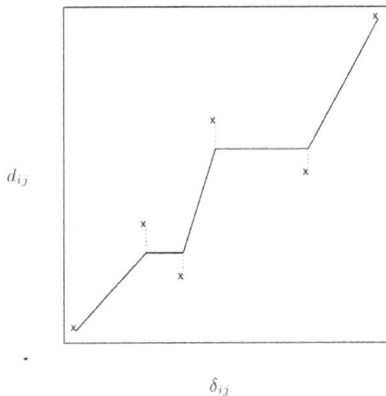

FIGURE 3.4: Example of monotonic regression

Metrical scaling

Classical scaling could be described as metrical scaling since, in contrast to non-metrical scaling, the fitted and original distances are expressed in the same metric. However, the term metrical scaling usually seems to be reserved for something which may most naturally be thought of as related to non-metrical (ordinal) scaling in another way. In classical scaling, we supposed that the distances were Euclidean distances. In ordinal scaling, we made use only of the rank order of the δ_{ij}s. This was tantamount to assuming that we had to make a monotonic transformation of the δ_{ij}s to turn them into Euclidean distances. In metrical scaling, we assume that they can be transformed into Euclidean distances by some other parametric transformation. In some fields, there may be good reasons for supposing that such transformations exist, but we are not aware of any convincing arguments for introducing them in social science applications. However, we mention two special cases because they are closely linked to classical scaling. *Interval scaling* refers to the case where it is supposed that a linear transformation will turn the δ_{ij}s into Euclidean distances. Instead of fitting a monotonic regression to the distances to obtain the disparities, we would now fit a linear regression. The disparities would then become the points on the regression line instead of points on the monotonic regression curve. The formula for stress remains the same except that the $\hat{d}_{ij}$s would now be obtained from the least squares regression line. In the special case of *ratio scaling*, when the regression goes through the origin, we are back to the situation we faced in classical scaling because multiplying the δ_{ij}s by a constant does not change the metric — if they were Euclidean before they will be Euclidean afterwards and vice versa. The difference here lies in the function which is being minimised. The Kruskal's stress formula applied in this case aims to achieve the closest degree of proportionality between the given distances and those fitted. Classical scaling aims to achieve the closest fit in a least squares sense. The two methods will often give very similar results and we shall use ratio scaling in one of the examples below.

3.4   Comments on computational procedures

Given the number of dimensions, K, the aim of MDS is to find a configuration in K dimensions such that the stress criterion used is minimised.

Most ordinal MDS computer packages start with an initial configuration in K dimensions, and then iteratively improve the configuration by moving the points short distances in such a manner so as to reduce the stress slightly on each iteration. When further changes to the configuration do not reduce the stress (or not by more than some pre-specified tolerance level), the procedure ends and that configuration is the MDS solution. Typically, the method of steepest descent is used. Kruskal and Wish (1978) give the analogy of a blind-folded parachutist trying to find the lowest point in a terrain by following the gradient downhill.

Unfortunately, it is possible that a *local* minimum rather than the *global* minimum will be found. Repeating the process with different starting config-urations to see whether the same minimum is found is one way of checking for this, but there is no absolute guarantee that there may not be some even smaller minimum lurking in a region of the space which has not been explored.

The MDS solution achieved depends on

i) the choice of initial configuration, and

ii) the stress criterion used.

For example, the program PROXSCAL (available in SPSS), with which many of the calculations in this chapter were done, arrives at a solution which min-imises a stress function with d_{ij} replaced with δ_{ij} in the denominator of the formula for Kruskal's stress type I. There are other variants of stress which measure the differences between the distances and the disparities in slightly different ways.

Full discussion of such computational issues is outside the scope of this book, but the reader should be aware that different packages may give slightly different solutions. If the solutions are very different, this suggests that either there is no strong structure in the data, or that at least one of the solutions is a local rather than a global optimum, or that complete convergence has not been achieved for one or both solutions.

3.5 Assessing fit and choosing the number of dimensions

There are a number of ways of assessing the fit of a MDS solution. One method involves comparing the stress obtained for the solution with the guidelines shown in Table 3.3. These were developed by Kruskal (1964) and are based on empirical experience rather than theoretical criteria. These should always be used flexibly with an eye on the interpretability of the solution to which they lead.

TABLE 3.3: Guidelines for assessing fit using stress

Stress (Kruskal's type I)	Assessment of fit
0.20	poor
0.05	good
0.00	perfect

Another method that may be used to choose the number of dimensions is to examine a scree plot in which the stress is plotted against the number of dimensions. A similar plot was used in the chapter on cluster analysis (Section 2.7) to choose the number of clusters in the K-means clustering method, where the number of clusters is plotted against the total within-cluster dissimilarity. As the number of dimensions increases the stress decreases, but there is a trade-off between improving fit and reducing the interpretability of the solution. In the scree plot, we look for an "elbow" which is the point at which increasing the number of dimensions has little further effect on the stress. Again there is a strong subjective element in using this method, but experience shows that it often works well. See, for example, Figure 3.5.

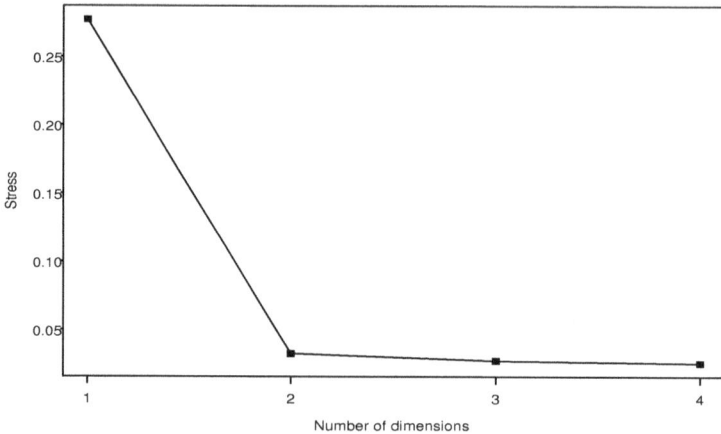

FIGURE 3.5: Scree plot of stress by number of dimensions, colour data

There are also a number of useful diagnostic plots. In the case of ordinal scaling, the plots involve all pairs of δ_{ij}, d_{ij} and $\hat{d}_{ij}$, that may be examined to evaluate the fit of a MDS solution.

i) Plot of d_{ij} (the inter-point distance in the configuration) versus $\hat{d}_{ij}$ (the disparity or fitted value of d_{ij} obtained from the monotonic regression on δ_{ij}). If the MDS solution is a good fit, this plot should show a linear relationship with a 45 degree slope and only a small amount of scatter about the line. If little smoothing of the d_{ij}s was necessary to produce the $\hat{d}_{ij}$s, then they should be in almost the same rank order and close in value since they are measured on the same scale. See, for example, Figure 3.7.

ii) Plot of d_{ij} (the inter-point distance in the configuration) versus δ_{ij} (the observed distance or dissimilarity or similarity). If the solution is a good fit, d_{ij} and δ_{ij} should have approximately the same (or the reverse) rank order and this plot should show a monotonic curve, either increasing (for dissimilarities) or decreasing (for similarities). See, for example, Figure 3.8.

iii) Plot of $\hat{d}_{ij}$ (the disparity or fitted value of the inter-point distance, d_{ij}) versus δ_{ij} (the observed distance or dissimilarity or similarity). The $\hat{d}_{ij}$s are the "smoothed" versions of the d_{ij}s constructed to have the same rank order as δ_{ij} (for dissimilarities) or reverse rank order (for similarities). If a large amount of smoothing were required to achieve a monotonic curve (i.e. if the solution were a poor fit), this plot would show a number of large horizontal steps where the smoothing took place. When the fit is good there will only be small steps. See, for example, Figure 3.9.

For metrical scaling, the $\hat{d}_{ij}$s are made to be proportional to the δ_{ij}s. Therefore, the plots involving δ_{ij} are redundant, leaving only the plot of d_{ij} versus $\hat{d}_{ij}$ to be examined.

3.6 A worked example: dimensions of colour vision

We now illustrate these ideas and methods on an example which was originally analysed by other means before the development of multidimensional scaling methods.

An experiment was conducted where subjects were asked to look at a screen which had two circular opaque glass windows. These windows were lit from two projectors behind the screen. Different colour filters could be inserted in the projectors. Fourteen colour filters were used, transmitting light of wavelengths $434m\mu$ to $674m\mu$. Each stimulus was combined with each other stimulus in a random order. The subjects were then asked to rate the degree of "qualitative similarity" between each pair of colour filters on a five-point scale. Further details, and the original analysis, will be found in Ekman (1954). The similarity matrix constructed by Ekman is given in Table 3.4. An ordinal MDS of these similarities was carried out.

This is a case where we might guess in advance that a one-dimensional solution would be possible because the difference in wavelength between two colours is a continuous metric measuring how far apart the colours are. However, the scree plot given in Figure 3.5 shows that there is a big reduction in stress in passing from one to two dimensions, so there must be other factors which come into play when making subjective assessments of colour. The "elbow" at two dimensions indicates that there is little reduction in stress after two dimensions. Therefore, we select a two-dimensional solution. This solution has stress of 0.03 (3%) which according to Kruskal's guidelines is a good fit.

In the two-dimensional configuration (Figure 3.6), the points appear on a curve to give a "horseshoe" effect — a common phenomenon. At one extreme, are the violets (colours 1 and 2) and at the other are the reds

TABLE 3.4: Similarities between colours based on subjective judgements

Colour	1	2	3	4	5	6	7	8	9	10	11	12	13	14
1	–													
2	0.86	–												
3	0.42	0.50	–											
4	0.42	0.44	0.81	–										
5	0.18	0.22	0.47	0.54	–									
6	0.06	0.09	0.17	0.25	0.61	–								
7	0.07	0.07	0.10	0.10	0.31	0.62	–							
8	0.04	0.07	0.08	0.09	0.26	0.45	0.73	–						
9	0.02	0.02	0.02	0.02	0.07	0.14	0.22	0.33	–					
10	0.07	0.04	0.01	0.01	0.02	0.08	0.14	0.19	0.58	–				
11	0.09	0.07	0.02	0.01	0.02	0.02	0.05	0.04	0.37	0.74	–			
12	0.12	0.11	0.01	0.01	0.01	0.02	0.02	0.03	0.27	0.50	0.76	–		
13	0.13	0.13	0.05	0.02	0.02	0.02	0.02	0.02	0.20	0.41	0.62	0.85	–	
14	0.16	0.14	0.03	0.04	0.01	0.01	0.01	0.02	0.23	0.28	0.55	0.68	0.76	–

(colours 11-14). As we go round the horseshoe, we encounter the colours in strict order of wavelength. However, it appears that subjects were making more subtle judgements in that reds are seen as closer to violets than to greens (colours 6-8), even though reds and greens are closer in terms of their wavelengths. Reference back to Table 3.4 confirms that this is not an accidental artefact of the MDS solution. There is clearly some other aspect of the perception of colour influencing the subject's comparisons than is conveyed by wavelength alone.

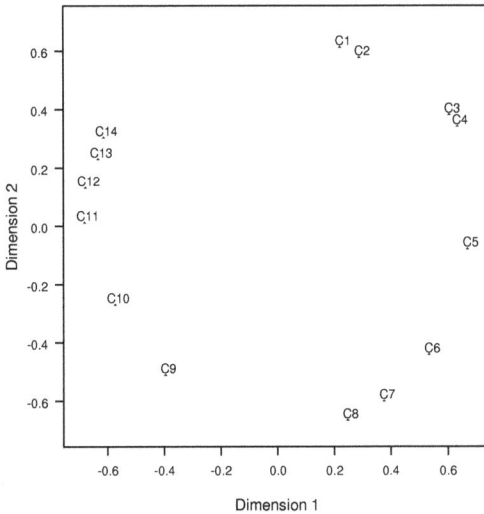

FIGURE 3.6: Two-dimensional configuration plot from an ordinal MDS of colour data

The three diagnostic plots are typical of what one finds with a reasonably good fit. On Figure 3.7, the points lie close to the 45 degree line; the curve in Figure 3.8 shows marked monotonicity, and Figure 3.9 has horizontal steps of short length reflecting the near monotonicity shown by the previous figure.

Before leaving this example, it is interesting to return to the one-dimensional solution plotted in Figure 3.10. The colours do not appear in order of their wavelength though there is a clear separation between the "blue" end of the spectrum (colours with low numbers) and the "red" end (colours with high numbers). Within those two groups, however, there seems to be some inversion of the order one would have expected. The fit, of course, was not good in this case. The stress was 0.28, which on Kruskal's criterion, indicates a poor fit.

The clear conclusion of our analysis is that colour perception involves more than is conveyed by the wavelength of the light. To return to the title of Ekman's paper, there appear to be two dimensions of colour vision.

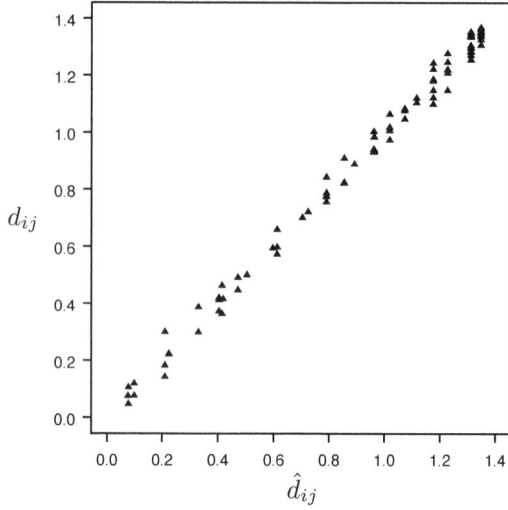

FIGURE 3.7: Plot of d_{ij} (inter-point distance in the configuration) versus $\hat{d}_{ij}$ (fitted value of d_{ij}) from a two-dimensional ordinal MDS of colour data

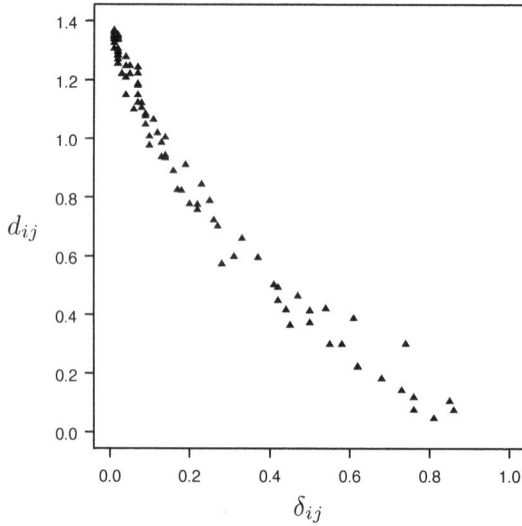

FIGURE 3.8: Plot of d_{ij} (inter-point distance in the configuration) versus δ_{ij} (observed similarity) from a two-dimensional ordinal MDS of colour data

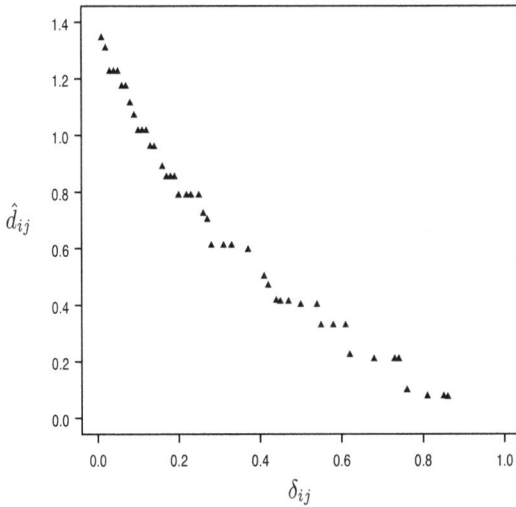

FIGURE 3.9: Plot of $\hat{d}_{ij}$ (fitted value of d_{ij}) versus δ_{ij} (observed similarity) from a two-dimensional ordinal MDS of colour data

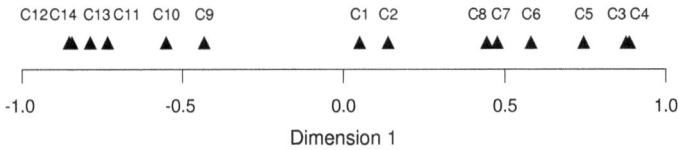

FIGURE 3.10: MDS solution for colour data in one dimension

3.7 Additional examples and further work

In this section, we give five further examples to illustrate the methods. We shall not carry out an exhaustive analysis on any of them, but focus on particularly interesting features which the individual examples show. You are invited to use these examples to explore the other options available in the various software packages. Two of the examples have already been considered in the chapter on cluster analysis, and our main interest in these cases will be to compare the two methods when applied to the same data.

Economic and demographic indicators for 25 countries

Table 3.5 shows the values of five economic and demographic indicators for a sample of 25 countries. The data refer to 1990 and they come from the United Nations Statistical Yearbook of 1997. The indicators are annual percentage population growth rate (Increase), life expectancy in years (Life), infant mortality rate per 1000 (IMR), total fertility rate (TFR), and Gross Domestic Product per capita in US dollars (GDP).

TABLE 3.5: Economic and demographic indicators for 25 countries, 1990, UN Statistical Yearbook of 1997

Country	Increase	Life	IMR	TFR	GDP
Albania	1.2	69.2	30	2.9	659.91
Argentina	1.2	68.6	24	2.8	4343.04
Australia	1.1	74.7	7	1.9	17529.98
Austria	1.0	73.0	7	1.5	20561.88
Benin	3.2	45.9	86	7.1	398.21
Bolivia	2.4	57.7	75	4.8	812.19
Brazil	1.5	64.0	58	2.9	3219.22
Cambodia	2.8	50.1	116	5.3	97.39
China	1.1	66.7	44	2.0	341.31
Colombia	1.7	66.4	37	2.7	1246.87
Croatia	−1.5	67.1	9	1.7	5400.66
El Salvador	2.2	63.9	46	4.0	988.58
France	0.4	73.0	7	1.7	21076.77
Greece	0.6	75.0	10	1.4	6501.23
Guatemala	2.9	62.4	48	5.4	831.81
Iran	2.3	67.0	36	5.0	9129.34
Italy	−0.2	74.2	8	1.3	19204.92
Malawi	3.3	45.0	143	7.2	229.01
Netherlands	0.7	74.4	7	1.6	18961.90
Pakistan	3.1	60.6	91	6.2	385.59
Papua New Guinea	1.9	55.2	68	5.1	839.03
Peru	1.7	64.1	64	3.4	1674.15
Romania	−0.5	66.6	23	1.5	1647.97
USA	1.1	72.5	9	2.1	21965.08
Zimbabwe	4.4	52.4	67	5.0	686.75

Ratio MDS was applied to these data. Since the data are in the form of an data matrix, the first stage of a MDS is to convert the data to a distance matrix showing the pairwise distances between countries. Since the variables differ greatly in terms of their variances, the variables are first standardised to have a variance of 1. Euclidean distances are then computed. Since we apply ratio MDS, the fitted distances will be proportional to the actual distances. You should try ordinal scaling and compare the results.

One aim of an MDS of these data might be to determine whether countries can be placed on a scale of development based on these five indicators. Therefore, the one-dimensional solution is of particular interest. Developed countries are generally characterised by low growth rate, high life expectancy, low infant mortality, low fertility and high GDP. If countries can be located on a single dimension of development, developed countries should be placed at one extreme with less developed countries (characterised by high growth rate, low life expectancy, high infant mortality, high fertility and low GDP) placed at the other extreme.

The stress (Kruskal type I) value for the one-dimensional solution was 0.17, suggesting a poor fit. The locations of the countries on a single dimension are given in Table 3.6. We find that the countries lie approximately where we would expect. At one extreme, we have the less developed mainly African and Asian countries, while at the other we have European countries, the USA and Australia. Since the one-dimensional fit is poor, however, you should go on to examine a two-dimensional solution to see whether a second dimension improves the fit and adds any new insight into the structure of the data.

TABLE 3.6: Coordinate for each country from a one-dimensional ratio MDS of Economic and demographic indicators (arranged in increasing order)

Country	Coordinate
Malawi	−2.027
Benin	−1.616
Cambodia	−1.414
Zimbabwe	−1.302
Pakistan	−1.133
Bolivia	−0.798
Papua New Guinea	−0.783
Guatemala	−0.706
El Salvador	−0.344
Peru	−0.277
Iran	−0.167
Brazil	−0.112
Colombia	0.036
China	0.188
Albania	0.220
Argentina	0.327
Romania	0.786
Greece	0.921
Australia	1.049
USA	1.105
Netherlands	1.158
Austria	1.164
Croatia	1.167
France	1.230
Italy	1.328

The stress value for the two-dimensional solution is 0.05, indicating a much better fit than the one-dimensional solution. Figure 3.11 shows the plot of d_{ij} versus $\hat{d}_{ij}$. The strong linear relationship between the distances in the configuration and the smoothed distances is a further indication that the data are well represented in two dimensions. As noted in Section 3.5, with ratio MDS, the other two diagnostic plots, involving δ_{ij}, are redundant since δ_{ij} and $\hat{d}_{ij}$ have been made to be proportional.

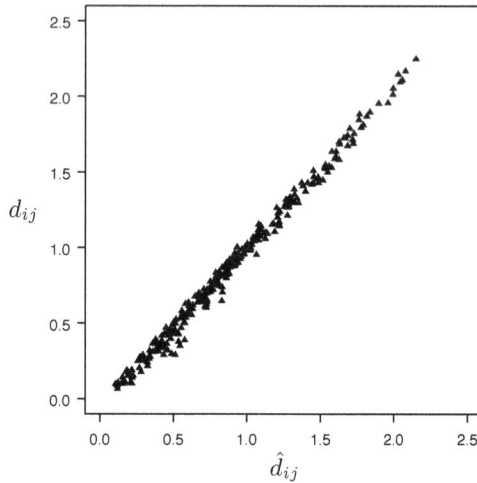

FIGURE 3.11: Plot of d_{ij} versus $\hat{d}_{ij}$ from a two-dimensional ratio MDS of economic and demographic indicators

The two-dimensional configuration is shown in Figure 3.12. The location of countries on dimension 1 is almost the same as in the one-dimensional solution. Dimension 1 could be interpreted as a measure of overall development. On the second dimension, Romania and Croatia stand out from the other countries. If you look at the profiles of these countries in Table 3.5, you can see that they both have characteristics associated with developed countries (low growth rate, moderately high life expectancy, fairly low infant mortality and very low fertility), which places them on the left-hand side of the first dimension together with other developed countries. However, they have very low GDPs compared to other developed countries. Those countries located at the other end of the second dimension generally have high GDPs. Thus, the second dimension is largely a function of GDP.

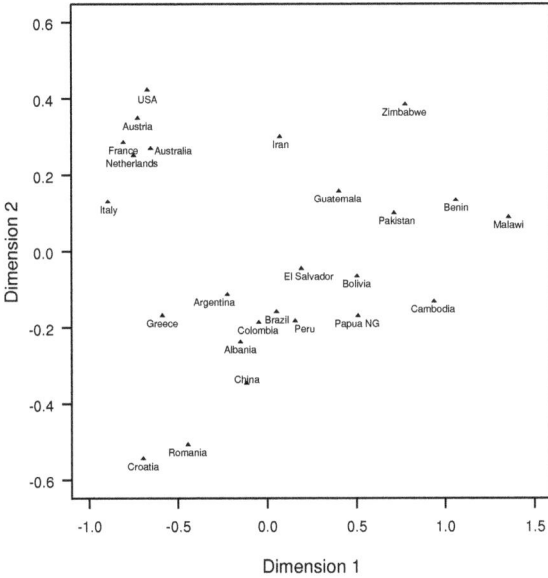

FIGURE 3.12: Configuration of countries from a two-dimensional ratio MDS of economic and demographic indicators

Persian archers

In Chapter 2, similarities between pairs of 24 archers (Table 2.13) were analysed using cluster analysis. The data are described in Section 2.9. The similarities may also be analysed using ordinal MDS.

From the scree plot in Figure 3.13, there is a suggestion of an elbow at two dimensions, indicating that a two-dimensional solution may be adequate, but three or four dimensions might improve the representation of the dissimilarities between the archers. The configuration for the two-dimensional solution is plotted in Figure 3.14.

Archaeologists want to know how the bas-reliefs were carved. Were they the work of a single sculptor, several independent sculptors, or of one or more teams of sculptors? Figure 3.14 shows five archers (20 to 24) clustered together to the left of centre near the bottom; eight archers (1-8) are spread out upwards and slightly diagonally on the left; the remaining archers (9 to 19) are spread out on the right. Roaf (1983), pp. 14-16, as we noted in Chapter 2, concluded that there could have been three teams of sculptors. One working on the top section of the staircase (1 to 8), another on the centre section (9 to 19) and a third on the bottom section (20 to 24), these last five being so similar that they could be the work of a single sculptor. Within this broad clustering into groups, adjacent archers on the staircase tend to be close to each other in the configuration.

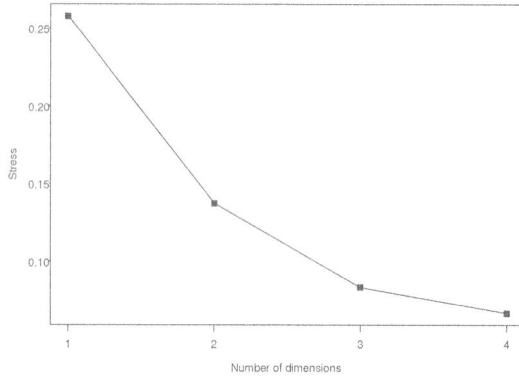

FIGURE 3.13: A scree plot for an ordinal MDS of data on 24 Persian Archers

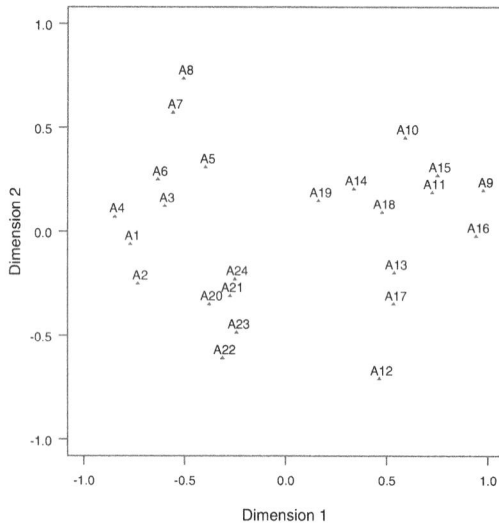

FIGURE 3.14: 24 Persian archers plotted in the two-dimensional space found through ordinal MDS

You may suggest explanations of why archers 1 to 8 are strung out in a line in Figure 3.14, why archer 2 appears relatively close to archers 20 to 24, and why archer 12 is distant from the others. Then turn to Figure 3.15

where lines have been added joining points (archers) with a similarity of 15 or more. Such additions to the plot of the MDS solution can clarify whether the relative positions of individual points in the configuration reflect their true similarities. Points close together on the map but with low similarities will be major contributors to the stress.

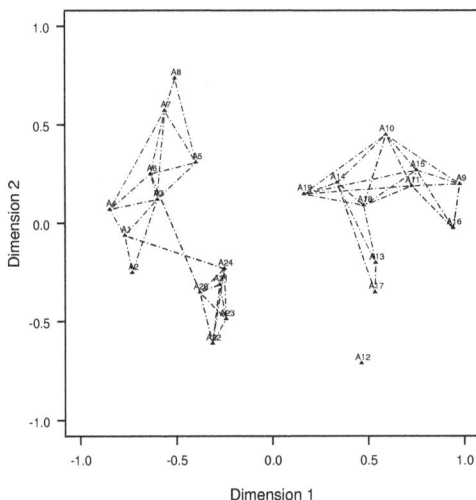

FIGURE 3.15: 24 Persian archers plotted in the two-dimensional space found through ordinal MDS, with lines drawn between pairs of archers with similarity ≥ 15

Dialect words in 25 English villages

The dataset on which this example is based was given in Section 2.5 where it formed the basis for demonstrating the various techniques of cluster analysis. We showed there that there was a fairly clear cluster structure in which the villages in each cluster were close together geographically, as one would have expected. Given that the villages can be represented on a map in two dimensions, it is natural to ask whether one would obtain a similar map if linguistic similarity were used as a measure of distance. We would then be able to see whether the pattern of villages on the linguistic map was similar to their geographical situations. If this turned out to be the case, we would infer that the result of easier interchange between villages close together led to them having more words in common. But major topographical features, like rivers, roads and railways, might make for greater similarity along the main lines of communication. There are no mountain barriers in that part of England, but a river like the Trent might well prove a barrier to easy communication.

Bearing these points in mind, you should carry out an ordinal MDS on the similarities in two dimensions. The stress is 0.14 which is not a particularly good fit, but given the particular interest of the two-dimensional plot in this case, it is given in Figure 3.16. This should be compared with the map in Figure 2.6. The orientation is not the conventional one with north at the top of the diagram. The most northerly village on the map is V4 which occurs on the extreme right of the figure. The orientation will therefore be approximately correct if we rotate the figure anti-clockwise through 90 degrees. In that case, the Huntingdon village (V22) will be on the right, as for the conventional view. Rotating the figure has thus produced something fairly close to the map given in Chapter 2. This is shown in Figure 3.17.

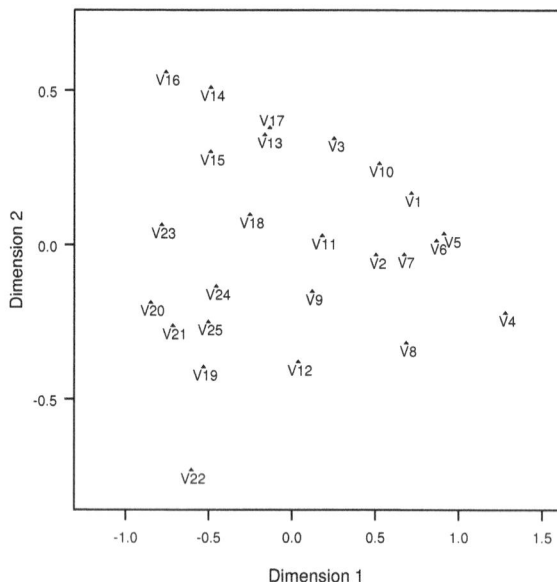

FIGURE 3.16: Two-dimensional representation of 25 English villages

A careful comparison of the "map" provided by your analysis with the true map will show a fairly good, but by no means exact, correspondence. This suggests that geographical factors play a major role in explaining the distribution of dialect words. It must be remembered, of course, that the measure of linguistic similarity we have used is based on a fairly small sample (60) of words.

In view of the relatively poor fit of the two-dimensional map, it is worth looking at the diagnostic plot of d_{ij} versus $\hat{d}_{ij}$, given in Figure 3.18. Although the fit is not as good as in some of the other examples, there is a broad correspondence between the d_{ij} and the $\hat{d}_{ij}$.

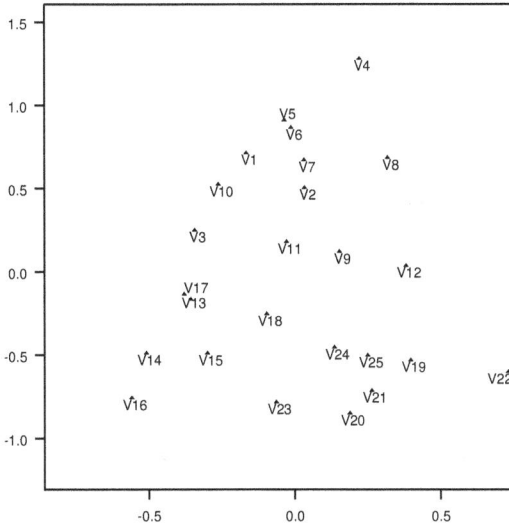

FIGURE 3.17: The points on Figure 3.16 in a more conventional orientation

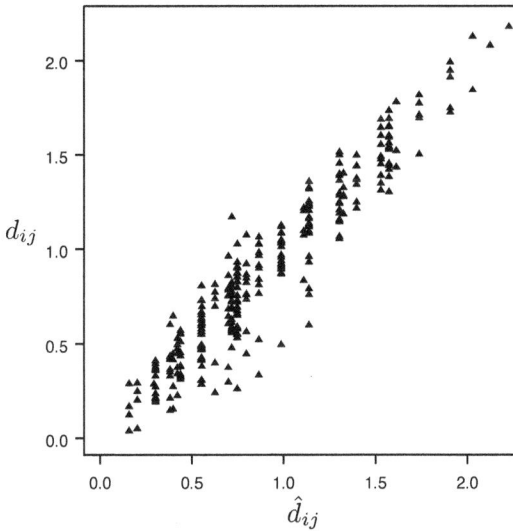

FIGURE 3.18: Plot of d_{ij} versus $\hat{d}_{ij}$ for the dialect data

Acoustic confusion of letters of the alphabet

In psychological experiments on memory, subjects may be asked to listen to and remember letters of the alphabet in some sequence. There is a risk that they may fail to give the right letters, not because of a failure of memory, but because they did not hear them clearly. Conrad (1964) reports the results of an experiment to investigate acoustic confusion in identifying letters of the alphabet. Three hundred post office employees wrote down the letters they thought they heard when letters were spoken against a background noise at a rate of one every five seconds.

Morgan (1973) calculated the similarities given in Table 3.7 by averaging the number of times the first letter was confused with the second, and the number of times the second was confused with the first, each letter being presented a total of 1440 times. The object of using MDS is to discover what led to letters being confused with each other. Figure 3.19 shows the minimum values of Kruskal's stress type I for one- through six-dimensional solutions obtained from an ordinal MDS. There is unfortunately no clear elbow, and it is not until you come to the four-dimensional solution that the stress falls below 0.1. A two-dimensional solution will not be adequate, but that does not mean that it will be of no use.

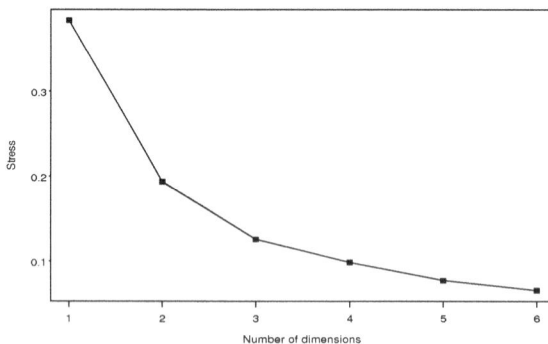

FIGURE 3.19: A scree plot of stress against the number of dimensions for the acoustic data

The configuration of letters for the two-dimensional solution is shown in Figure 3.20. The tightest cluster consists of c, t, p, b, e, d, v, g, and q, u. Referring back to Table 3.7, you can see that these letters including q and u all do have relatively high similarities with each other. You can easily see that most members of this group share the "ee" sound, and it is presumably that fact which leads to them frequently being confused. It is not so obvious why q and u also come in this group, although q and u share the "oo" sound.

TABLE 3.7: Similarities between letters (average of number of times each was confused with the other), acoustic data

	letter	w	g	c	q	p	t	b	d	e	u	v	h	f
1	w	–												
2	g	6	–											
3	c	6	41	–										
4	q	8	142	73	–									
5	p	8	185	385	446	–								
6	t	10	128	274	265	786	–							
7	b	35	182	203	137	237	227	–						
8	d	27	151	90	106	235	201	322	–					
9	e	18	242	129	118	283	287	379	418	–				
10	u	30	222	78	153	95	40	139	101	190	–			
11	v	21	172	81	61	125	72	290	252	174	426	–		
12	h	18	5	22	32	27	32	18	17	53	28	18	–	
13	f	13	3	8	0	13	18	8	5	15	6	4	81	–
14	s	7	7	20	10	7	15	4	9	23	8	4	194	824
15	x	3	3	11	3	7	7	3	5	25	15	1	191	483
16	l	38	6	2	7	6	2	9	6	11	3	3	16	41
17	j	13	20	16	19	26	14	35	24	10	31	25	23	13
18	k	21	5	11	20	45	19	13	12	16	25	23	43	37
19	m	25	25	18	10	33	15	21	16	72	28	12	18	35
20	n	39	34	26	12	29	20	23	27	112	35	31	55	40
21	a	83	39	11	11	16	20	27	28	26	38	26	104	19
22	o	77	27	5	9	13	40	14	10	27	25	20	50	28
23	i	22	13	8	13	13	15	15	14	114	55	9	4	7
24	r	9	10	3	10	1	10	3	5	19	5	6	8	18
25	y	16	12	12	5	4	9	8	7	12	8	11	4	12
26	z	97	5	8	14	34	10	26	14	10	12	21	53	121

	letter	s	x	l	j	k	m	n	a	o	i	r	y	z
1	w	7	3	38	13	21	25	39	83	77	22	9	16	97
2	g	7	3	6	20	5	25	34	39	27	13	10	12	5
3	c	20	11	2	16	11	18	26	11	5	8	3	12	8
4	q	10	3	7	19	20	10	12	11	9	13	10	5	14
5	p	7	7	6	26	45	33	29	16	13	13	1	4	34
6	t	15	7	2	14	19	15	20	20	40	15	10	9	10
7	b	4	3	9	35	13	21	23	27	14	15	3	8	26
8	d	9	5	6	24	12	16	27	28	10	14	5	7	14
9	e	23	25	11	10	16	72	112	26	27	114	19	12	10
10	u	8	15	3	31	25	28	35	38	25	55	5	8	12
11	v	4	1	3	25	23	12	31	26	20	9	6	11	21
12	h	194	191	16	23	43	18	55	104	50	4	8	4	53
13	f	824	483	41	13	37	35	40	19	28	7	18	12	121
14	s	–												
15	x	575	–											
16	l	60	13	–										
17	j	40	8	74	–									
18	k	41	15	68	222	–								
19	m	44	11	115	46	82	–							
20	n	49	15	76	106	144	846	–						
21	a	42	9	203	161	246	151	360	–					
22	o	44	7	101	87	101	339	89	594	–				
23	i	24	5	86	14	13	83	77	20	54	–			
24	r	20	11	193	18	27	65	52	36	56	292	–		
25	y	15	6	123	118	31	69	58	28	22	164	194	–	
26	z	120	78	47	150	80	48	58	26	53	7	30	41	–

This example shows that even when there is a rather poor fit, some meaning can still be extracted from the analysis. You may investigate solutions in three or more dimensions to see whether further meaningful groupings occur.

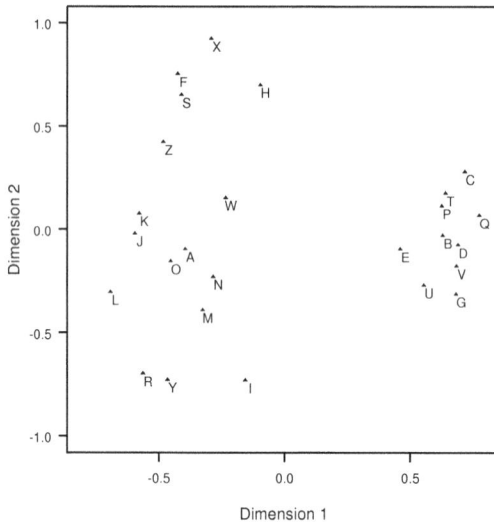

FIGURE 3.20: Two-dimensional configuration of acoustic data from an ordinal MDS

Perturbed distances between pairs of cities

This example is based on the example in Section 3.2 on the distances between pairs of cities in Europe and Asia. Instead of the true distance matrix in Table 3.1, suppose that we observe the perturbed distance matrix in Table 3.8, which is obtained by adding independent random errors to the true one. Similar perturbations may occur in real applications due to measurement error.

We apply metric MDS to the perturbed distance matrix with varying numbers of projection dimensions and then compare the distance matrices implied by the estimated configurations with the true distance matrix in Table 3.1. Here, the true configuration is almost two-dimensional (with a slight deviation due to the curvature of the Earth). In Table 3.9, we show the average squared differences between the estimated and true pairwise distances for projection dimensions one through five, where the average is taken over the 45 pairs of cities. The average squared difference between the perturbed and true distances is 28765.

TABLE 3.8: Perturbed distances between ten cities in air miles

	London	Berlin	Oslo	Moscow	Paris	Rome	Beijing	Istanbul	Gibraltar	Reykjavik
London	–									
Berlin	445	–								
Oslo	747	459	–							
Moscow	1383	1302	1209	–						
Paris	529	618	994	1664	–					
Rome	956	606	1359	1459	952	–				
Beijing	4886	4127	4544	3569	5079	4967	–			
Istanbul	1647	1305	1676	796	1118	771	4533	–		
Gibraltar	1238	1441	1805	2314	949	1018	5977	2009	–	
Reykjavik	1285	1477	682	2144	1105	2260	4849	2671	1912	–

TABLE 3.9: Average squared differences between the estimated and true pairwise distances

Dimension	1	2	3	4	5
Average squared difference	315814	16065	18489	18555	18555

We observe that the one-dimensional solution recovers the true distance matrix poorly. This is not surprising as it introduces a large bias into the estimated configuration by missing the information in the additional dimension of the true configuration. The two- to five-dimensional solutions all better recover the true distance matrix than the input distance matrix. This is because these solutions reduce the noise in the input distance matrix while keeping the bias low. Among these solutions, the two-dimensional solution achieves the largest noise reduction without substantially increasing the bias and thus has the smallest average squared difference.

We further note that the four- and five-dimensional solutions are very close to each other, as the fifth dimension in the latter almost degenerates. This is due to the fact that their distance matrices are already very close to the "projection" of the input distance matrix onto the space of Euclidean matrices, i.e. the Euclidean distance matrix that best approximates the input distance matrix. Further increasing the projection dimension makes little difference to the resulting distance matrix.

This example shows that when the input distance matrix is noisy and the true configuration is approximately low-dimensional, a low-dimensional solution may provide a representation that better recovers the true distance matrix. It also shows that choosing the right number of dimensions is critical in MDS when our goal is to obtain a good representation of the data for subsequent analysis. In that case, we need to choose a dimension that is not too low or too high to balance the bias and variance of the MDS solution.

3.8   Further topics

Multidimensional unfolding

Multidimensional Unfolding (MDU) is a task closely related to MDS. It concerns simultaneously learning and visualising a low-dimensional representation of two groups of objects given only the between-group distances or dissimilarities. For example, MDU is commonly used in marketing research to study the relationship between customer preferences and product positioning (e.g. Ho, Chung, and Lau 2010). In this example, the two groups of objects may be the customers and the brands. An MDU procedure takes the distances between customer-brand pairs as input. It outputs a joint display of customers and brands on a single map, where the distances may be constructed based on customers' shopping behaviours. For another example, MDU can be used to study the voting behaviours of politicians and the political positioning of legislative proposals, in which case the two groups of objects are the politicians and the legislative proposals, and the distance between a politician and a legislative proposal is determined by the corresponding vote.

MDU can be viewed as a special case of MDS with missing values in the input distance matrix. Let n and m be the numbers of objects of the two groups. The input distance matrix for MDU is an $n \times m$ matrix $\{\delta_{ij}\}$. We can view the MDU problem as an MDS problem with $(n+m)$ objects and an $(n+m) \times (n+m)$ input distance matrix of the blockwise form

$$\left(\begin{array}{c|c} \{\mathrm{NA}\}_{n\times n} & \{\delta_{ij}\}_{n\times m} \\ \hline \{\delta_{ji}\}_{m\times n} & \{\mathrm{NA}\}_{m\times m} \end{array} \right),$$

where the diagonal blocks $\{\mathrm{NA}\}_{n\times n}$ and $\{\mathrm{NA}\}_{m\times m}$ represent two matrices with missing entries. Therefore, the MDU problem can be solved using procedures for MDS with missing data, for which metric and non-metric procedures similar to the ones described earlier in this chapter are available. However, the solutions from these procedures for MDU can sometimes be unstable due to the large proportion of missing data in the input distance matrix. We refer interested readers to Chapters 14–16 of Borg and Groenen (2005) for more details about MDU.

Latent space models

The MDS methods described earlier in this chapter are descriptive. Alternatively, one can solve the same problem by assuming a statistical model, typically known as a latent space model (Oh and Raftery 2001). In such a model, each object is assumed to be represented by an ideal point in a (multidimensional) latent Euclidean space, and the distance or dissimilarity between two objects is a random variable whose distribution depends on the distance between their ideal points and possibly additional model parameters. A popular application of latent space models is social network analysis (Hoff, Raftery, and Handcock 2002), where each object is an actor in the social network, and the dissimilarity between two objects is determined by the presence or absence (or intensity) of the link between the actors.

Compared with the descriptive MDS methods, latent space models allow us to draw statistical inferences about the estimated configuration, such as determining the dimension of the configuration and constructing confidence regions for individual ideal points. Both Bayesian and frequentist methods have been developed to perform statistical inference for latent space models; see, e.g. Oh and Raftery (2001) and Hoff, Raftery, and Handcock (2002). It is also worth noting that the model-based approach is not without limitations. Specifically, the validity of the inference results relies on the correct specification of the model assumptions, and violating the assumptions can lead to a substantial bias. In addition, the model-based approach tends to be more computationally demanding, especially for Bayesian procedures. These factors should be considered when choosing between model-based and non-model-based methods for MDS.

3.9   Further reading

Borg, I. and Groenen, P. J. F. (2005). *Modern Multidimensional Scaling* (2nd ed.). New York: Springer-Verlag.

Cox, T. F. and Cox, M. A. A. (2001). *Multidimensional Scaling* (2nd ed.). London: Chapman and Hall/CRC.

Everitt, B. S. and Rabe-Hesketh, S. (1997). *The Analysis of Proximity Data*. London: Arnold.

Kruskal, J. B. and Wish, M. (1978). *Multidimensional Scaling*. Series Quantitative Applications in the Social Sciences, Number 11. London, Sage Publications.

4

Correspondence Analysis

4.1 Aims of correspondence analysis

Correspondence analysis (CORA) is an exploratory technique for analysing multi-way frequency tables, that is, cross-classifications of two or more categorical variables. We will focus on the analysis of two-way tables, but the analysis of multi-way tables depends on much the same set of ideas and will be discussed briefly later in the chapter. Like MDS, correspondence analysis aims to convert a table of numbers into a plot of points in a small number of dimensions — usually two. The term *correspondence analysis* derives from the French *Analyse Factorielle de Correspondances* which is the term used by Benzecri and others who developed the technique. However, the basic idea is found much earlier in attempts to scale the categories of contingency tables.

The usual way to begin the analysis of a two-way table would be to perform a chi-squared test of association between the row and column variables. If a significant association were found, the nature of the association could be explored by examining row and/or column percentages. However, when the number of categories is large, perhaps hundreds, comparisons of row (column) percentages across columns (rows) is difficult. The aim of CORA is to represent the raw data in a low-dimensional space so that it is easier to identify the key features of the data. CORA can be used to explore questions such as the following.

   i) Are there row categories which have similar distributions over the column categories?

   ii) Are any of the column categories similar with respect to their distributions over the row categories?

   iii) Are the row/column categories ordered with respect to their distributions across the column/row categories? If so, are the categories fairly evenly spaced?

   iv) Questions i) and ii) are concerned with the extent to which row/column distributions vary across the column/row categories. Further questions arise concerning the extent to which a cell given by a row and column category contributes to the overall association.

DOI: 10.1201/9781003483342-4

Form of data input

Frequency tables may arise in a number of ways. Most commonly, the row and column variables have mutually exclusive categories, in which case the table is called a *contingency table*. These variables may be nominal or ordinal. Indeed, as noted above, CORA may be used to explore whether a variable that is suspected to be ordinal may be treated as such. Table 4.1 shows a cross-tabulation of the political party voted for in the 1992 British general election by the main reason for voting for the chosen party (among those who voted). The data are from the 1992 British General Election Study (Heath et al. 1993). Both variables are nominal and have mutually exclusive categories.

TABLE 4.1: Voting preference by reason, British General Election Study 1992

Reason	Party					Total
	Conservative	Labour	Lib Dem	Other	Refused	
Always vote that way	244	405	48	39	18	754
Best party	933	542	305	127	46	1953
My party had no chance	59	74	87	31	5	256
Total	1236	1021	440	197	89	2963

The categories of the row and/or column variables do not have to be mutually exclusive. For example, the data in Table 4.2 are from a survey on leisure activities in Norway (Clausen 1998). Respondents were asked whether they had engaged in any of the ten activities in the previous year. Since each respondent may have engaged in more than one of the activities, the categories are not mutually exclusive. Other examples of this type of frequency table are found in market research, where a number of product brands are rated on a series of attributes.

TABLE 4.2: Leisure activities by occupation, Survey of Level of Living 1995, Norway

Activity	Occupation						Total
	Manual	Low NM*	High NM	Farmer	Student	Retired	
Sport event	301	497	208	50	254	187	1497
Cinema	261	550	250	27	339	157	1584
Dance/disco	361	534	204	59	324	216	1698
Cafe/restaurant	463	766	334	72	350	601	2586
Theatre	89	350	195	12	143	167	956
Classical concert	23	182	124	10	60	110	509
Pop concert	117	298	145	11	184	56	811
Art exhibition	104	379	219	21	152	213	1088
Library	130	352	153	17	272	264	1188
Church service	168	370	187	51	162	424	1362
Total	2017	4278	2019	330	2240	2395	13279

*NM denotes non-manual

4.2   Carrying out a correspondence analysis: a simple numerical example

To demonstrate how CORA is performed, consider a simple 3×3 table. As often happens in explaining techniques in multivariate analysis, it is easiest to grasp the idea if it is illustrated in a very simple example. We shall do that here, but it must be remembered that the full power of the technique can only be appreciated on much larger tables. In this particular case, it is to be expected that CORA will tell us little more than could be learnt from careful inspection of the table.

A cross-classification of attitude to abortion in the US and years of education is given in Table 4.3. The data are from the General Social Surveys of 1972-1974 and appear in Haberman (1978), p.264.

TABLE 4.3: Attitude to abortion by education in the US, 1972-74: cell frequencies

		Attitude			
		Positive	Mixed	Negative	Total
Education	≤ 8	101	120	320	541
	9-12	599	341	756	1696
	≥ 13	475	161	308	944
	Total	1175	622	1384	3181

The chi-squared statistic for a test of independence between rows and columns is 157.58 on 4 degrees of freedom, which indicates that there is a significant association between education and attitude to abortion in the US.

Row profiles and row masses

To investigate this association further, we might look at the distribution of respondents across the attitude categories for each education category, that is the row proportions. We refer to the sets of row proportions as *row profiles*. The row profiles for the data in Table 4.3 are shown in Table 4.4. Also shown are the *row masses* (overall proportion in each row) and the *centroid* or *average row profile* (overall proportion in each column). We could also examine the distribution across education categories for each attitude category. To do so, we would calculate the column profiles, column masses and the average column profile. For now, we will consider only the row profiles but will discuss the role of column profiles later.

It is clear that there are marked differences in the three row profiles. The proportion with a positive attitude tends to increase as we move down the table; that is, attitude tends to become more positive for more years of education. However, note that we are only able to describe the pattern in these simple terms because the categories are ordered. In general, the categories will not be ordered and part of the purpose of the analysis will be to see whether there is an ordering which helps to make sense of the table. For our immediate purpose, we shall not make use of the ordering information.

TABLE 4.4: Attitude to abortion by education in the US, 1972-74: row profiles

		Attitude		
	Positive	Mixed	Negative	Row mass
Education ≤ 8	0.187	0.222	0.591	0.170
9-12	0.353	0.201	0.446	0.533
≥ 13	0.503	0.171	0.326	0.297
Centroid or Average row profile	0.369	0.196	0.435	

For a table with three columns, the row profiles can be represented as points in two-dimensional space, because the proportions must add to 1. In Figure 4.1, they are represented by points inside an equilateral triangle, where the centre of the triangle corresponds to equal proportions of responses in each category, and a point nearer a vertex (a corner of the triangle) corresponds to a higher proportion in that category.

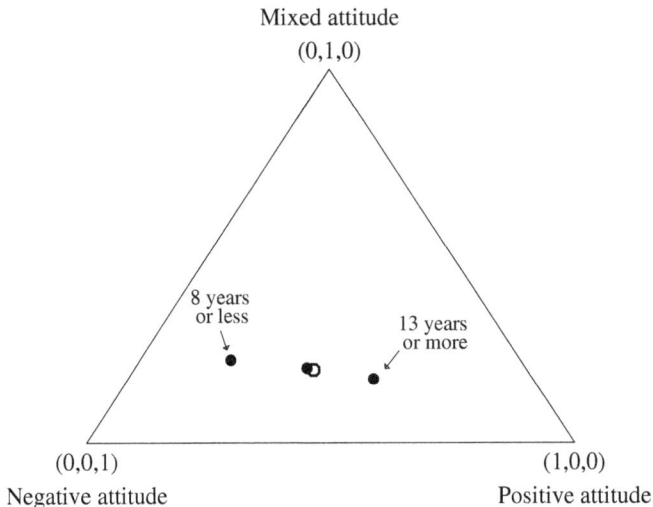

FIGURE 4.1: The row profiles for the US education and attitude to abortion data. The open circle represents the centroid or average row profile and the solid dots represent the row profiles for the three educational groups.

In Figure 4.1, the average row profile shows approximately equal proportions of positive and negative responses with a lower proportion of mixed responses. The profile for the highest education group (≥ 13 years) has a higher proportion of positive responses, while the profile for the lowest education group (≤ 8 years) has a higher proportion of negative responses (and a slightly higher proportion of mixed responses). The profile for the 9-12 year group is practically the same as that for the average row profile.

If there had been no association between attitude to abortion and education level, the row profiles would have been identical and the solid dots in Figure 4.1 would all have coincided with the average row profile (represented by the open circle). Their distance apart and pattern tell us something about the nature of the association. As a first step towards exploring this, we could calculate distances between each pair of row profiles, and between each row profile and the centroid. One possibility is to use the Euclidean distance, which is equal to the square root of the sum of squared differences between the profile values. For example, the Euclidean distance between the row profiles for ≤ 8 and 9-12 years of education is

$$\sqrt{(0.187 - 0.353)^2 + (0.222 - 0.201)^2 + (0.591 - 0.446)^2} = 0.221.$$

However, in CORA, each dimension is weighted inversely by the corresponding coordinate of the average row profile, so that column categories with a higher relative frequency do not dominate those with a lower relative frequency. The weighted Euclidean distance between the row profiles for ≤ 8 and 9-12 years of education is thus

$$\sqrt{\frac{(0.187 - 0.353)^2}{0.369} + \frac{(0.222 - 0.201)^2}{0.196} + \frac{(0.591 - 0.446)^2}{0.435}} = 0.354.$$

This measure of distance is often referred to as the chi-squared distance because weighting the squared difference between two profile values by the average profile value is analogous to weighting the squared difference between observed and expected values by the expected value. The full set of squared chi-squared distances is given in Table 4.5. In the last row of Table 4.5 are the squared chi-squared distances between row i and the average row profile or centroid, which we denote by d_i^2.

TABLE 4.5: Squared chi-squared distances between row profiles, and between row profiles and the centroid, attitude to abortion by education, US, 1972-74

	Row		
	1	2	3
1	0	–	–
2	0.125	0	–
3	0.445	0.099	0
Centroid	0.149	0.001	0.079

At this stage, we have a situation reminiscent of cluster analysis and MDS where the first step was to calculate a distance matrix. We could, indeed, proceed to carry out either kind of analysis on distance tables calculated from frequency tables. A cluster analysis, for example, might identify clusters of rows which had very similar profiles and this would suggest that those categories should be close together in some sense. MDS would provide a plot of points representing the rows and, again, it might be possible to infer something about the association from the pattern of the points. However, in CORA, our attention is focused more specifically on the nature of the association and the contribution which the various row and column categories make to it.

Inertia

In CORA, the term *inertia* is used to describe the measure of scatter or "variance" in the row (or column) profiles about the centroid. The total inertia is defined as

$$\sum_{i=1}^{I}(\text{mass for row } i) \times d_i^2,$$

where I is the number of rows in the table. The term *inertia*, like the more familiar *degrees of freedom*, comes from mechanics; the analogy on which its use is based arises from the formula for inertia which is a mass multiplied by a squared distance.

From Tables 4.4 and 4.5, the total inertia for the US abortion data is

$$(0.170 \times 0.149) + (0.533 \times 0.001) + (0.297 \times 0.079) = 0.050.$$

It can be shown that the total inertia is related to the chi-squared statistic (X^2) divided by the grand total, n, that is

$$\text{Inertia} = \frac{X^2}{n}.$$

This result provides an interesting alternative way of looking at the chi-squared statistic. It now appears as a measure of the variation of the row profiles. If we interchange the rows and columns of the table the value of chi-squared remains the same, and it then follows that its value can also be represented as a measure of the variation of the column profiles.

The inertia measures the variation between rows, which are multidimensional objects and can thus be represented as points in space, as we have seen above. Equally, it measures variation between columns. CORA depends on the fact that the inertia can be decomposed in another way, with each part measuring the variation in a single dimension. Because of the relationship between inertia and the chi-squared statistic, this is equivalent to decomposing the chi-squared statistic. If it turns out that most of the variation takes place in a small number of dimensions, two for example, it will be possible to picture the variation, and so perhaps interpret it in a meaningful way.

A two-dimensional representation

In the US education and attitude to abortion example, the row profiles may be represented in three-dimensional space because there are three rows and three columns. Actually, they may be represented perfectly in two dimensions, as in Figure 4.1, because any element in a row, say, may be obtained by subtracting the other two elements from the row total. However, CORA is most useful in much larger tables where the number of rows and columns is much greater than three. In such cases, it is desirable to reduce the dimensionality of the row profiles so that they may be plotted in two-dimensional, or at most three-dimensional, space. The question is how to find the coordinates of points representing the row profiles in two dimensions, and then to assess how good a representation of the original data these provide. In geometrical terms, the aim of CORA is to find a plane that is as close as possible to all the points and which also reproduces, as accurately as possible, the chi-squared distances between them. The row profiles are projected onto this plane to obtain points which represent the profiles in two dimensions.

Before describing how to do this, we look first at the interpretation of two-dimensional plots using the row profiles for the US education and attitude to abortion data as shown in Figure 4.2. From Figure 4.2 we see that most of the variation between the row categories takes place in dimension 1 and, as we might expect, this dimension corresponds to "years of education". The row categories are thus given a metric with values –0.82, –0.07, and 0.60 which are spaced at roughly equal intervals. However, there is a second dimension separating the middle group, 9–12 years, from the extremes. The variation in this dimension is much less.

If this kind of analysis is to be carried out for larger tables, we need some way of carrying out the decomposition in a manner which produces the successive dimensions algebraically. We now briefly outline the way that this is done without going into the mathematical details, which depend on what is known as the singular value decomposition of a matrix.

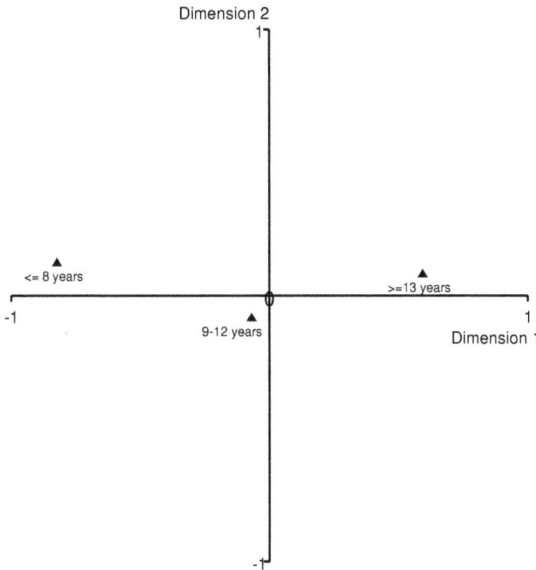

FIGURE 4.2: Row profiles in two dimensions for attitude to abortion by education, US, 1972-74

4.3 Carrying out a correspondence analysis: the general method

Pearson residuals

In correspondence analysis, the chi-squared statistic is partitioned in the manner described above. For the mathematics of the decomposition, it turns out to be more convenient to work, not with the profiles themselves, but with closely related quantities which we shall call *Pearson residuals*. We begin with an $I \times J$ matrix of observed frequencies, where I and J are the numbers of rows and columns, respectively, and convert this to a matrix of Pearson residuals. These Pearson residuals are deviations between the observed frequencies and those expected under the model of independence. Denote the observed frequency for row i and column j of the table by O_{ij}, the total for row i by O_{i+} and the total for column j by O_{+j}. Denote the matrix of Pearson residuals by **C**. The elements of **C** are

$$c_{ij} = \frac{O_{ij} - E_{ij}}{\sqrt{E_{ij}}},$$

where

$$E_{ij} = \frac{O_{i+}O_{+j}}{n} \qquad (i = 1, \ldots, I; \ j = 1, \ldots, J).$$

(Under the hypothesis of independence, each element of $\mathbf{C}$ has approximately the same distribution with zero mean and unit variance and thus the Pearson residuals are, in a sense, put on an equal footing.) If the row profiles are the same, the elements of each row of $\mathbf{C}$ will be zero. The size and pattern of the deviations from zero therefore tells us about the nature of the association. The matrix $\mathbf{C}$ for the US education and attitude to abortion data is given in Table 4.6.

TABLE 4.6: Pearson residuals for attitude to abortion by education, US, 1972-74

		Attitude		
		Positive	Mixed	Negative
Education	≤ 8	-6.99	1.38	5.52
	9-12	-1.10	0.51	0.67
	≥ 13	6.76	-1.74	-5.07

The singular value decomposition theorem tells us that we can write the typical element of $\mathbf{C}$ as

$$c_{ij} = \sum_{k=1}^{K} \lambda_k^{1/2} u_{ik} v_{jk} \qquad (i = 1, \ldots, I; \ j = 1, \ldots, J),$$

where K is the smaller of $I - 1$ and $J - 1$. The λ_ks are known as *eigenvalues* and their square roots, that is $\sqrt{\lambda_k}$ or $\lambda_k^{1/2}$, are the singular values. These are mathematical terms which are used here as convenient labels, but for our purposes, it is not necessary to know anything about their technical role in the derivation of the decomposition. The u_{ik}s and v_{jk}s may be thought of as scores attached to the rows and columns. In the simple analysis we carried out on the attitude to abortion data, we found scores for the row categories. In that case, $K=2$, so there was no dimensional reduction, and there were two scores for each category. In the general case, each row category is represented by a point in K dimensions with coordinates $(u_{i1}, u_{i2}, \ldots, u_{iK})$, and the column categories by points with coordinates $(v_{j1}, v_{j2}, \ldots, v_{jK})$. Usually, however, we wish to represent row and column categories by points in a low-dimensional (preferably two) space.

The best approximation to c_{ij} in two dimensions is

$$c_{ij} \simeq \lambda_1^{1/2} u_{i1} v_{j1} + \lambda_2^{1/2} u_{i2} v_{j2},$$

where λ_1 and λ_2 are the two largest eigenvalues. Therefore, to present a graphical display in two dimensions, we could plot (u_{i1}, u_{i2}) and (v_{j1}, v_{j2}).

Usually, the coordinates are standardised in some way. We will use the following standardisation (implemented in SPSS), where the standardised row coordinate for dimension k is calculated as

$$u_{ik}^* = \frac{u_{ik}\lambda_k^{1/4}}{\sqrt{O_{i+}/n}}.$$

The u_{ik}s are multiplied by the inverse of the square roots of the row masses to ensure that row categories with high relative frequencies do not dominate rows with small frequencies. The u_{ik}s are further multiplied by $\lambda^{1/4}$ so that more weight is attached to coordinates corresponding to the most important dimensions than to coordinates for less important dimensions. The v_{jk}s are transformed in a similar way to obtain standardised column coordinates, v_{jk}^*.

The eigenvalues of $\mathbf{C}$ are the principal way of judging the importance of the various dimensions. Associated with each dimension is an eigenvalue which represents the scatter of profiles about the centroid on that dimension, that is, the contribution to X^2 associated with that dimension. An alternative way of calculating the total contribution to X^2 is to take the sum of the eigenvalues. Eigenvalues may be compared across dimensions to assess the relative importance of each dimension in explaining X^2. The eigenvalues are ordered such that $\lambda_1 \geq \lambda_2 \geq \ldots \geq \lambda_K$. The dimensions are therefore constructed so that the first dimension explains the largest portion of X^2, or equivalently of the inertia, the second dimension explains the largest portion of the remaining inertia, and so on.

The proportion of X^2, or of inertia, explained by dimension k is

$$\frac{\lambda_k}{\lambda_1 + \lambda_2 + \ldots + \lambda_K}.$$

For the US education and attitude to abortion data, the proportion of inertia explained by the first dimension is $0.049/0.05 = 99\%$. The second dimension explains only 1% of inertia. Another way to look at the contribution of the first dimension is to estimate the value of X^2 if only the first dimension is considered. This is calculated as $\lambda_1 \times n = 0.049 \times 3181 = 151.0$ (compared with 157.58 for the full table). Most of the variation in the row profiles can therefore be expressed in one dimension.

Dual scaling

It is worth noting, at this point, a convergence between two different approaches to the analysis of association in contingency tables. In the approach which we have been following here, the category scores arise out of the decomposition of the table; but there was no thought at the outset of trying to assign scores to the categories. The other approach starts with what seems to be a quite different objective where the focus is on scaling both individuals and categories, and it sometimes goes under the name of dual scaling. We know a great deal about investigating the correlation structure of continuous variables. It might be possible to utilise this knowledge for contingency tables if there were some ways of turning categorical data into continuous data. Dual scaling, also known as optimal scaling, asks whether there is some optimum way of assigning scores to individuals and categories so that the structure can be explored in terms of regression and correlation. We cannot go into the details here but the method turns out to be equivalent to correspondence analysis as we have described it. For some purposes, the scaling approach is the more natural way to approach some of the questions we outlined at the beginning of the chapter. For example, question iii) in Section 4.1 asked whether the categories were ordinal. If this were so, one might expect to be able to assign scores to the categories so that they formed an increasing (or decreasing) sequence. Our analysis shows that, in general, there can be no unambiguous answer to that question because the method assigns K sets of scores to the row categories and each will give a different ranking. However, as we saw with the abortion data, if one dimension is dominant we can reasonably treat the scores to which that leads as converting a nominal scale to an interval scale.

4.4   The biplot

The process by which row profiles are represented geometrically (Section 4.2) may be repeated for column profiles. It follows from the general representation of the singular value decomposition of $\mathbf{C}$ given above that the scatter of column profiles about the average column profile is equal to the scatter of row profiles about the average row profile. Therefore, the total inertia can be derived by considering either row profiles or column profiles. Also, the dimensionality of the row profiles is the same as that for the column profiles, even when the numbers of rows and columns are unequal. The maximum number of dimensions needed to represent either row profiles or column profiles is $K = \min(I - 1, J - 1)$. In terms of dimension reduction, the best-fitting line or the best-fitting plane to the row profiles explains the same proportion of inertia as the best-fitting line or the best-fitting plane to the column profiles.

Two-dimensional plots representing row or column profiles may be examined to identify whether any row or column categories have similar profiles. Row or column categories that have similar profiles will appear in close proximity on the plot. This could be useful to determine whether any row or column categories could be combined in subsequent analysis. However, also of interest is how row and column categories interact with one another in contributing to the overall association. This aspect can be explored by means of *biplots*, which are plots of the points (u_{i1}^*, u_{i2}^*) and (v_{j1}^*, v_{j2}^*) on the same diagram. The purpose of a biplot can be seen by going back to the decomposition of the matrix of standardised residuals given above. This shows how the row and column score contribute to the overall size of the residual.

Recall that u_{ik}^* is the coordinate on dimension k representing row category i, and v_{jk}^* is the coordinate on dimension k representing column category j. The product $u_{ik}^* v_{jk}^*$ represents the joint contribution of row i and column j to the residual arising from dimension k. This is often spoken of as the "association" of column i and column j, but it must be distinguished from the overall association between the row categories and the column categories measured by X^2 with which we started the analysis. It is more accurately spoken of as a contribution to the overall association arising from a particular row and column.

In that sense, a large positive value for $u_{ik}^* v_{jk}^*$ indicates a positive relative association between row i and column j on dimension k. A large positive value is obtained if u_{ik}^* and v_{jk}^* are both large and positive or both large and negative, that is, the points for these categories appear close together on the biplot and far from zero on dimension k.

A large negative value for $u_{ik}^* v_{jk}^*$ indicates a negative relative association between row i and column j on dimension k. A large negative value is obtained if one of u_{ik}^* and v_{jk}^* is large and positive, and the other is large and negative, that is the points for these categories appear far apart on the biplot with neither point close to a value of zero on dimension k.

A value close to zero for $u_{ik}^* v_{jk}^*$ indicates no association between row i and column j on dimension k. A value close to zero is obtained if one or both of u_{ik}^* and v_{jk}^* is close to zero on dimension k.

There are two types of biplot that may be used in CORA: asymmetric plots and symmetric plots. It is the symmetric plot that we have been discussing above and will illustrate below. This is more generally useful.

Symmetric plots

The association between a row category and a column category may be assessed according to the proximity of their profile points on the biplot. However, these proximities must be interpreted with caution. If the point for row category 1 is closest to the point for column category 2, we cannot say anything about the *magnitude* of their interaction in an absolute sense. We can only interpret it in *relative* terms. That is, we can say, for example, that individuals in row category 1 are relatively more likely (compared to the average row profile) to be in column category 2. It could be the case that, overall, there are very few individuals in column category 2; all we can say from the symmetric plot is that individuals in row category 1 are more likely to be in column category 2 than are individuals in the other row categories.

The coordinates of points in a symmetric biplot are scaled so that row or column points for rows or columns with high masses (marginal frequencies) do not dominate. They are further scaled (as described in Section 4.3) so that more weight is attached to coordinates corresponding to the most important dimensions than to coordinates for less important dimensions.

This procedure is illustrated on the biplot in Figure 4.3 showing the row and column profiles for the US abortion data. The coordinates for "≥ 13 years" (row category 3) and "positive" (column category 3) are both large and positive on dimension 1, giving a large positive value for $u_{31}^* v_{31}^*$. Thus, respondents with 13 or more years of education are relatively positively associated with having a positive attitude to abortion. The large negative coordinate for "$\leq$ 8 years" (row category 1) on dimension 1 and the large positive coordinate for "positive" leads to a large negative value for $u_{11}^* v_{31}^*$. Respondents with 8 or fewer years of education are less likely to have a positive attitude than are more educated respondents.

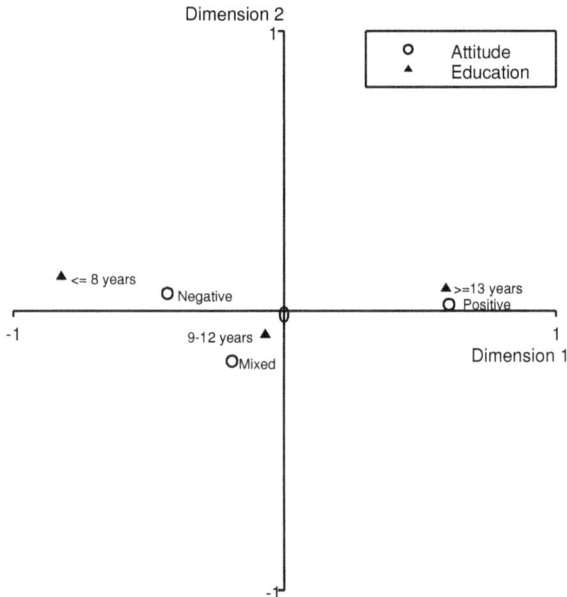

FIGURE 4.3: Biplot for attitude to abortion by education, US, 1972-74

Note that the coordinates for "9-12 years" and "mixed" on dimension 1 are both close to zero. Having 9-12 years of education is not associated with any attitude category, and having a mixed attitude is not associated with any education category. The latter can be seen by looking at the row profiles (Table 4.4). The proportion with a mixed attitude varies little across education categories.

Symmetric biplots are of fundamental importance for the interpretation of frequency tables and we shall therefore give two further examples. Biplots from CORA of the data in Table 4.1 and Table 4.2 are shown in Figure 4.4 and Figure 4.5 respectively.

From Figure 4.4, we can say that voting Labour is relatively associated with voting for a party because the respondent "always votes that way". We reach this conclusion because these particular row and column categories are in close proximity on the plot. Voting Conservative is relatively associated with voting for the "best party". Individuals who refused to state who they voted for are closer to Conservative voters in terms of their distribution across the main reason for voting categories. Conservative and Labour voters are relatively unlikely to vote for those parties because they thought their party had no chance of winning.

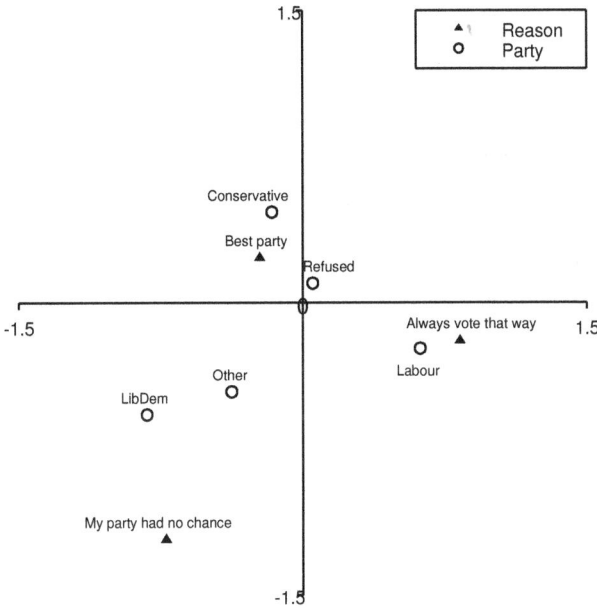

FIGURE 4.4: Biplot for cross-tabulation of party preference by main reason for party choice, British General Election Study 1992

It is important to emphasise the use of the word *relatively* in describing the association between a row category and a column category. From the type of plot shown above (a symmetric biplot), we cannot say anything about the *absolute* level of association. We can only say that a pair of row-column categories that are close together are more strongly associated than a pair of categories that are further apart.

CORA is more useful for analysing large contingency tables such as Table 4.2. From Figure 4.5, we can see, for example, that being retired is relatively associated with going to church, and students and low nonmanual workers are relatively more likely to go to the cinema and pop concerts than other occupation groups. Clausen (1998) also places an interpretation on the two dimensions. Dimension 1 separates the young (students) from the old (retired), while dimension 2 separates arts activities (e.g. going to a classical concert) from light entertainment (e.g. going to a disco).

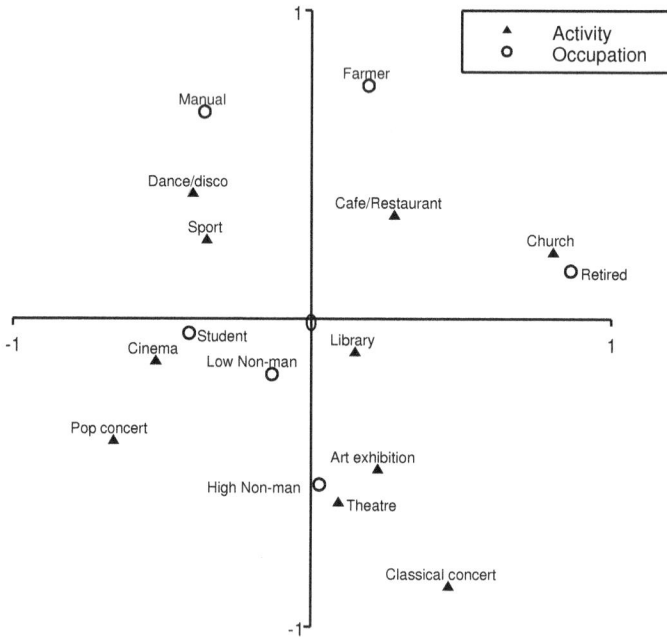

FIGURE 4.5: Biplot for cross-tabulation of leisure activities by occupation, Survey of Level of Living 1995, Norway

Asymmetric plots

In an asymmetric biplot, the row profiles are compared not with the column profiles but with the column vertices. (Alternatively, column profiles and row vertices are plotted simultaneously.) The point of doing this can be illustrated using the case of the 3×3 table. The column vertices are the following points in three-dimensional space: (1,0,0), (0,1,0) and (0,0,1), where the first, second and third coordinates are the proportions in columns 1, 2, and 3 respectively. These points represent extreme cases in which all members of a row fall into one column category. The corner points of the triangular region shown in Figure 4.1 are the column vertices. Suppose that the row profile point for row 1 in the table was very close to the (1,0,0) vertex, (0.9, 0.04, 0.06). This would indicate that the majority of individuals in row 1 are in column 1. In other words, there is a large positive association between row category 1 and column category 1. On the other hand, suppose that the row profile point for row 1 was very close to one of the other vertices. This would indicate a high negative association between row category 1 and column category 1, since individuals in row category 1 are relatively unlikely to be in column category 1. For a higher dimension table, CORA is used to obtain an approximate lower dimensional representation of the row profiles and column vertices, preferably in one- or two-dimensional space. This low-dimensional plot is interpreted in the same way.

The main problem with the asymmetric map is that usually the row profiles are fairly close to the centroid (average profile). The addition of the column vertices to the plot alters the scale of the map so that the row profiles tend to appear very close together and are almost indistinguishable. However, if the scatter of the row profiles about the centroid is large (meaning the inertia is large), an asymmetric map may be useful. Also, there will be cases where it is unclear whether to view a cross-tabulation in terms of row profiles or column profiles; for example, if either the row or the column variables may be regarded as dependent variables. In this case, row and column profiles are of equal substantive interest and a symmetric plot may be more appropriate.

4.5   Interpretation of dimensions

It is sometimes possible to interpret or "label" the dimensions obtained from a CORA. We do so by examining the position of row/column categories along each dimension and thinking about what row/column categories that appear close together have in common, and what distinguishes those that appear far apart. For example, in his analysis of the Norwegian leisure data, Clausen (1998) found that on the first dimension, light entertainment activities were grouped together and appeared far away from a cluster of arts activities. Biplots provide a visual display of such groupings of row/column categories. However, when interpreting a dimension, it is important to pay particular attention to those points which contribute the most to the inertia or scatter of points along that dimension.

Suppose we wish to interpret dimension k with respect to the row profiles. We can partition each point's contribution to the total inertia into its contributions to the inertia on each dimension in the CORA solution. The amount of inertia along dimension k explained by row point i is

$$\frac{(\text{mass for row } i) \times u_{ik}^{*\,2}}{\sqrt{\lambda_k}} = u_{ik}^2.$$

Thus points corresponding to rows with a high row mass and with a large coordinate on dimension k will contribute the most to the inertia on dimension k. The amount of inertia explained by a given column point is calculated in a similar way. Points with relatively large contributions are most important to that dimension and provide the key to its interpretation. These values are examined together with the sign of the corresponding coordinates to interpret dimension k.

We illustrate this process using the US education and attitude to abortion data. Table 4.7 shows the row masses and row coordinates in dimensions 1 and 2 (plotted in Figure 4.3), and the contribution of each row point to the inertia on each dimension in the two-dimensional solution. The contributions to inertia are expressed as proportions of the total inertia on that dimension. Table 4.8 shows the same quantities for the column categories. Starting with the row categories, we see that "≤ 8 years" and "≥ 13 years" explain similar proportions of the inertia on dimension 1. These two categories have coordinates which are opposite in sign. Thus we might label dimension 1 "level of education". Turning to the column categories (Table 4.8), we find that the "positive" category makes the largest contribution to inertia in dimension 1, followed by "negative". The corresponding coordinates in dimension 1 are opposite in sign, leading to this dimension being labelled "direction of attitude". Considering the interpretation of dimension 1 with respect to both row and column categories suggests that a high level of education is associated with a more positive attitude to abortion. In this case, we reached the same interpretation of the dimensions simply by examining the biplot in Figure 4.3. In general, however, row/column points with the largest coordinates will not always make the largest contributions to inertia since they may correspond to row/column categories with small relative frequencies. It is therefore important to examine both the coordinates of row/column points and the contributions of points to inertia in order to interpret dimensions.

TABLE 4.7: Coordinates and contribution to inertia of row points for attitude to abortion by education, US, 1972-74

Education	Row mass	Coordinate		Contribution to inertia	
		Dim 1	Dim 2	Dim 1	Dim 2
≤ 8	0.170	−0.821	0.123	0.516	0.314
9-12	0.533	−0.069	−0.084	0.012	0.455
≥ 13	0.297	0.595	0.080	0.473	0.230

TABLE 4.8: Coordinates and contribution to inertia of column points for attitude to abortion by education, US, 1972-74

Attitude	Column mass	Coordinate		Contribution to inertia	
		Dim 1	Dim 2	Dim 1	Dim 2
Positive	0.369	0.606	0.022	0.609	0.022
Mixed	0.196	−0.191	−0.180	0.032	0.773
Negative	0.435	−0.428	0.062	0.359	0.206

4.6 Choosing the number of dimensions

As with MDS, the aim of CORA is to balance goodness-of-fit with parsimony when choosing the number of dimensions. The aim is to choose as few dimensions as possible, as this makes the task of interpretation easier. At the same time, the dimensions we choose to interpret should explain a reasonable amount of inertia. A commonly used tool is the scree plot, similar to the one used in MDS. The inertia for each dimension is plotted and the plot is examined for an "elbow", that is, the point after which there is little decrease in inertia. The scree plot for the Norwegian leisure activities data is shown in Figure 4.6. The maximum number of dimensions needed to represent the data is $\min(10-1, 6-1) = 5$. The elbow at three (or possibly four) suggests that two (possibly three) dimensions are sufficient to represent the data.

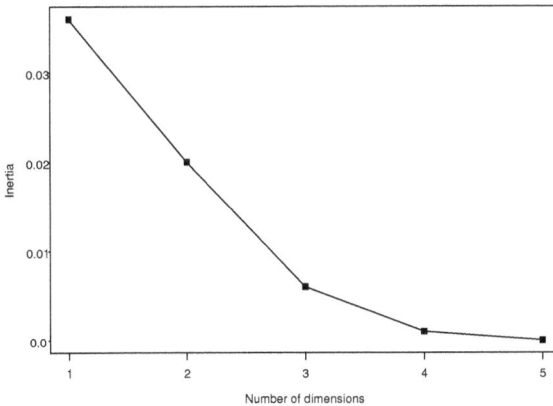

FIGURE 4.6: Scree plot for Norwegian leisure activities data

Another way of determining the number of dimensions is to examine the cumulative proportion of inertia explained by the dimensions. For example, in the case of the Norwegian data, the first two dimensions explain 90% of inertia, while the first three explain 99%. The proportion of total inertia explained by the first k dimensions can be thought of as a measure of overall goodness-of-fit of the k-dimensional solution. We can also examine how well each row/column category is represented in k dimensions. Once again, we start by considering row categories. The total inertia of row point i is

$$\sum_{k=1}^{K} \lambda_k \times \text{(amount of inertia on dimension } k \text{ explained by point } i\text{)}$$
$$= \sum_{k=1}^{K} \sqrt{\lambda_k} \times \text{(mass for row } i\text{)} \times u_{ik}^{*\,2}.$$

The contribution of dimension k to the inertia of row point i is then

$$\frac{\sqrt{\lambda_k} \times \text{(mass for row } i\text{)} \times u_{ik}^{*\,2}}{\sum_{k=1}^{K} \sqrt{\lambda_k} \times \text{(mass for row } i\text{)} \times u_{ik}^{*\,2}}.$$

For example, using the information in Table 4.7 and the fact that the inertia on dimension 1 is 0.049, the total inertia of row point 1 (≤ 8 years of education) for the US abortion data is

$$(0.049 \times 0.516) + (0.001 \times 0.314) = 0.026,$$

and the contribution of dimension 1 to the inertia of row point 1 is

$$\frac{(0.049 \times 0.516)}{0.026} = 0.99.$$

The contribution of dimension k to the inertia of column points may be calculated in a similar way. The contributions of dimensions 1 and 2 to the inertia of row and column points for the US education and attitude to abortion data are shown in Table 4.9. These quantities measure how well each row and column point is described by each dimension. In this case, since dimension 1 is highly dominant, each row and column point is extremely well represented by the first dimension alone. Here, a maximum of two dimensions is required to represent the row/column profiles, so the sum of the contributions across dimension 1 and 2 equals 1. In general, however, when a larger table is analysed the number of dimensions required to achieve a perfect fit will be large, and we will wish to assess the fit of a solution in considerably fewer dimensions.

TABLE 4.9: Contributions of dimensions to inertia of points for attitude to abortion by education, US, 1972-74

		Dim 1	Dim 2	Total
Education	≤ 8	0.999	0.001	1.000
	9-12	0.949	0.051	1.000
	≥ 13	0.999	0.001	1.000
Attitude	Positive	1.000	0.000	1.000
	Mixed	0.968	0.032	1.000
	Negative	0.999	0.001	1.000

4.7 Example: confidence in purchasing from European Community countries

We now illustrate the use of CORA with a more realistic example where the dimensions of the two-way cross-classification are large. The data are from the 1995 Eurobarometer Survey in which respondents from 15 countries in the European Community (EC) were asked about the confidence with which they would purchase various products or services from other EC countries (Reif and Marlier 1995). Using the responses to these questions, three binary items were constructed where 1 indicates confidence in buying a given product/service from another EC country and 0 indicates lack of confidence.

   i) Confidence in buying food and/or wine

   ii) Confidence in buying household electrical appliances

   iii) Confidence in buying medical services and/or financial services

Since each item is binary, there are $2^3 = 8$ possible combinations of responses (response patterns) across the three items. A categorical variable was created with a category for each response pattern as described in Table 4.10.

TABLE 4.10: Description of response patterns on three binary indicators of confidence in purchasing from other EC countries (1=Confident, 0=Not confident)

Confident about purchasing ...	Response pattern
None of three types of product/service	000
Food/wine only	100
Household electrical only	010
Medical/financial services only	001
Any except medical/financial services	110
Any except household electrical appliances	101
Any except food/wine	011
Any of the products/services	111

A cross-tabulation of the response patterns on the three items by country of residence is shown in Table 4.11. The chi-squared statistic for this table is 3513.0 on 98 degrees of freedom, providing evidence of an association between confidence in shopping in the EC and country. To explore this association further, a CORA was carried out. We begin by examining the row profiles, that is the distribution of respondents across response patterns for each country (see Table 4.12). It can be seen that Spain and Greece have very similar row profiles. For example, respondents in these countries are the most likely to feel confident about purchasing any of the three types of products or services. Overall, there appears to be a division between Southern European countries

(Italy, Greece, Portugal, and Spain) and Northern European countries. If respondents in Southern Europe feel confident about purchasing only one of the three products or services, it is relatively unlikely to be food/wine. In contrast relatively high proportions of respondents in Northern Europe (particularly Belgium, Germany, and the UK) feel confident about buying food/wine only.

TABLE 4.11: Cross-tabulation of confidence in buying products/services from EC by country of residence, 1995

	Response pattern for three products/services								
Country	000	100	010	001	110	101	011	111	Total
Austria	203	140	59	27	194	88	39	262	1012
Belgium	115	278	11	9	156	158	2	293	1022
Denmark	68	211	21	13	247	79	4	358	1001
Finland	115	177	44	20	209	85	29	363	1042
France	165	165	89	23	203	61	31	281	1018
Germany	243	600	46	48	272	380	27	483	2099
Greece	138	26	56	26	66	29	114	551	1006
Ireland	74	244	20	14	163	130	8	410	1063
Italy	144	81	118	117	118	102	141	318	1139
Luxembourg	76	51	10	8	30	40	6	155	376
Netherlands	51	155	13	10	145	243	12	385	1014
Portugal	281	63	103	41	85	44	56	379	1052
Spain	177	56	65	36	86	74	49	476	1019
Sweden	75	211	23	15	166	182	22	381	1075
UK	102	369	15	17	251	192	9	402	1357
Total	2027	2827	693	424	2391	1887	549	5497	16295

TABLE 4.12: Row profiles for EC purchasing data

	Response pattern for three products/services								
Country	000	100	010	001	110	101	011	111	Row mass
Austria	0.201	0.138	0.058	0.027	0.192	0.087	0.039	0.259	0.062
Belgium	0.113	0.272	0.011	0.009	0.153	0.155	0.002	0.287	0.063
Denmark	0.068	0.211	0.021	0.013	0.247	0.079	0.004	0.358	0.061
Finland	0.110	0.170	0.042	0.019	0.201	0.082	0.028	0.348	0.064
France	0.162	0.162	0.087	0.023	0.199	0.060	0.030	0.276	0.062
Germany	0.116	0.286	0.022	0.023	0.130	0.181	0.013	0.230	0.129
Greece	0.137	0.026	0.056	0.026	0.066	0.029	0.113	0.548	0.062
Ireland	0.070	0.230	0.019	0.013	0.153	0.122	0.008	0.386	0.065
Italy	0.126	0.071	0.104	0.103	0.104	0.090	0.124	0.279	0.070
Luxembourg	0.202	0.136	0.027	0.021	0.080	0.106	0.016	0.412	0.023
Netherlands	0.050	0.153	0.013	0.010	0.143	0.240	0.012	0.380	0.062
Portugal	0.267	0.060	0.098	0.039	0.081	0.042	0.053	0.360	0.065
Spain	0.174	0.055	0.064	0.035	0.084	0.073	0.048	0.467	0.063
Sweden	0.070	0.196	0.021	0.014	0.154	0.169	0.020	0.354	0.066
UK	0.075	0.272	0.011	0.013	0.185	0.141	0.007	0.296	0.083
Column mass	0.124	0.173	0.043	0.026	0.147	0.116	0.034	0.337	

When a cross-tabulation has a large number of rows and columns (as in the case of Table 4.11), it can be difficult to pick out all of the important patterns in the data. CORA is particularly useful for analysing such tables. In the EC purchasing example, 7 dimensions are required to provide an exact representation of the row or column profiles (since the minimum of $(15-1)$ and $(8-1)$ equals 7). However, we hope that considerably fewer will be necessary to provide a good approximation. Table 4.13 shows the inertia on each dimension and the proportion of total inertia explained by each dimension. The first dimension is highly dominant accounting for almost 63% of the total inertia. The second and third dimensions each explain over 10% of inertia. We will consider the two-dimensional solution which accounts for 76.3% of inertia, although in practice it would be advisable to also examine the three-dimensional solution. We leave this as an exercise for the reader.

TABLE 4.13: Percentage of inertia explained by each dimension for the EC purchasing data

Dimension	Inertia	Inertia explained %	Cumulative %
1	0.135	62.6	62.6
2	0.030	13.7	76.3
3	0.026	11.8	88.1
4	0.016	7.4	95.5
5	0.006	2.9	98.4
6	0.002	1.1	99.4
7	0.001	0.6	100.0
Total	0.216		

The coordinates on each dimension and the contribution to inertia of each dimension are shown for rows and columns in Table 4.14 and Table 4.15, respectively. To aid interpretation, countries and response patterns have been ordered according to their position on the first dimension. The countries with the largest contributions to the inertia on dimension 1 are (Italy, Greece, Portugal, Spain) with positive coordinates and (UK, Belgium, Germany) with negative coordinates (see Table 4.14). Thus, in general, dimension 1 contrasts Southern European countries with Northern European countries. Turning to the column points (Table 4.15), we find that the response patterns with large contributions to inertia on dimension 1 are (011,010,001) with positive coordinates and (100,101) with negative coordinates. The response patterns with large positive coordinates indicate confidence in purchasing a combination of one or two items, neither of which include food/wine; the categories with large negative coordinates correspond to confidence in purchasing selected items which include food/wine. Thus dimension 1 distinguishes between respondents who feel confident in purchasing food/wine from another EC country and those who do not.

TABLE 4.14: Coordinates and contribution to inertia of row points for the EC purchasing data

Country	Coordinate		Contribution to inertia	
	Dim 1	Dim 2	Dim 1	Dim 2
UK	−0.618	−0.033	0.087	0.001
Belgium	−0.596	−0.085	0.061	0.003
Germany	−0.552	−0.404	0.107	0.122
Netherlands	−0.472	0.504	0.038	0.092
Ireland	−0.405	0.326	0.029	0.040
Denmark	−0.402	0.198	0.027	0.014
Sweden	−0.365	0.256	0.024	0.025
Finland	−0.042	0.071	0.000	0.002
Luxembourg	0.160	0.305	0.002	0.012
France	0.202	−0.427	0.007	0.066
Austria	0.217	−0.427	0.008	0.066
Spain	0.657	0.436	0.073	0.069
Portugal	0.912	−0.227	0.146	0.019
Greece	1.032	0.887	0.179	0.282
Italy	1.057	−0.676	0.213	0.186

TABLE 4.15: Coordinates and contribution to inertia of column points for the EC purchasing data

Response pattern	Coordinate		Contribution to inertia	
	Dim 1	Dim 2	Dim 1	Dim 2
100	−0.759	−0.306	0.272	0.094
101	−0.617	−0.039	0.120	0.001
110	−0.327	−0.152	0.043	0.020
111	0.181	0.528	0.030	0.547
000	0.507	−0.334	0.087	0.081
001	1.021	−0.985	0.074	0.147
010	1.089	−0.664	0.137	0.109
011	1.607	−0.083	0.237	0.001

We now consider the interpretation of dimension 2. Starting with row categories, we find that Greece (and to a lesser extent Spain) are contrasted with (Italy, France, Austria, Germany) (see Table 4.14). At this stage, it is not clear why countries should be grouped in this way, but we will return to these groupings after considering the interpretation of dimension 2 with respect to column categories. Of the column categories, dimension 2 is dominated by the group of respondents who feel confident in buying any of the three types of products or services (response pattern 111 with a contribution to inertia on dimension 2 of 0.547 — see Table 4.15). These respondents are contrasted with a group who would be confident about buying financial/medical services only

(001) or household electrical goods only (010). This dimension may be loosely interpreted as a measure of the degree of confidence in purchasing goods or services from other EC countries. Response patterns located at the negative end of this dimension indicate confidence in buying selected products only, while the 111 pattern with a positive score indicates confidence in buying any products or services.

Next, we consider the association between row and column categories by considering the location of row and column points jointly. Figure 4.7 shows a symmetric biplot for the two-dimensional solution. The Southern European countries lie on one side of dimension 1, together with the response patterns which indicate lack of confidence in buying food/wine from other European countries; Northern European countries tend to lie at the other extreme of dimension 1, along with patterns indicating confidence in buying food/wine. This distinction may be partly explained by Spain, Portugal, and Italy being wine-producing countries; respondents in these countries may be expressing the attitude that it is unnecessary to purchase wine from elsewhere. Dimension 2 has been interpreted as an indicator of the degree of confidence in purchasing from other EC countries. Located at the positive end of this dimension towards the 111 response pattern are Greece and Spain; Italy is located at the negative end of dimension 2 indicating less confidence in purchasing elsewhere.

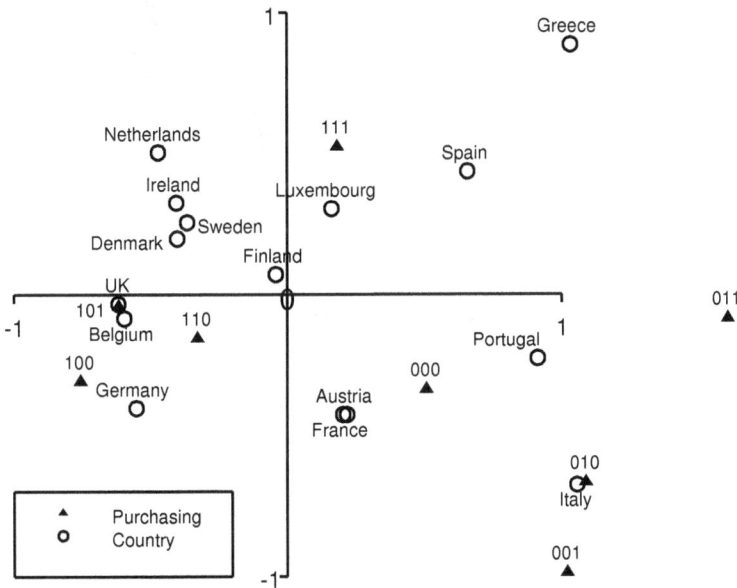

FIGURE 4.7: Biplot for cross-tabulation of confidence in purchasing from EC country by country of residence

To further evaluate the goodness-of-fit of the two-dimensional solution, we examine the contributions of dimensions 1 and 2 to the inertia of row and column points (Table 4.16). While many row and column categories are well represented in one dimension (e.g. Belgium, UK, 100 and 011), others are not (e.g. Finland, Luxembourg and 111). Finland and Luxembourg are not well represented in dimension 2 either, so their position in the two-dimensional solution relative to other countries and response patterns must be interpreted with caution. It is not surprising that some categories are not adequately represented in two dimensions since 24% of inertia remains unexplained by the first two dimensions. The addition of a third dimension may improve the representation of those categories which are poorly represented by two dimensions. In contrast, we find that response pattern 111 is well represented by a second dimension (also reflected in the high contribution of this column category to the inertia on dimension 2 — see Table 4.15).

TABLE 4.16: Contributions of dimensions to inertia of points for the EC purchasing data

		Contribution of dimension		
		Dim 1	Dim 2	Total
Country	Austria	0.174	0.314	0.489
	Belgium	0.914	0.009	0.922
	Denmark	0.375	0.042	0.417
	Finland	0.019	0.025	0.044
	France	0.128	0.268	0.395
	Germany	0.661	0.166	0.827
	Greece	0.697	0.241	0.938
	Ireland	0.671	0.203	0.875
	Italy	0.637	0.122	0.759
	Luxembourg	0.077	0.132	0.210
	Netherlands	0.357	0.190	0.547
	Portugal	0.751	0.022	0.773
	Spain	0.743	0.153	0.896
	Sweden	0.658	0.152	0.811
	UK	0.948	0.001	0.949
Purchasing	000	0.452	0.092	0.543
	100	0.872	0.066	0.938
	010	0.792	0.138	0.930
	001	0.504	0.220	0.724
	110	0.347	0.035	0.382
	101	0.579	0.001	0.581
	011	0.811	0.001	0.812
	111	0.198	0.790	0.988

4.8 Correspondence analysis of multi-way tables

So far, we have considered using CORA for the analysis of two-way tables. In one sense, therefore, we have been dealing only with the bivariate problem — one variable for the rows and one for the columns. Extensions to simple CORA, called multiple or joint CORA, have been developed for cross-classifications of more than two variables. All of them involve turning the multi-way table into a two-way table. We shall show, by means of an example, one way in which this may be done and then briefly outline a way of looking at the two-way table which generalises immediately to tables of any dimension.

A method where one variable is treated as dependent

We consider again the US education and attitude to abortion data, but now suppose we were interested in the association between attitude to abortion, education level and a third variable, religion. We could create a single new variable, called an *interactive* variable, with categories for each combination of education level and religion. Note that we could have combined education and attitude, or religion and attitude, in the same way to form a two-way table. However, it makes more substantive sense in this particular case to treat attitude as a dependent variable which is determined by combinations of the other variables. The new education/religion variable is cross-classified with attitude to abortion to create a two-way table (Table 4.17). The two-dimensional symmetric biplot from a CORA of these data is shown in Figure 4.8.

TABLE 4.17: Attitude to abortion by education and religion, US, 1972-74

| Education | Religion | Attitude | | | |
		Positive	Mixed	Negative	Total
≤ 8	Northern Protestant	49	46	115	210
9-12	Northern Protestant	293	140	277	710
≥ 13	Northern Protestant	244	66	100	410
≤ 8	Southern Protestant	27	34	117	178
9-12	Southern Protestant	134	98	167	399
≥ 13	Southern Protestant	138	38	73	249
≤ 8	Catholic	25	40	88	153
9-12	Catholic	172	103	312	587
≥ 13	Catholic	93	57	135	285
	Total	1175	622	1384	3181

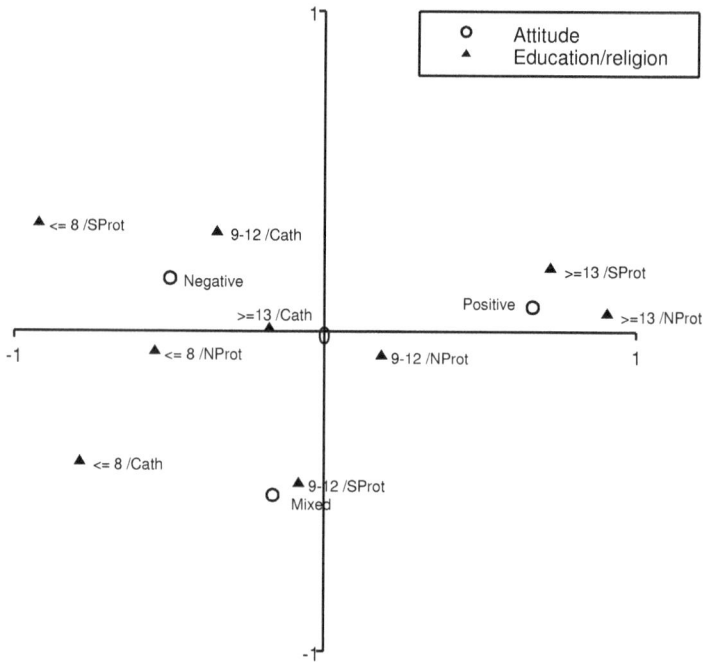

FIGURE 4.8: Symmetric biplot from cross-tabulation of education-religion by attitude to abortion, US, 1972-74

Looking first at the scores derived for the row and column categories, we notice that attitudes move from "negative" to "positive" through "mixed" along dimension 1, whereas dimension 2 separates those with a mixed view from those who have a definite view one way or the other. For the education/religion categories, dimension 1 corresponds roughly to the "amount of education" variable with less than nine years on the left and 13 or more on the right. There is no such clear grouping for religion, suggesting that it is the interaction of education and religion rather than religion itself which counts.

This becomes clearer if we look at the biplot (Figure 4.8). We can see that Protestants with ≥ 13 years of education have a relatively positive attitude to abortion, while Catholics with the same level of education tend to have a negative attitude. Those with less education, of both religions, seem to have a more negative or mixed attitude.

Direct analysis of the "data matrix"

The starting point of our discussion was the two-way contingency table which is very familiar from any elementary statistics course. However, there is much to be learnt if we go back a stage and see the table as the first step in summarising a data matrix. We introduced the idea of a data matrix in Chapter 1 and have shown in Chapters 2 and 3 how to construct distance matrices from it on the way to cluster analysis or MDS. The data expressed in a two- (or multi-) way contingency table can also be set out in a data matrix.

In the standard data matrix, the rows correspond to objects (individuals, sample members, etc.) and the columns to variables. The entry in any cell of the table is the value of the variable indicated at the head of the column for the object in the row. In the examples we have so far considered, the value of the variable was indicated by a single number — a 1 or 0 in the case of binary variables or a scale value for a metrical variable. For categorical variables, we use several numbers to indicate the category into which an individual falls. This is done by what is known as an indicator vector. Thus for example, if we consider the "religion" variable above listed in the order Northern Protestant, Southern Protestant and Catholic, the indicator vector would consist of 0s and 1s with 1 indicating the category into which the object falls. Thus a Southern Protestant would be recorded as (010) and a Catholic as (001). This notation extends immediately to any number of categories; (001000) would indicate an object which falls into the third category of a six-category variable. Instead of having a single number in each cell of the data matrix, we therefore have a vector. (We could economise on notation and delete one element of the vector because if we know all elements but one, we can deduce what the other must be. This is exactly what we do if we have binary data where in the data matrix we only record whether or not the object is in the first category. This halves the size of the data matrix and means that we only have to deal with single numbers rather than vectors. That advantage is lost when we move to more than two categories, and it turns out to be more convenient to use the full indicator variables as we have defined them.)

A typical row of a data matrix for six variables might appear as follows:

$$100 : 01000 : 10 : 010 : 0001 : 01000$$

The colons are included only for clarity and the rows would normally be written without colons or spaces. Such a matrix is sometimes called the *super-indicator matrix*.

The data matrix, written in this fashion, is a two-way array and can be subjected to correspondence analysis exactly like any other two-way table. The great merit of looking at the general problem in this way is that the two-way table is not now special in any way but can be included within the framework of a form of multiple correspondence analysis.

It is clear that an analysis carried out on such a data matrix will yield scores for both columns (categories of the variables) and rows (objects). The whole analysis can thus be viewed as a scaling exercise for both objects and categories.

We have described the data matrix as it arises when the categories are mutually exclusive. However, this representation may also be used quite generally. If an individual gives responses in more than one category, there will be a 1 entered in each column for which a response is given. In the "mutually exclusive case", the sum of every row is p, the number of variables. This is because there is precisely one 1 for each variable. In the general case, the row totals will be greater than p. In both cases, the column sums are the marginal totals for each category of each variable. For example, the first column refers to the first category of the first variable and its sum is simply the number of times an object falls into that category.

Apart from this property concerning the column sums, the data matrix looks very different to the contingency table, and it is not obvious that correspondence analysis of the two arrays will lead to essentially the same result. The fact that it does results from properties of the singular value decompositions of the two matrices which lie beyond the scope of this book.

4.9 Additional examples and further work

We give two further examples for you to work through, reminding you that we have already suggested some further analysis to be carried out on the example in Section 4.7. Both of the following examples demonstrate the usefulness of CORA as a way of identifying key patterns in fairly large two-way tables. For large tables such as these, a large number of dimensions is required in order to represent the data perfectly, but it turns out that in each case a two-dimensional solution provides a very good representation. We therefore focus on the interpretation of the first two dimensions.

As you work through these examples, you are encouraged to carry out some analyses of your own. For instance, you should assess the goodness-of-fit of the one- or two-dimensional solutions by examining scree plots and the contribution of each dimension to the inertia of row and column points. You might also examine three-dimensional solutions. Some suggestions for further analysis are given below.

Newspaper readership by occupation in the UK

Table 4.18 shows a cross-tabulation of newspaper readership by occupation in the UK. The data are from the 1999 Eurobarometer Survey (Melich, 1999). Respondents were asked which of the following newspapers they read most regularly:

1. Daily Mirror/Daily Record/Sunday Mirror
2. Sun/News of the World

3. Daily or Sunday Mail

4. Daily or Sunday Express/Sunday People

5. Times or Sunday Times

6. Daily or Sunday Telegraph

7. Guardian/Observer/Independent/Independent on Sunday

8. Other

9. None

Monday to Saturday and Sunday versions of the same newspaper have been grouped, as have two broadsheet newspapers (the Guardian and the Independent). For readers not familiar with British newspapers, we give a brief description here. First, a distinction used to be made between tabloids and broadsheets. Newspapers in categories 1-4 above were tabloids, while 5-7 were broadsheets. A further distinction can be made based on political ideology. Newspapers 1, 2 and 7 were generally considered centre or left-of-centre, while 3, 4 and 6 were right-of-centre. The Times (5) was generally thought to be somewhere between the Guardian/Independent (7) and the Telegraph (6).

TABLE 4.18: Cross-tabulation of newspaper read most regularly by occupation, UK, 1999

	Newspaper									
Occupation	Mirr	Sun	Mail	Expr	Times	Teleg	Guard	Other	None	Total
Self-employed	8	16	12	6	6	4	7	2	6	67
Manager	9	14	19	8	17	10	18	12	9	116
Other white collar	20	20	20	9	7	10	6	12	8	112
Manual worker	90	103	34	28	10	4	7	30	28	334
Keeping house	51	64	19	9	9	4	6	24	29	215
Unemployed	22	30	6	1	2	3	5	12	11	92
Retired	60	50	52	31	8	19	7	48	31	306
Student	15	17	10	5	10	3	7	11	6	84
Total	275	314	172	97	69	57	63	151	128	1326

A chi-squared test provides evidence of an association between newspaper readership and occupation ($X^2 = 204.23$, degrees of freedom $= 56$, $p < 0.001$). CORA may be carried out to obtain a low-dimensional representation of the two-way table. Seven dimensions are needed to provide a perfect representation. However, you will find that the first dimension explains 64.1% of the total inertia while the second explains a further 23.3%; therefore the first two dimensions explain 87.4% of inertia. There is a sharp decline in the inertia explained after the second dimension, suggesting that two dimensions should be adequate. As a further check of the goodness-of-fit of the two-dimensional solution, you should look at the contributions of each dimension to the inertia of the newspaper and occupation points.

Tables 4.19 and 4.20 show the coordinates and contribution to inertia from the two-dimensional solution for row (occupation) and column (newspaper) points respectively. To assist interpretation of the first dimension, occupations and newspapers have been ordered according to their location on dimension 1. You should begin with an interpretation of the first dimension. From Table 4.19, you will see that Manager (with a large positive coordinate) and Manual (with a large negative coordinate) make the largest contributions to inertia on dimension 1. Therefore, dimension 1 might be labelled "social class". Table 4.20 shows that the Times and Guardian/Independent (with large positive coordinates) and the Mirror and Sun (with large negative coordinates) make the largest contributions to inertia on dimension 1. The Telegraph also makes a moderate contribution to inertia and has a large, positive coordinate. Thus, this dimension distinguishes between left-of-centre tabloids and broadsheets. The occupation and newspaper categories are plotted on a biplot in Figure 4.9. On dimension 1, you can see that the broadsheets are located close to Manager; the Mirror and Sun are located close to Manual and Unemployed and, somewhat surprisingly, Keeping house.

TABLE 4.19: Coordinates and contribution to inertia of row points for the UK newspaper data

	Coordinate		Contribution to inertia	
Occupation	Dim 1	Dim 2	Dim 1	Dim 2
Manual	−0.508	−0.145	0.207	0.028
Keeping house	−0.412	−0.300	0.088	0.077
Unemployed	−0.403	−0.459	0.036	0.077
Retired	0.031	0.707	0.001	0.609
Other white collar	0.410	0.356	0.045	0.056
Student	0.468	−0.399	0.044	0.053
Self-employed	0.640	−0.339	0.066	0.031
Manager	1.359	−0.387	0.514	0.069

Among occupation categories (Table 4.19), dimension 2 is dominated by Retired which is quite distinct from the other categories, though closest to Other white collar. Turning to the newspapers (Table 4.20), a contrast can be seen between right-of-centre papers (Express, Mail, Telegraph) and more left-of-centre papers (Guardian/Independent, Times, Sun). This suggests that retired people are relatively more likely to read right-of-centre newspapers.

Some further observations can be made from the biplot in Figure 4.9. The self-employed and students have similar profiles with respect to their newspaper preferences, as do manual workers and the unemployed. Newspapers which have similar profiles across occupation categories are the Express and Mail (right-of-centre tabloids), Sun and Mirror (left-of-centre tabloids), and Times and Guardian/Independent (centre to left-of-centre broadsheets). The Telegraph appears quite distinct from the other newspapers.

TABLE 4.20: Coordinates and contribution to inertia of column points for the UK newspaper data

	Coordinate		Contribution to inertia	
Newspaper	Dim 1	Dim 2	Dim 1	Dim 2
Mirror	−0.476	−0.021	0.150	0.000
Sun	−0.444	−0.387	0.148	0.188
None	−0.186	−0.039	0.011	0.001
Other	−0.017	0.400	0.000	0.096
Express	0.111	0.587	0.003	0.133
Mail	0.379	0.463	0.059	0.147
Telegraph	0.969	0.687	0.128	0.107
Times	1.160	−0.728	0.223	0.146
Guardian/Independent	1.356	−0.851	0.278	0.182

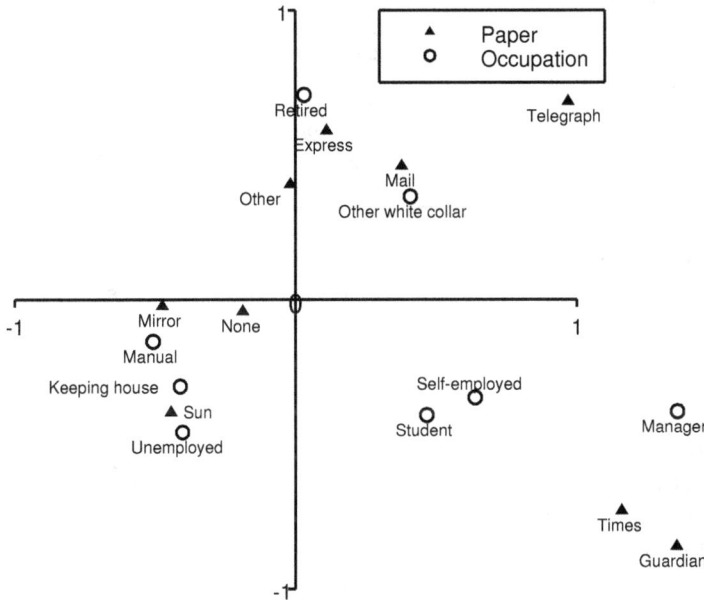

FIGURE 4.9: Biplot for cross-tabulation of newspaper readership by occupation

Contraceptive method choice in Indonesia

Table 4.21 shows a cross-tabulation of choice of contraceptive method by age in Indonesia. The data are from the Indonesia Demographic and Health Survey of 1997 (CBSI 1998) and consist of 26833 women who were married at the time of the survey.

A chi-squared test provides very strong evidence of an association between contraceptive method and age (X^2=2644.98, degrees of freedom = 42, $p <$ 0.001). A correspondence analysis may be carried out to determine whether the patterns of contraceptive method choice across age groups may be represented in a small number of dimensions. Table 4.21 may be perfectly represented in six dimensions. However, you will find that the first dimension explains 73.0% of inertia and the first and second dimensions together explain 98.7%. This suggests that we should focus on the two-dimensional solution.

TABLE 4.21: Cross-tabulation of contraceptive method choice by age, Indonesia, 1997

Method								

Method	15-19	20-24	25-29	30-34	35-39	40-44	45-49	Total
Pill	138	641	1009	902	852	491	219	4252
IUD	16	139	281	421	511	476	295	2139
Injectable	186	1037	1326	1126	837	398	133	5043
Implant	44	203	339	327	289	126	45	1373
Other modern	1	7	23	42	65	39	15	192
Sterilisation	0	1	24	128	220	266	198	837
Traditional	7	64	142	205	231	185	124	958
None	674	1789	2262	1958	1928	1579	1849	12039
Total	1066	3881	5406	5109	4933	3560	2678	26833

Tables 4.22 and 4.23 show the coordinates and contributions of points to the inertia of each dimension for the two-dimensional solution, for row (method) and column (age) points respectively. Starting with the interpretation of dimension 1 with respect to contraceptive method (Table 4.22), Sterilisation and IUD (with negative scores) and Injectable (with a positive score) make the largest contribution to the inertia on this dimension. Dimension 1 contrasts clinical methods (Sterilisation and IUD) with hormonal methods (Injectable, Implant, Pill). From Table 4.23, you can see that the first dimension also contrasts young women with older women. The clinical methods tend to be used for longer durations (or permanently in the case of sterilisation) by older women who have achieved their desired number of children, while hormonal methods are typically seen as shorter-term methods and are predominantly used by younger women who wish to space rather than limit births.

Turning to the interpretation of the second dimension, the contraceptive method category "None" accounts for a large part of the inertia on this dimension. Notice also that "None" has the only negative coordinate on this dimension. Dimension 2 contrasts non-users (with a negative score) with users of any method (each method has a positive score). In Table 4.23, very young women (age 15-19) and the oldest age group are contrasted with women in their thirties and early forties. If you look at a biplot, you should find that the oldest women are located quite close to the "non-use" category on the second dimension since these categories have negative scores of a similar magnitude; reasons for high non-use among the 44-49 age group might include a low perceived risk of conception and conservative attitudes to family planning. The relatively high level of non-use for 15-19 year-olds could reflect early childbearing for women who marry at a very young age.

TABLE 4.22: Coordinates and contribution to inertia of row points for the Indonesia contraception data

Method	Coordinate		Contribution to inertia	
	Dim 1	Dim 2	Dim 1	Dim 2
Pill	0.319	0.305	0.060	0.093
IUD	−0.692	0.571	0.142	0.163
Injectable	0.697	0.133	0.340	0.021
Implant	0.471	0.416	0.042	0.056
Other modern	−0.556	1.140	0.008	0.058
Sterilisation	−1.660	0.519	0.320	0.053
Traditional	−0.540	0.582	0.039	0.076
None	−0.168	−0.413	0.047	0.481

TABLE 4.23: Coordinates and contribution to inertia of column points for the Indonesia contraception data

Age (years)	Coordinate		Contribution to inertia	
	Dim 1	Dim 2	Dim 1	Dim 2
15-19	0.229	−1.053	0.008	0.277
20-24	0.563	−0.316	0.171	0.091
25-29	0.484	−0.031	0.176	0.001
30-34	0.189	0.278	0.025	0.093
35-39	−0.161	0.394	0.018	0.180
40-44	−0.696	0.291	0.240	0.071
45-49	−0.953	−0.655	0.363	0.288

To assess further the fit of the one- and two-dimensional solutions, you should examine the contribution of each dimension to the inertia of the method and age group points.

4.10 A further topic: Goodman's RC model

The correspondence analysis methods discussed above can be seen as descriptive approaches to analysing and visualising the relationship between two or more categorical variables, which do not involve any parametric model assumptions. Alternatively, one can solve the same problem using a probabilistic approach based on Goodman's Row-and-Column (RC) model (Goodman 1979; Goodman 1981; Goodman 1991), which is closely related to the factor models discussed in Chapters 7–9. This approach finds low-dimensional representations of row and column profiles by estimating a statistical model. An advantage of this model-based approach to correspondence analysis is that standard statistical inference tools, such as maximum likelihood estimation, can be used to estimate the row and column scores and construct confidence intervals. In what follows, we briefly introduce the RC model and connect it to the descriptive methods discussed above. The statistical inferences about the RC model are omitted here, and interested readers can refer to Chapter 6, Kateri (2014) for the details, including the computational implementations.

Consider an $I \times J$ matrix of observed frequencies, where I and J are the numbers of rows and columns, respectively, and denote the observed frequency for the cell in row i and column j of the table by O_{ij}. The RC model assumes that O_{ij}, $i = 1, ..., I$, $j = 1, ..., J$, are independent random variables, each following a Poisson distribution. The expected value of O_{ij}, denoted by E_{ij}, satisfies

$$E_{ij} = \exp\left(\alpha_i + \beta_j + \sum_{k=1}^{R} \lambda_k^{1/2} u_{ik} v_{jk}\right),$$

where α_i, β_j, λ_k, u_{ik}, and v_{jk}, $i = 1, ..., I$, $j = 1, ..., J$, $k = 1, ..., R$, are unknown model parameters, and R is a pre-specified value that can be chosen by model selection. The row and column scores u_{ik} and v_{jk} satisfy standardisation constraints

$$\sum_{i=1}^{I} u_{ik} = 0, \quad \sum_{j=1}^{J} v_{jk} = 0, \quad \sum_{i=1}^{I} u_{ik}^2 = 1,$$

$$\sum_{j=1}^{J} v_{jk}^2 = 1, \quad \sum_{i=1}^{I} u_{ik} u_{ik'} = 0, \text{ and } \sum_{j=1}^{J} v_{jk} v_{jk'} = 0$$

for $k \neq k'$. We intentionally let λ_k, u_{ik} and v_{jk} follow the same notations as those in the singular value decomposition of the Pearson residuals in correspondence analysis due to their similarities explained below.

Given the standardisation constraints on the row and column scores, the parameters α_i and β_j together capture the average row and column profiles. If $\sum_{k=1}^{R} \lambda_k^{1/2} u_{ik} v_{jk} = 0$ for all i, j, then the expected count has the factorisation $E_{ij} = \exp(\alpha_i) \exp(\beta_j)$, which implies independence of the row and column variables. The term $\sum_{k=1}^{R} \lambda_k^{1/2} u_{ik} v_{jk}$ can be interpreted as the cell's deviation from the expected count under the independence model, mimicking the singular value decomposition of Pearson residuals in correspondence analysis (the row and column scores derived from singular value decomposition satisfy similar standardisation constraints). When R is set to two or three, the row and column scores can be plotted as in correspondence analysis to visualise the dependence structure among the row and column categories.

The RC model can be extended to analyse multi-way frequency tables; see, for example, Chapter 6, Kateri (2014).

4.11 Further reading

Greenacre, M. and Blasius, J. (Eds.) (1994). *Correspondence Analysis in the Social Sciences*. San Diego: Academic Press.

Greenacre, M. and Blasius, J. (Eds.) (2006). *Multiple Correspondence Analysis and Related Methods*. New York: Chapman and Hall/CRC.

Michailidis, G. and De Leeuw, J. (1998). The Gifi system of descriptive multivariate analysis. *Statistical Science*, 13, 307–336.

Chapter 6, Kateri, M. (2014). *Contingency Table Analysis. Methods and Implementation Using R*, Aachen: Springer.

5

Principal Components Analysis

5.1 Introduction

The main aim of principal components analysis (PCA) is to replace p continuous correlated variables by a much smaller number of uncorrelated variables which contain most of the information in the original set. This greatly simplifies the task of understanding the structure of the data since it is much easier to interpret two or three uncorrelated variables than 20 or 30 that have a complicated pattern of interrelationships. In order to translate this objective into a practical method, we have to be more precise about what it is to retain "most of the information".

The central idea is based on the concept of the proportion of the *total variance* (the sum of the variances of the p original variables) that is accounted for by each of the new variables. PCA transforms the set of correlated variables $(y_1, \ldots, y_p)$ to a set of uncorrelated variables $(z_1, \ldots, z_p)$ called principal components, in such a way that z_1 explains the maximum possible of the total variance, z_2 the maximum possible of the remaining variance, and so on. The full set of p principal components fully explains the total variance:

$$\sum_{j=1}^{p} \text{var}(z_j) = \sum_{i=1}^{p} \text{var}(y_i).$$

However, if it turns out that the first few principal components account for a large enough part of the total variance, most of the variation in the ys being explained by the first few zs, then the remaining principal components can be discarded without too great a loss of information. It is usual to standardise the ys to unit variance before carrying out PCA so that each y-variable makes the same contribution to the total variance, and thus

$$\sum_{i=1}^{p} \text{var}(y_i) = p.$$

Another aim of PCA is to interpret the underlying structure of the data in terms of the most important principal components. Often the principal

DOI: 10.1201/9781003483342-5

components may be identified with some quantity of substantive interest. For example, the first principal component is frequently found to be positively correlated with each of the *y*s so that it can be interpreted as a measure of what is common to all the variables. Suppose we had measurements on the height, footsize, span, weight, waist and hip measurements of adult men. Because these six variables are all positively correlated with each other the first component, a measure of size, will be positively correlated with all of them. Sometimes components contrast one subset of the variables with another subset. Such components may be interpreted by thinking about what each subset of variables has in common. So, for the body measurements described above, the second component could be a contrast between the subset (height, footsize, span) and the subset (weight, waist, hip). The first component distinguishes between men by size, small men at one end and large at the other end of the scale. The second component has men who are relatively fat for their size at one extreme and men who are relatively thin at the other. The examples in Sections 5.5, 5.6, and 5.9 provide further illustrations of contrast effects and their interpretation.

PCA is analogous to correspondence analysis (CORA), but the two methods are applied to different types of data. PCA is a method for reducing the dimensionality of a set of correlated continuous variables, while CORA is a method for reducing the dimensionality of a cross-tabulation of associated categorical variables. In CORA, the dimensions are derived in order of importance in the sense that the first dimension explains the largest proportion of Pearson's chi-squared statistic or, equivalently, inertia. In PCA, the components are also derived in order of importance but in terms of the proportion of variance explained. PCA is also similar to a multidimensional scaling (MDS) analysis of variables. Both methods aim to find good low-dimensional representations of the original data. However, PCA aims to approximate the sample covariance matrix, while MDS aims to approximate a distance or dissimilarity matrix of the variables.

We also note that PCA is closely related to factor analysis which we shall introduce in Chapter 7. Factor analysis, in particular the exploratory version of factor analysis, can be viewed as a statistical-model-based alternative to PCA. See Chapter 7 for a comparison between the two approaches.

5.2   Some potential applications

i) Suppose we have examination scores in different subjects for a set of individuals. We would like to combine these in some way to obtain an overall measure of academic ability. From a PCA of the data, we find that the first component is positively correlated with each examination score. We interpret this component as a measure of general ability. However, the second component also explains a large amount of the variance in examination scores and contrasts science subjects with humanities subjects. From this, we conclude that ability may not be captured adequately by a single variable; there is a difference between ability in science subjects and ability in humanities subjects.

ii) Suppose we are interested in constructing a measure of deprivation. We have several indicators that could be thought of as deprivation measures. Each measures some slightly different aspect of deprivation and we somehow want to extract what is common to all of them to get at the core of the concept. What do we do? One approach would be to carry out a PCA on the set of indicators which would play the role of the y-variables. If the first component explains a large proportion of the total variation of the original deprivation measures, it could be used in place of the original variables as a measure of deprivation.

iii) We may wish to find a good low-dimensional representation of our raw data before carrying out further analyses (such as cluster analysis or multiple regression) by reducing a large number of highly correlated variables to a few independent principal components (whilst sacrificing as little information as possible).

5.3 Illustration of PCA for two variables

Before dealing with PCA itself, we will give a simple example to show why transformations of the data might achieve the object we desire. Suppose we have observations on two variables, y_1 and y_2, which have the same scale and origin of measurement. Suppose also that they are highly correlated; this means that if we were to plot them they would lie close to the 45 degree line through the origin. Next we transform them to two new variables $z_1 = (y_1 + y_2)/\sqrt{2}$ and $z_2 = (y_2 - y_1)/\sqrt{2}$. Because y_1 and y_2 are highly correlated, it is clear that the second variable will have small values and will vary little; z_1 on the other hand, will have a much larger variance. The difference in the ys, therefore tells us very little about the variation between individuals whereas the sum tells us much more. In that sense, z_1 contains most of the information in the two variables about the variation among individuals. We now develop this idea into a full PCA using an example with two variables. Suppose that we have two y-variables, each having a variance 1, and that the correlation between them is 0.90. A scatterplot is shown in Figure 5.1.

Finding the principal components for two variables involves an orthogonal rotation of the axes. An orthogonal rotation is one where the axes are kept at right angles to one another. The first principal component will be in the direction of greatest variance. This is found by minimising the sum of the squared perpendicular distances from the observations to the first component. Once the first component is positioned, the second component is fixed since it must be orthogonal (at right angles) to the first. The dashed lines in Figure 5.2 represent the two principal components for the data in Figure 5.1.

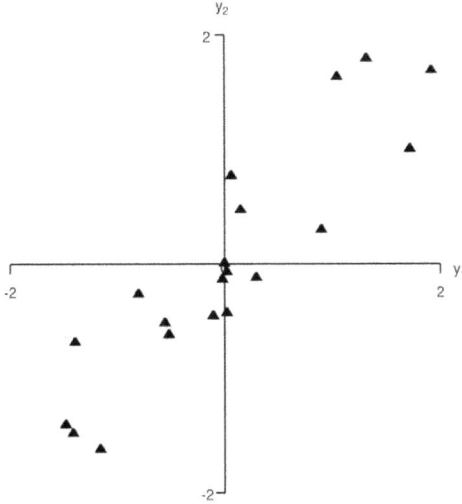

FIGURE 5.1: Scatterplot of two variables, y_1 and y_2, with equal variance and correlation 0.90

If the variances of y_1 and y_2 are equal and the correlation between them is positive, as in Figure 5.2, then the first component will always lie at a 45 degree angle to the y_1 and y_2 axes. If y_1 and y_2 had unequal variance, the first component would lie closer to the axis with greater variance. Having found the principal components, we could plot our observations taking the components as our new axes. A scatterplot of the observations with the components as axes is shown in Figure 5.3.

Note that the proportion of total variance explained by the first component will depend on the degree of correlation between y_1 and y_2. The higher the correlation, the closer the observations will lie to the first component and the greater the proportion of variance explained by the first component. The variance of the first principal component is 1.90 and that of the second is 0.10. The original total variance of 2 is now apportioned unequally with the much greater part allocated to the first component.

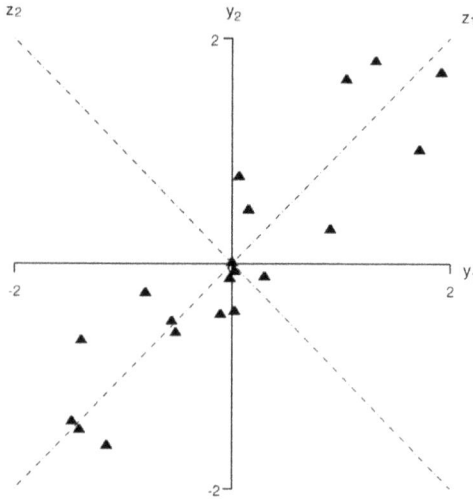

FIGURE 5.2: The principal components (z_1 and z_2) for two correlated variables (y_1 and y_2) with equal variance

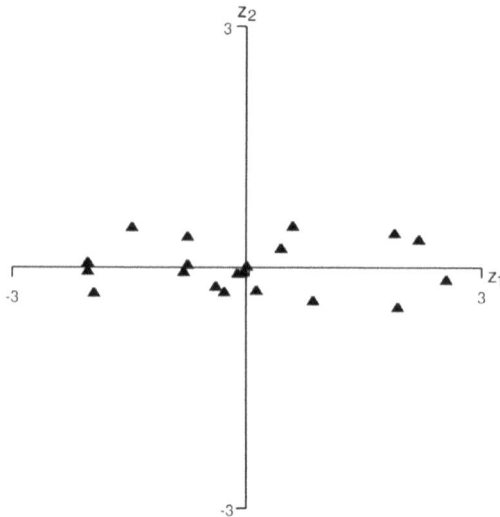

FIGURE 5.3: Scatterplot of the data in Figure 5.1 taking the principal components as axes

Although it is less easy to visualise, we can imagine the same approach being applied with three y-variables. The scatterplot for three positively correlated variables such as height, footsize, and span will show a cluster of points with a shape similar to a slightly flattened rugby football. The first principal component will then lie along the major axis of the ball, the second — at right angles — along the largest diameter through the centre, and the third mutually at right angles to the first two.

In the example given above, the principal components could be found, approximately, by eye. This is sufficient to show that the answer we get will depend on the variances of the ys. In Figure 5.1, the variances were both 1, but we noted that it was easy to see what would happen if the variance of y_1, say, had been much greater than that of y_2. This would be equivalent to stretching the horizontal axis, the effect of which would be to reduce the slope of the line representing the first principal component. This makes the first component more like y_1. In general, the larger the variance of any variable, the more dominant its role. However, in many applications, especially in social sciences, the units of measurement are arbitrary and so any analysis whose results depend on these arbitrary scalings is practically meaningless. We can arrange that all the variables carry equal weight by first standardising them so that they each have unit variance. In that case, the analysis depends only on the correlation matrix. Unless the original scale of the variables is meaningful, the first step in a PCA is, therefore, to compute the correlation matrix. The (ik)th element of the correlation matrix is the correlation between y_i and y_k as defined below. We note that a similar scaling problem occurs in the K-means clustering method, as discussed in Chapter 2, where the result we get also depends on how the variables are scaled and, thus, a standardisation step is typically needed.

Formulae for means, variances, covariances and correlations

Consider a sample of n observations on variables y_i and y_k, say. The sample means are

$$\bar{y}_i = \frac{1}{n} \sum_{t=1}^{n} y_{ti} \text{ and } \bar{y}_k = \frac{1}{n} \sum_{t=1}^{n} y_{tk}.$$

The sample variances are

$$s_i^2 = \frac{1}{(n-1)} \sum_{t=1}^{n} (y_{ti} - \bar{y}_i)^2 \text{ and } s_k^2 = \frac{1}{(n-1)} \sum_{t=1}^{n} (y_{tk} - \bar{y}_k)^2.$$

The standard deviations, s_i and s_k are the square roots of the variances. The covariance between y_i and y_k is

$$\text{cov}(y_i, y_k) = \frac{1}{(n-1)} \sum_{t=1}^{n} (y_{ti} - \bar{y}_i)(y_{tk} - \bar{y}_k).$$

The correlation between y_i and y_k is

$$\text{corr}(y_i, y_k) = \frac{\text{cov}(y_i, y_k)}{s_i s_k}.$$

5.4   An outline of PCA

PCA transforms a set of correlated variables (ys) into a set of uncorrelated components (zs). The principal components are linear combinations of the ys which we write as:

$$
\begin{aligned}
z_1 &= a_{11}y_1 + a_{21}y_2 + \cdots + a_{p1}y_p, \\
z_2 &= a_{12}y_1 + a_{22}y_2 + \cdots + a_{p2}y_p, \\
&\;\;\vdots \\
z_p &= a_{1p}y_1 + a_{2p}y_2 + \cdots + a_{pp}y_p.
\end{aligned}
$$

Each component is a weighted sum of the ys, where the $a_{ij}s$ are the *weights*, or *coefficients*.

It is clear that there have to be some constraints on the $a_{ij}s$. Otherwise we could make the variance of any z as large as we pleased simply by making the $a_{ij}s$ large enough. We can see what is required by going back to the treatment for two variables. We arrived at the principal components by rotating the axes while keeping them at right angles (orthogonal). In the general case, we have to find the equivalent algebraic formulation for orthogonal rotation. It turns out that this requires that the $a_{ij}s$ must satisfy the following conditions:

$$\sum_{i=1}^{p} a_{ij}^2 = 1 \qquad (j = 1, 2, \ldots, p),$$

and

$$\sum_{i=1}^{p} a_{ij} a_{ik} = 0 \qquad (j \neq k; \; j = 1, \ldots, p; \; k = 1, \ldots, p).$$

Another way of describing what these conditions do is to say that they leave the relative positions or the configuration of the points unchanged.

An important consequence of the orthogonality condition is that, as stated in Section 5.1, the total variance of the zs is equal to the total variance of the ys, that is

$$\sum_{j=1}^{p} \text{var}(z_j) = \sum_{i=1}^{p} \text{var}(y_i).$$

This means that the total variance does not change; rather variance is re-distributed among the variables. The zs are derived in decreasing order of importance such that z_1 has maximum variance and, therefore, explains the largest proportion of the total variance. The first principal component may thus be thought of as the best one-dimensional summary of the data. The second component, z_2, is derived so that it has the second largest variance, subject to the constraints

$$\sum_{i=1}^{p} a_{i2}^2 = 1 \quad \text{and} \quad \sum_{i=1}^{p} a_{i1}a_{i2} = 0,$$

so that it is orthogonal to (uncorrelated with) z_1. The first two components provide the best two-dimensional summary of the data. Subsequent compo-nents are derived in decreasing order of variance, each component being un-correlated with the previous components.

The mathematical problem that this procedure poses, therefore, is to find a method of determining the a_{ij}s so that the components have the required properties. Although this seems a formidable problem, it is easily solved be-cause it turns out to be equivalent to a well-known problem in matrix algebra concerned with finding what are called the eigenvalues and eigenvectors of a matrix — the matrix in this case being either the covariance matrix or, more commonly, the correlation matrix. There are standard algorithms which deter-mine the weights a_{ij} and the variances of the principal components. The latter are usually denoted by $(\lambda_1, \lambda_2, \dots, \lambda_p)$ and are listed from largest to smallest.

Choosing the number of principal components

As for CORA and MDS, one aim of PCA is to be able to plot the data in one, two or three dimensions without losing too much information. When choosing the number of components, the aim is to retain as small a set as possible but, at the same time, to have a sufficient number to provide a good representation of the original data. The variance of component j is the eigenvalue λ_j. Since the components are derived in order of variance, $\lambda_1 \geq \lambda_2 \cdots \geq \lambda_p$. If the ys are standardised so that the correlation matrix is analysed, the sum of the variances of the ys will be equal to p. Thus the sum of the eigenvalues, the total variance of the zs, will also equal p.

The proportion of total variance explained by component j is

$$\frac{\lambda_j}{\lambda_1 + \lambda_2 + \cdots + \lambda_p}.$$

The proportion explained by the first k components together is

$$\frac{\lambda_1 + \lambda_2 + \cdots + \lambda_k}{\lambda_1 + \lambda_2 + \cdots + \lambda_p}.$$

In practice, these proportions are often expressed as percentages.

There are a number of criteria that may be used to decide how many components should be retained:

i) Retain the first k components which explain a "large" proportion of the total variation, say 70-80%.

ii) If the correlation matrix is analysed, retain only those components with eigenvalues greater than 1. The logic behind this rule of thumb is that a component with an eigenvalue of 1 explains the same amount of variation as one of the original ys. However, Jolliffe (1972) suggests that retaining components with eigenvalues greater than 0.7 is better than the cut-off at 1.

iii) Examine a scree plot. This is a plot of the eigenvalues versus the component number. The idea is to look for the "elbow" which corresponds to the point after which the eigenvalues decrease more slowly. Adding components after this point explains relatively little more of the variance. See Figure 5.4 for an example of a scree plot.

iv) Consider whether the component has a sensible and useful interpretation.

Interpretation

The weight given to variable i on component j is a_{ij}. The relative sizes of the a_{ij}s reflect the relative contributions made by each variable to the component. To interpret a component, we examine the pattern in the a_{ij} values for that component.

Often, the coefficients are rescaled so that the coefficients for the most important components (i.e. the ones that explain the most variation) are larger than those for less important components. These rescaled coefficients, called *component loadings*, are the coefficients for reconstructing the ys from the zs as explained in Section 5.8 and are calculated as

$$a_{ij}^* = \sqrt{\lambda_j} a_{ij} \qquad (i = 1, \ldots, p; \ j = 1, \ldots, p).$$

When the correlation matrix of the ys is analysed, a_{ij}^* may be interpreted as the correlation coefficient between variable i and component j. This is especially useful for interpretation.

5.5 Examples

Children's personality traits

To demonstrate the interpretation of components, consider the following example. In a study of children's personality traits, a sample of children were scored on the eight variables listed in Table 5.1 which also gives the loadings, a_{ij}^*, for the first two components of the correlation matrix. The first two components explain 77% of the total variance of the scores.

TABLE 5.1: Component loadings for the first two principal components, children's personality trait data

Personality trait	Variable	a_{i1}^*	a_{i2}^*
Mannerliness	y_1	0.68	0.58
Approval seeking	y_2	0.60	0.59
Initiative	y_3	0.65	-0.52
Guilt	y_4	0.65	-0.59
Sociability	y_5	0.61	0.57
Creativity	y_6	0.71	-0.61
Adult role	y_7	0.69	-0.49
Cooperativeness	y_8	0.67	0.61
Proportion of total variance		44%	33%

For the first component, the loadings are all fairly large, positive, and of a similar magnitude. Each variable is positively correlated with the first component. The first component might therefore be interpreted as some overall measure of personality: a child who scores highly on each trait would have a high score on this component, while a child who has a low score on each trait would have a low score on this component.

The coefficients for the second component have a bipolar structure. Mannerliness (y_1), approval seeking (y_2), sociability (y_5), and cooperativeness (y_8) all have relatively high positive loadings, while the other traits have relatively high negative loadings. To interpret this component, we need to think about what the variables in each subset have in common, and how they differ from the other subset of variables. The variables with positive loadings measure how well a child relates to other people. The variables with negative loadings are more internal to the individual and might be thought of as measures of the child's independence. Therefore, the second component may be interpreted as a contrast between these two aspects of personality. Children with a high score on this component will tend to get along relatively well with others but to show relatively less independent behaviour. In contrast, children with a low score on this component will tend to be relatively more independent in their behaviour but not to be so sociable. Children who are located towards the middle of the second component will have roughly equal scores for (y_1, y_2, y_5, y_8) as for (y_3, y_4, y_6, y_7).

The above structure is quite common in a PCA of a correlation matrix. If the original variables, the ys, are all positively correlated with one another, the loadings for the first component will have the same sign (either positive or negative) and will tend to be of a similar magnitude. Subsequent components will be contrasts between subsets of variables.

Subject marks

Table 5.2 shows the pairwise correlation coefficients between subject scores for a sample of 220 boys. The data are taken from Lawley and Maxwell (1971), p. 66.

TABLE 5.2: Pairwise correlation coefficients between subject marks

	Gaelic	English	History	Arithmetic	Algebra	Geometry
Gaelic	1.00					
English	0.44	1.00				
History	0.41	0.35	1.00			
Arithmetic	0.29	0.35	0.16	1.00		
Algebra	0.33	0.32	0.19	0.59	1.00	
Geometry	0.25	0.33	0.18	0.47	0.46	1.00

It is always good practice to inspect the correlation matrix before embarking on any analysis. This may reveal anomalous entries and enable us to check the plausibility of the data. From such an inspection, one may be able to anticipate the results of a PCA because the analysis only makes explicit what is already implicit in the correlations. In this case, each subject score is positively correlated with each of the scores in the other subjects, indicating that there is a general tendency for those who do well in one subject to do well in others. The highest correlations are between the three mathematical subjects and, to a slightly lesser extent, between the three humanities subjects, suggesting that there is more in common within each of these two groups than between them.

A PCA was carried out on this correlation matrix and produced the eigenvalues shown in Table 5.3. Only the first two eigenvalues are greater than 1. Also shown are the percentage of variance explained by each component and the cumulative percentage of variation explained by the first k components. For example, the first two components explain just over 64% of the total variation.

TABLE 5.3: Variance explained by each principal component, subject marks data

Component	Variance explained		
	Variance (Eigenvalue)	%	Cumulative %
1	2.73	45.48	45.48
2	1.13	18.81	64.29
3	0.62	10.26	74.55
4	0.60	10.05	84.59
5	0.52	8.71	93.30
6	0.40	6.70	100.00

Figure 5.4 shows the scree plot. There is an elbow at the third component. The third and subsequent components have similar eigenvalues which means that they each explain a similar but small proportion of the total variance. From Table 5.3 and Figure 5.4, we would conclude that the first two components should provide an adequate representation of the ys.

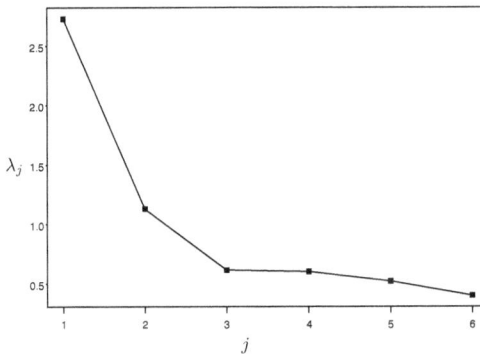

FIGURE 5.4: Scree plot showing eigenvalue by number of principal component, subject marks data

The loadings for the first two components are shown in Table 5.4 and are plotted in Figure 5.5. The loadings of the first component are all positive and fairly large. Remembering that they can be interpreted as correlations between the subject scores and the component, we infer that the first component represents something which is common to performance on all subjects. This could be what we usually describe as general academic ability. The second component is a contrast between humanities subjects and mathematical subjects. Boys who are better at humanities subjects than at mathematics will score highly on this component and conversely. This component measures the contrast between the two types of ability. Figure 5.5 makes the same point graphically with the three mathematics loadings appearing as a tight cluster separated from those for the humanities on the vertical dimension.

TABLE 5.4: Loadings for first two principal components, subject marks data

Subject	a_{i1}^*	a_{i2}^*
Gaelic	0.66	0.44
English	0.69	0.29
History	0.52	0.64
Arithmetic	0.74	-0.42
Algebra	0.74	-0.37
Geometry	0.68	-0.35

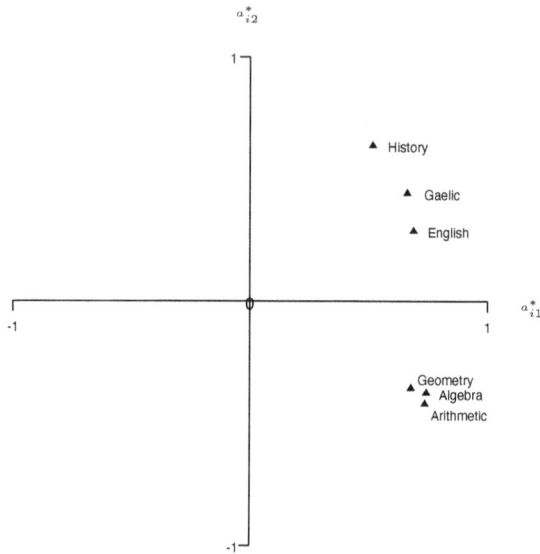

FIGURE 5.5: Plot of loadings for the first two components, subject marks data

Employment in 26 European countries

The percentage of the total workforce employed during 1979 in each of the nine industries listed below was recorded for 26 European countries (see Hand et al. 1994).

y_1: Agriculture   % employed in agriculture
y_2: Mining   % employed in mining
y_3: Manufacture   % employed in manufacturing
y_4: Power   % employed in power supply industries
y_5: Construction   % employed in construction
y_6: Service   % employed in service industries
y_7: Finance   % employed in finance
y_8: Social   % employed in social and personal services
y_9: Transport   % employed in transport and communications

There are very large differences in the variances of these variables (ranging from 0.14 to 241.70) and, as we wish to give the variables equal weight, the correlation matrix will be analysed. The lower part of the correlation matrix is shown in Table 5.5. From this, we can see that the percentage employed in agriculture has a fairly strong and negative correlation with the percentage employed in all other industries apart from mining. The percentage employed in manufacturing is quite highly, positively correlated with the percentage in construction. Also, the percentage employed in social and personal services is positively correlated with the percentages in service industries and the percentage in transport and communications. This matrix does not conform to the pattern of positive correlations observed in the previous examples, and so we would not expect the first principal component to have all positive loadings.

TABLE 5.5: Correlation matrix for European employment data

	y_1	y_2	y_3	y_4	y_5	y_6	y_7	y_8	y_9
y_1	1.00								
y_2	0.04	1.00							
y_3	−0.67	0.45	1.00						
y_4	−0.40	0.41	0.39	1.00					
y_5	−0.54	−0.03	0.49	0.06	1.00				
y_6	−0.74	−0.40	0.20	0.20	0.36	1.00			
y_7	−0.22	−0.44	−0.16	0.11	0.02	0.37	1.00		
y_8	−0.75	−0.28	0.15	0.13	0.16	0.57	0.11	1.00	
y_9	−0.57	0.16	0.35	0.38	0.39	0.19	−0.25	0.57	1.00

The percentage explained by each of the nine principal components is given in Table 5.6. The first two components account for just over 60% of the total variation. The first three components have eigenvalues greater than 1, and the eigenvalue for component 4 is 1. The scree plot (Figure 5.6) shows some suggestion of an elbow at component 3, indicating that between two and four components might be considered. The loadings for the first three components are shown in Table 5.7.

TABLE 5.6: Variance explained by each component, European employment data

Component	Variance	Variance explained %	Cumulative %
1	3.49	38.75	38.75
2	2.13	23.67	62.42
3	1.10	12.21	74.63
4	1.00	11.05	85.68
5	0.54	6.04	91.71
6	0.38	4.26	95.97
7	0.23	2.51	98.48
8	0.14	1.52	100.00
9	0.00	0.00	100.00

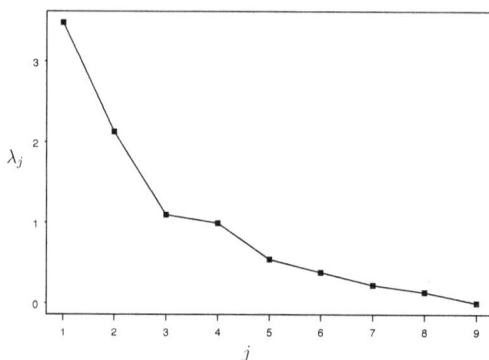

FIGURE 5.6: Scree plot of eigenvalue versus number of components, European employment data

TABLE 5.7: Loadings for first three components, European employment data

Industry	a_{i1}^*	a_{i2}^*	a_{i3}^*
Agriculture	−0.98	0.08	−0.05
Mining	0.00	0.90	0.21
Manufacture	0.65	0.52	0.16
Power	0.48	0.38	0.59
Construction	0.61	0.07	−0.16
Service	0.71	−0.51	0.12
Finance	0.14	−0.66	0.62
Social and personal services	0.72	−0.32	−0.33
Transport	0.69	0.30	−0.39

Notice that the first component is almost perfectly negatively correlated with agriculture (y_1), and has low correlations with mining (y_2) and finance (y_7). The other variables have moderate to high positive correlations with the first component. The first component may be interpreted as distinguishing countries with agricultural economies from those with industrial economies. The second component is positively correlated with mining (y_2), manufacturing (y_3), power (y_4), and transport (y_9) and is negatively correlated with service (y_6), finance (y_7) and social and personal services (y_8). This component may be interpreted as a contrast between countries with relatively large and relatively small service sectors. These contrasts are clearly brought out in Figure 5.7. There is no obvious interpretation of the third component, which confirms our hope based on the scree plot that two components would be adequate.

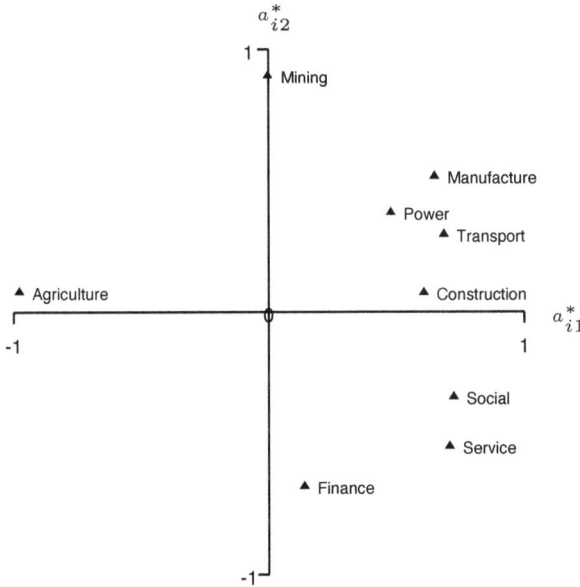

FIGURE 5.7: Plot of loadings for first two components, European employment data

5.6 Component scores

Suppose we wish to compute an individual's score on a particular component from the PCA of the standardised data. We have

$$z_j = a_{1j}y_1 + a_{2j}y_2 + \cdots + a_{pj}y_p,$$

where $y_1, y_2, \ldots, y_p$ are all standardised to have a mean of zero and a variance of 1, and where z_j has variance λ_j. See, for example, the scores for z_1 and z_2 plotted in Figure 5.3.

However, it is more usual to standardise the component scores to have unit variance, so that $\tilde{z}_j = z_j / \sqrt{\lambda_j}$ has variance 1. Hence,

$$\tilde{z}_j = \tilde{a}_{1j} y_1 + \tilde{a}_{2j} y_2 + \cdots + \tilde{a}_{pj} y_p,$$

where

$$\tilde{a}_{ij} = \frac{a_{ij}}{\sqrt{\lambda_j}} = \frac{a_{ij}^*}{\lambda_j}.$$

These $\tilde{a}_{ij}$s are referred to as the *component score coefficients*.

For example, since in our PCA of the subject marks data we have interpreted the first component as a measure of overall ability, a boy's ability could be estimated by computing component scores, $\tilde{z}_1$, using the coefficients in the first column of Table 5.8. Using the coefficients in the second column, we would be able to score individuals on the humanities versus mathematics dimension, $\tilde{z}_2$.

TABLE 5.8: Component score coefficients for the first two components, subject marks data

Subject	$\tilde{a}_{i1}$	$\tilde{a}_{i2}$
Gaelic	0.24	0.39
English	0.25	0.26
History	0.19	0.57
Arithmetic	0.27	−0.37
Algebra	0.27	−0.33
Geometry	0.25	−0.31

The coefficients for computing standardised scores $(\tilde{z}_j)$ on the first two components from the PCA of the European employment data are given in Table 5.9. For example, a country's score on component 1 would be calculated as

$$\tilde{z}_1 = -0.28 y_1 + \cdots + 0.20 y_9,$$

where y_1 is the *standardised* (to zero mean and unit variance) percentage employed in agriculture, etc. The scores for the first two components have been calculated for each of the 26 countries. Figure 5.8 shows a scatterplot of the scores on the first component versus the scores on the second component. On the first component, Turkey and, to a lesser extent, Yugoslavia, stand out from the other countries. This would suggest that Turkey and Yugoslavia had a more agricultural economy than the other countries in 1979. The second component separates the capitalist Western countries (negative scores) from the former communist Eastern countries (positive scores). Since countries with high percentages in service industries (service, finance and social) would have a low score on this component, which suggests that the capitalist Western countries had a larger service sector than the communist Eastern countries.

TABLE 5.9: Component score coefficients for the first two components, European employment data

Industry	$\tilde{a}_{i1}$	$\tilde{a}_{i2}$
Agriculture	−0.28	0.04
Mining	−0.00	0.42
Manufacture	0.19	0.24
Power	0.14	0.18
Construction	0.17	0.04
Service	0.20	−0.24
Finance	0.04	−0.31
Social and personal services	0.21	−0.15
Transport	0.20	0.14

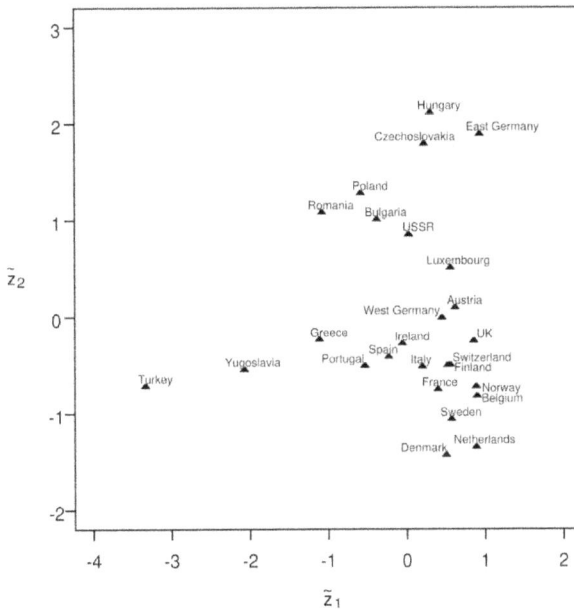

FIGURE 5.8: Twenty six European countries plotted using standardised principal component scores, European employment data

5.7 The link between PCA and multidimensional scaling, and between PCA and correspondence analysis

The link between PCA and multidimensional scaling

A basic PCA only requires the correlation matrix, but to calculate the component scores for the n individuals (e.g. the 26 European countries), it is also necessary to know the values of the p (standardised) variables given in the $n \times p$ data matrix $\mathbf{Y}$ as defined in Chapter 1:

$$
\mathbf{Y} = \begin{pmatrix}
y_{11} & y_{12} & \cdots & y_{1p} \\
y_{21} & y_{22} & \cdots & y_{2p} \\
\vdots & & & \vdots \\
y_{n1} & y_{n2} & \cdots & y_{np}
\end{pmatrix}.
$$

Multidimensional scaling can be carried out using just the distances between individuals — but it gives no information about the variables from which the distances might have been constructed. Given the data matrix, either one or both techniques could be applied to produce a plot of the individuals in, say, two dimensions.

Standard metrical MDS, like ordinal MDS, uses an iterative procedure to find a solution that minimises some stress criterion. PCA uses an algebraic procedure to maximise the variances of the components. In general, the results will be different (although maybe not very different when the structure is strong). However, where squared Euclidean distances have been calculated from a data matrix, *classical* MDS (sometimes called *principal coordinate analysis*) gives the PCA solution exactly.

We demonstrate their relationship below by carrying out a PCA of the correlation matrix for the economic and demographic data for 25 countries analysed in Section 3.7. Table 5.10 shows the proportion of variance explained by each of the five components. The first component is dominant, explaining 80% of the total variance.

TABLE 5.10: Variance explained by each component, economic and demographic data

Component	Variance	Variance explained %	Cumulative %
1	4.01	80.28	80.28
2	0.57	11.38	91.66
3	0.25	5.05	96.71
4	0.09	1.92	98.62
5	0.06	1.38	100.00

Table 5.11 shows the loadings for the principal components. The first component is highly correlated with all the variables, negatively with y_1, y_3, and y_4 and positively with y_2 and y_5. A country with a high rate of population increase, low life expectancy, high infant mortality rate, high fertility rate and low GDP would have a low score on this component. A country with a low rate of population increase, high life expectancy, low infant mortality rate, low fertility rate, and high GDP would have a high score. Therefore, the first component may be interpreted as a measure of overall development. The

second component is positively correlated with GDP and, to a lesser degree, rate of population increase and fertility rate. Life expectancy and infant mortality rate do not contribute to this component. Countries with a low rate of population increase, low fertility rate, and low GDP would have a low score on this component.

TABLE 5.11: Loadings for all five components, economic and demographic country data

Index	a_{i1}^{*}	a_{i2}^{*}	a_{i3}^{*}	a_{i4}^{*}	a_{i5}^{*}
Population increase	−0.86	0.39	0.32	0.09	−0.06
Life expectancy	0.95	0.03	0.24	0.05	0.19
Infant mortality rate	−0.95	−0.01	−0.21	0.20	0.13
Fertility rate	−0.95	0.19	−0.01	−0.22	0.12
GDP	0.76	0.62	−0.21	0.01	0.00

The standardised scores for each of the 25 countries on the first two components have been plotted in Figure 5.9. Note the close resemblance between this plot and the two-dimensional metrical MDS configuration for these data (see Figure 3.12). (To aid comparison, the horizontal axis has been reversed so that $\tilde{z}_1$ *decreases* from left to right.)

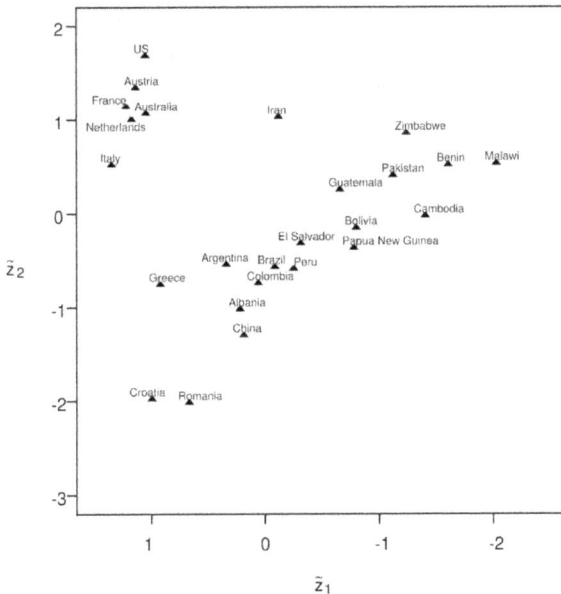

FIGURE 5.9: Plot of scores on the first two principal components, economic and demographic data. (Note, the horizontal axis has been reversed.)

The link between PCA and correspondence analysis

The usual derivation of PCA starts with the correlation matrix. An alternative derivation, giving the same results, involves the singular value decomposition (SVD) of the $n \times p$ standardised data matrix: $\mathbf{Y} = \{y_{ti}\}$, where y_{ti}, the value for individual (row) t on variable (column) i, has been standardised to zero mean and unit standard deviation. This parallels the SVD of the matrix of Pearson residuals in CORA described in Section 4.3.

From the SVD, we obtain the u_{tj} (corresponding to rows or individuals), the v_{ij} (corresponding to columns or variables) and the singular values, $\sqrt{(n-1)\lambda_j}$, such that

$$y_{ti} = \sum_j u_{tj} v_{ij} \sqrt{(n-1)\lambda_j},$$

where n is the sample size. So, it turns out that the square of the singular value is *proportional* to λ_j, the jth eigenvalue of the correlation matrix (and the variance explained by the jth principal component); $u_{tj}\sqrt{(n-1)}$ is $\tilde{z}_{tj}$, the score for individual t on principal component j; and $v_{ij}\sqrt{\lambda_j}$ is a_{ij}^*, the component loading of y_i on $\tilde{z}_j$. As with CORA, it is possible to plot individuals (rows) and variables (columns) in a single bi-plot, using the first two PCs.

The difference between CORA and PCA is that they start with different matrices. In CORA there is a symmetry: the rows and columns in the $I \times J$ matrix of Pearson residuals have the same status, whereas in the $n \times p$ matrix of standardised variables for PCA, the rows represent individuals and the columns represent variables.

5.8 Using principal component scores to replace the original variables

One use of PCA is to replace a larger set of p variables by a smaller set of q principal components. The first component, z_1, might be used alone as a univariate ($q = 1$) summary of the original variables, $y_1, \ldots, y_p$, either for use in further analyses or as an index. Indeed, the component score coefficients are sometimes used to score *new* individuals on such an index. The first two components might be used to plot the data — either scaled so that $\mathrm{var}(z_1) = \lambda_1$, $\mathrm{var}(z_2) = \lambda_2$, as in Figure 5.3, or with components, $\tilde{z}_1$ and $\tilde{z}_2$, standardised to unit variance as in Figures 5.8 and 5.9.

The question arises as to how much information is lost by replacing the p ys by their first q principal components; or more particularly, as to how well y_i can be reconstructed from $\tilde{z}_1, \ldots, \tilde{z}_q$ for $i = 1, \ldots, p$.

In Section 5.6, the (standardised) principal components are given as linear functions of the (standardised) original variables,

$$\tilde{z}_j = \tilde{a}_{1j}y_1 + \cdots + \tilde{a}_{pj}y_p \qquad (j = 1, \ldots, p).$$

These equations can be inverted to give

$$y_i = a_{i1}^* \tilde{z}_1 + \cdots + a_{ip}^* \tilde{z}_p \qquad (i = 1, \ldots, p),$$

where $a_{ij}^* = \lambda_j \tilde{a}_{ij} = \sqrt{\lambda_j} a_{ij}$ is the component loading introduced in Section 5.4. Remember, this loading is the correlation between y_i and z_j. Now suppose that we try to reconstruct y_i using only the first two components. The reconstructed value is

$$\hat{y}_i = a_{i1}^* \tilde{z}_1 + a_{i2}^* \tilde{z}_2.$$

This will be close to y_i if the remaining correlations or loadings, $a_{i3}^*, \ldots, a_{ip}^*$, are all close to zero. Equivalently, we can judge how well each y_i is reproduced from the first q components by seeing how close the *communality* is to one, where the communality is the sum of the first q squared loadings so that for y_i, the communality equals

$$a_{i1}^{*2} + \cdots + a_{iq}^{*2} \qquad (i = 1, \ldots, p).$$

We shall meet the communality again in the context of factor analysis. It is the square of the multiple correlation coefficient between y_i and $z_1, \ldots, z_q$.

Table 5.12 gives the communalities for the subject mark data, for one component and for two components. History, for example, would not be adequately summarised by just the first component, but the first two components together might be judged adequate. In Figure 5.5, the communalities for two components are represented by the squared distances from the origin to the plotted variables.

TABLE 5.12: Communalities for one and for two components for the subject marks data

Subject	One component	Two components
Gaelic	0.44	0.63
English	0.47	0.56
History	0.27	0.68
Arithmetic	0.54	0.72
Algebra	0.55	0.70
Geometry	0.46	0.58

5.9   Additional examples and further work

Social mobility in the UK

The correlations analysed are taken from Ridge (1974) and are based on information provided by 713 male or female married respondents to a survey carried out in 1949 by D.V. Glass and associates at the London School of Economics. The variables relate to the respondent, her or his spouse, father, father-in-law, and firstborn son and are described in Table 5.13. The correlations (which are all positive) are given in Table 5.14, and the loadings for the first six principal components are given in Table 5.15. You can see that many of the correlations are small and none of them are very large. Of interest is whether there are identifiable differences between the generations, and also to what extent the three measures (occupational status, further education and qualifications) are all indicators of a family's status.

TABLE 5.13: Descriptions of social mobility variables

Variable	Generation	Code	Description
y_1	1	HF/O	Husband's father's occupational status
y_2	1	WF/O	Wife's father's occupational status
y_3	2	H/FE	Husband's further education
y_4	2	H/Q	Husband's qualifications
y_5	2	H/O	Husband's occupational status
y_6	2	W/FE	Wife's further education
y_7	2	W/Q	Wife's qualifications
y_8	3	FB/FE	Firstborn's further education
y_9	3	FB/Q	Firstborn's qualifications
y_{10}	3	FB/O	Firstborn's occupational status

TABLE 5.14: Pairwise correlations between social mobility variables

Variable	y_1	y_2	y_3	y_4	y_5	y_6	y_7	y_8	y_9	y_{10}
y_1	1.00									
y_2	0.37	1.00								
y_3	0.23	0.23	1.00							
y_4	0.10	0.13	0.53	1.00						
y_5	0.43	0.38	0.35	0.24	1.00					
y_6	0.17	0.15	0.28	0.23	0.20	1.00				
y_7	0.13	0.10	0.28	0.38	0.14	0.47	1.00			
y_8	0.18	0.18	0.32	0.31	0.23	0.26	0.21	1.00		
y_9	0.08	0.10	0.25	0.35	0.11	0.12	0.19	0.50	1.00	
y_{10}	0.29	0.28	0.29	0.22	0.44	0.19	0.16	0.44	0.33	1.00

TABLE 5.15: Loadings for first six components, social mobility data

Variable		a_{i1}^*	a_{i2}^*	a_{i3}^*	a_{i4}^*	a_{i5}^*	a_{i6}^*
y_1	HF/O	0.50	0.56	0.15	0.07	0.06	0.61
y_2	WF/O	0.48	0.52	0.08	−0.01	0.60	−0.34
y_3	H/FE	0.68	−0.13	0.11	−0.51	−0.09	−0.06
y_4	H/Q	0.62	−0.40	0.06	−0.50	0.07	0.03
y_5	H/O	0.62	0.49	0.07	−0.13	−0.34	−0.08
y_6	W/FE	0.51	−0.24	0.52	0.44	−0.07	−0.15
y_7	W/Q	0.51	−0.41	0.50	0.21	0.06	0.11
y_8	FB/FE	0.65	−0.19	−0.44	0.27	0.03	−0.03
y_9	FB/Q	0.52	−0.35	−0.56	0.11	0.23	0.20
y_{10}	FB/O	0.65	0.22	−0.33	0.21	−0.35	−0.20
Variance		3.34	1.44	1.17	0.89	0.68	0.62
% variance explained		33.42	14.37	11.73	8.90	6.82	6.16

The first six eigenvalues (explained variance) are 3.34, 1.44, 1.17, 0.89, 0.68, and 0.61, suggesting that three or four principal components should be used. For component 1 (which explains 33% of the total variance), the loadings vary between 0.5 and 0.7, suggesting that this is a summary of the ten variables measuring the status of the family. The second component (which explains 14% of the total variance) contrasts variables y_1, y_2, y_5, and y_{10} (which give the occupational status of different family members) and variables y_4, y_7, and y_9 (relating to qualifications) with a lesser contribution from variables y_3, y_6, and y_8 (relating to further education). The third component (which explains a further 11% of the variance) contrasts the firstborn son (variables y_8, y_9, and y_{10}) with his mother (variables y_6 and y_7) and possibly other ancestors (the remaining variables). The fourth component might also be interpretable largely as a contrast between husband and wife, but this may be pushing the data beyond its limits as the eigenvalue for this component is less than one and successive components become less reliable.

In conclusion, the correlation structure suggests that all ten variables have something in common which could be referred to as family status, but that there are differences between occupational status, further education, and qualifications, and to a lesser extent, there are differences between the generations. We shall return to this example in Chapter 7.

Educational circumstances

The correlation matrix for nine variables relating to circumstances and test results in 1964 and 1968 of girls in their fourth year of secondary school in 1968 (Table 2.17) has already been subjected to cluster analysis in Chapter 2. If you carry out a PCA of these data, you will find that there is a dominant first component, accounting for 42.6% of the total variation. When examining the scree plot, you will notice an elbow at the second component, suggesting that only the first component is necessary. Although only the first two components have eigenvalues greater than one, components 3 and 4 have eigenvalues close to one, and by Jolliffe's criterion the first five components should be examined. Components 2, 3, and 4 have similar eigenvalues and account for 12.4%, 11.1% and 9.1% respectively of the total variance. It is therefore not clear how many components should be considered. The first three components are examined here, but you should look at the fourth to see whether it offers further insights.

You will find that the first component is positively correlated with all variables, reflecting the mainly positive correlations in the correlation matrix. The first component might therefore be interpreted as some general indicator of a girl's circumstances both at home and at school. Components 2 and 3 do not have a clear interpretation. However, some patterns emerge if we look at pairwise plots of the component loadings. Figures 5.10 and 5.11 show plots of the loadings for component 1 versus component 2, and component 1 versus component 3, respectively. From Figures 5.10 and 5.11, you can see that the loadings for variables y_1 and y_7 (parental circumstances in 1964 and 1968) are close together on all of the first three principal components. Similarly variables y_5, y_9, and y_6 (the two test scores and the type of school) have loadings close to each other on the first three principal components, but the remaining variables do not closely resemble each other or the two clusters of variables above. Compare Figures 5.10, 5.11, and 2.20, where Figure 2.20 shows the results of a cluster analysis. You will find that the results of the PCA tend to confirm the cluster analysis of these data. This illustrates how the different methods of summarising a correlation matrix can reinforce each other.

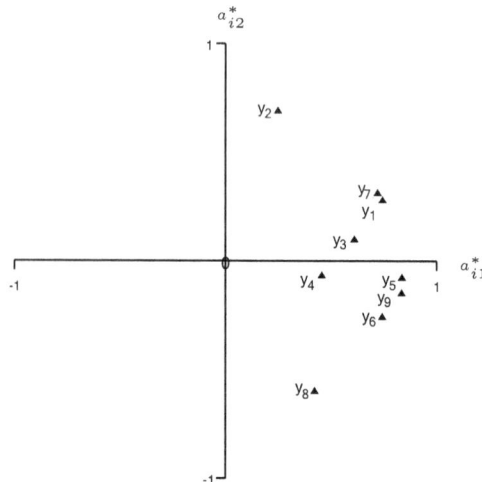

FIGURE 5.10: Plot of loadings for first two components, educational circumstances

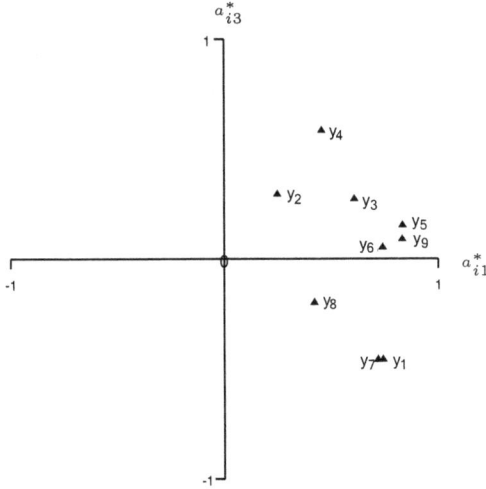

FIGURE 5.11: Plot of loadings for component 1 and component 3, educational circumstances

Television viewing in the UK

A sample of 7000 UK adults was asked whether they "really liked to watch" a range of ten television programmes (Ehrenberg 1977). The pairwise correlations between the ten variables measuring "liking to watch" the programmes are shown in Table 5.16. The programmes fall into two broad categories: sports programmes (World of Sport, Match of the Day, Grandstand, Professional Boxing, and Rugby Special) and current affairs programmes (24 Hours, Panorama, This Week, Today, and Line-Up).

TABLE 5.16: Pairwise correlations between liking to watch ten television programmes

	WoS	MoD	GrS	PrB	RgS	24H	Pan	ThW	Tod	LnU
World of Sport	1.00									
Match of the Day	0.58	1.00								
Grandstand	0.62	0.59	1.00							
Prof. Boxing	0.51	0.47	0.47	1.00						
Rugby Special	0.30	0.33	0.34	0.31	1.00					
24 Hours	0.14	0.12	0.14	0.12	0.12	1.00				
Panorama	0.19	0.13	0.18	0.17	0.15	0.52	1.00			
This Week	0.14	0.08	0.13	0.11	0.06	0.39	0.35	1.00		
Today	0.09	0.04	0.07	0.07	0.05	0.24	0.20	0.27	1.00	
Line-Up	0.08	0.05	0.08	0.09	0.10	0.27	0.20	0.19	0.15	1.00

The scree plot from a PCA of these data is shown in Figure 5.12. From this plot, you can see an elbow at the third component. Further, only the eigenvalues for the first two components are greater than one. You should also examine the proportions of variance explained by each component: the first component explains nearly 32% of the total variance and the first two components explain 50% of the variance. The third only accounts for an additional 9% of variance explained. All this points towards choosing the first two components to summarise the data.

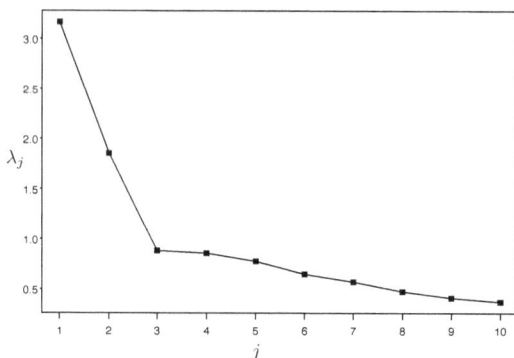

FIGURE 5.12: Scree plot of eigenvalue versus number of component, television viewing data

The loadings for the first two components are plotted in Figure 5.13. You can see that all variables are positively correlated with the first component. The first component might therefore be interpreted as a general measure of liking to watch television. The second component has a mixture of positive and negative loadings. If you refer to the description of the programmes given above, you will find that the current affairs programmes have positive loadings on the second component while the sports programmes have negative loadings. Thus this component contrasts liking to watch these two different types of programmes.

5.10   Measuring association between categorical variables

Measures of association are a building block of multivariate methods. The correlation coefficient may be the simplest measure of association, which measures the degree of association between two continuous variables. This notion of association plays a crucial role in the current chapter for principal component analysis, as well as other chapters that concern multiple continuous variables, including the linear factor models in Chapter 7.

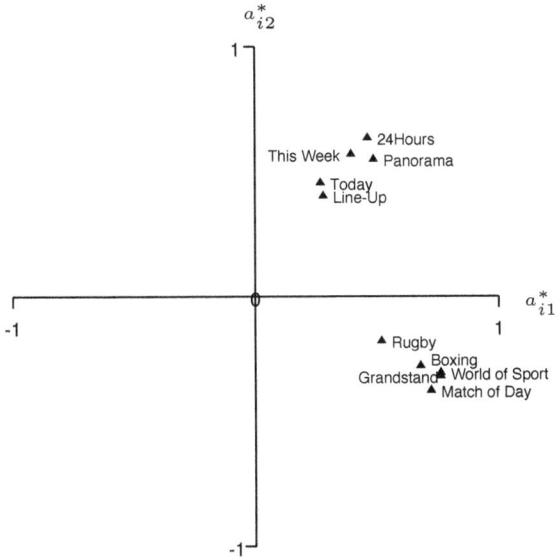

FIGURE 5.13: Plot of loadings for first two components, television viewing data

Correlation is not suitable for categorical variables. There are two different approaches for measuring the association between categorical variables; one is based on a model involving continuous latent variables, and the other is based on an independence test for categorical variables. We first demonstrate these two approaches using binary variables and then extend this discussion to variables with more categories.

Consider two binary random variables, y_1 and y_2, taking values 0 and 1. As illustrated in Figure 5.14, the latent variable approach assumes that there are continuous latent variables y_1^* and y_2^* underlying these two variables, respectively, and y_1 and y_2 are obtained by truncating y_1^* and y_2^* at thresholds τ_1 and τ_2, respectively. That is, $y_i = 1$ if $y_i^* \geq \tau_1$ and $y_i = 0$ otherwise, $i = 1, 2$. y_1^* and y_2^* cannot be observed, which is why they are called latent variables. The latent variables y_1^* and y_2^* are also called underlying variables or underlying latent variables. They are introduced for mathematical convenience and do not always have substantive interpretations, which somehow differs from the latent variables introduced in the models in Chapters 7–11. The latent variables (y_1^*, y_2^*) are assumed to follow a bivariate normal distribution, where each y_i^* marginally follows a standard normal distribution. The correlation between y_1^* and y_2^* is used to measure the association between the binary variables y_1 and y_2. This association measure is called the tetrachoric correlation. A tetrachoric correlation equal to zero implies the independence between the two latent variables, which further implies the independence between y_1 and y_2. Given a sample of observations for (y_1, y_2), the tetrachoric correlation is estimated together with the thresholds τ_1 and τ_2 by matching the model implied frequencies of $(y_1, y_2) = (0,0)$, $(0,1)$, $(1,0)$, and $(1,1)$ with the observed frequencies. This latent variable approach easily adapts to measuring the association between two original binary variables. Note that an ordinal variable with K categories can be obtained by truncating a continuous latent variable at $K-1$ thresholds. Therefore, similar to the binary data setting, the distribution of two ordinal variables can be implied by a pair of bivariate normal latent variables. The correlation between the two latent variables, known as the polychoric correlation, is used to measure the association between the ordinal variables. PCA can be performed based on an estimated tetrachoric or polychoric correlation matrix of multiple binary or ordinal variables. In this case, however, the analysis is essentially applied to the underlying latent variables instead of the raw variables, where, for example, the PCs are constructed by the latent variables. The underlying variable approaches for the factor analysis of binary and ordinal data in Chapters 8, 9, and 11 are also based on tetrachoric and polychoric correlations.

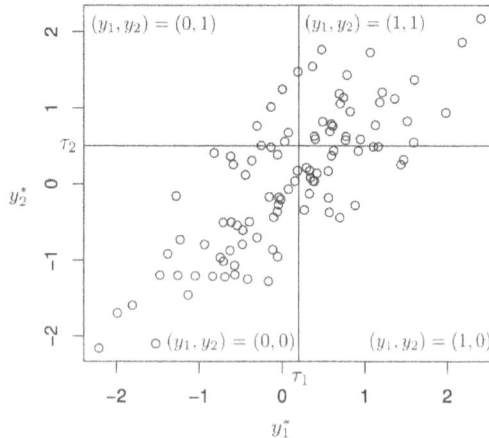

FIGURE 5.14: An illustration of the latent variable model for tetrachoric correlation

The second approach concerns testing the independence between two categorical variables. We start with two binary random variables, y_1 and y_2. Suppose we have a sample of observations for (y_1, y_2), for which n_{ij} denotes the observed frequency for $(y_1, y_2) = (i, j)$, $i, j = 0, 1$. Then we can calculate a sample odds ratio as $(n_{00}n_{11})/(n_{01}n_{10})$, which estimates the population odds ratio defined as the ratio between the odds of $y_1 = 1$ relative to $y_1 = 0$ when $y_2 = 1$ and the odds when $y_2 = 0$. An odds ratio of 2, for example, means that the odds of $y_1 = 1$ relative to $y_1 = 0$ is twice as high if $y_2 = 1$ as if $y_2 = 0$. The odds ratio larger than 1 means a positive association between the two binary variables, in the sense that y_1 is more likely to be 1 given y_2 is 1 than given y_2 is 0. Similarly, the odds ratio smaller than 1 means a negative association. The association between the two variables can be measured by testing whether the population odds ratio equals 1, based on the sample odds ratio and its sampling distribution.

Now consider categorical variables y_1 and y_2, where y_i has categories 1, ..., K_i, $i = 1, 2$. When $K_i > 2$, the association between y_1 and y_2 can no longer be measured by a single odds ratio. Still, an association measure can be constructed based on a test of independence using a sample of observations for (y_1, y_2). Let n denote the sample size and n_{ij} denote the observed frequency for $(y_1, y_2) = (i, j)$, $i = 1, ..., K_1$ and $j = 1, ..., K_2$. If y_1 and y_2 are independent, then we would expect $n_{ij} \approx e_{ij} = (n_{i1} + \cdots + n_{iK_2})(n_{1j} + \cdots + n_{K_1j})/n$. The association between the two variables is thus measured by the deviation of the observed frequencies from their expected values under the independence assumption. There are different ways to measure the deviation. A commonly used measure is Pearson's chi-squared statistic which takes the form

$$\sum_{i=1}^{K_1} \sum_{j=1}^{K_2} \frac{(n_{ij} - e_{ij})^2}{e_{ij}}.$$

This statistic is central for correspondence analysis in Chapter 4.

Finally, people are often interested in measuring the conditional association between two variables given some other variables. In particular, measures of conditional associations can be defined based on the concepts of correlation and odds ratio. These measures of conditional associations are crucial in all the models introduced in Chapters 6–14, especially the undirected graphical models in Chapter 12.

5.11   Further reading

Basilevsky, A. (1994). *Statistical Factor Analysis and Related Methods*. New York: Wiley.

Jolliffe, I. T. (1986). *Principal Components Analysis*. New York: Springer-Verlag.

6

Regression Analysis

6.1 Basic ideas

This chapter marks the transition from descriptive to model-based methods. We shall begin by treating regression descriptively and then introduce probability assumptions. In the last chapter all the variables had an equal status and PCA provided one way of looking at their interrelationships, beginning with the correlations. Regression looks at things differently, in an asymmetrical way, by asking how well one of the variables (usually denoted by y) can be predicted or estimated from the others (usually denoted by $x_1, x_2, \ldots, x_k$). Regression analysis is an important topic in its own right and it is very widely used in social science, including economics. It is also important as an ingredient of the techniques to which we come later in the book. Regression ideas lie at the heart of factor analysis and latent variable models in Chapters 7–9, structural equation models in Chapter 11, multilevel models in Chapter 13, and longitudinal models in Chapter 14.

We start with a simple artificial example involving only two variables. In PCA all variables were denoted by y and subscripts were used to distinguish one variable from another. Anticipating the asymmetry of regression analysis we now denote the variables x and y. You might like to think of x as the height and y as the weight of a man.

Figure 6.1 (a) is a scatterplot of y against x showing something close to a straight line relationship. We begin by asking which of the many possible lines through the points is best. The answer, as always in statistics, depends on what we want to use the line for! The acceptability of any particular line can be judged by how close the points are to it. One obvious possibility is to look at the perpendicular distances at right angles to the line as illustrated in Figure 6.1 (b).

We can use the sum of squares of these perpendicular distances as a measure of closeness of a line to the points. If we minimise the sum of squares we shall get the line which is closest to the set of points. This line is the first principal component as explained in Section 5.3. There the aim was to *maximise* the variation *along* the line, here we *minimise* the variation at *right angles* to it. But the two operations are equivalent because the total variation stays the same — so if we maximise one we minimise the other.

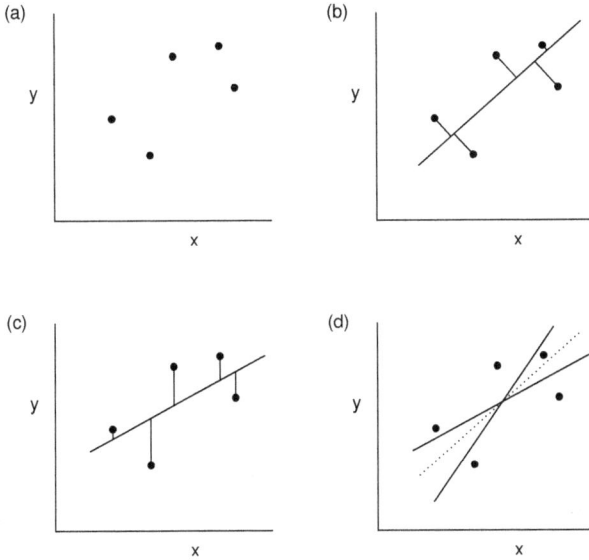

FIGURE 6.1: Diagram to illustrate two regression lines and principal component line

However, in regression analysis we are not interested in finding a composite variable to replace x and y but in how best to predict y *given* x. For this purpose we need to look at the scatterplot in a different way. This is done in Figure 6.1 (c).

A line now has to be judged by the *vertical* distances from the line as marked on Figure 6.1 (c). The best line for this purpose, in the least squares sense, is the one for which the sum of squares of the vertical distances is minimised. This line is known as the regression line of y on x.

We pause here to explain the rather strange use of the word *regression*. It was introduced by Francis Galton (1886) in the context of the inheritance of physical and mental characteristics which he illustrated by the relationship between the heights of fathers and of their sons. Galton fitted a line to predict a son's height from his father's. This showed that sons of tall fathers were, on average, taller than average but not, on average, as tall as their fathers, and conversely sons of short fathers were on average shorter than average but not, on average, as short as their fathers. For a given father's height the son's height was, on average, closer to the mean. Galton referred to this phenomenon as "regression towards mediocrity". (The puzzle which he went on to explain is why then the variance of sons' heights is still as large as the variance of fathers' heights.) The term *regression* has little meaning in most applications but it has stuck and there is now no chance of avoiding or changing it!

The focus of this chapter will be on multiple regression which is widely used in social applications. However, many of the ideas can be explained more simply in terms of what is called *simple linear regression*. Having established the main principles in this special case we shall then move on to the more practically important case of multiple regression.

6.2 Simple linear regression

Galton's regression of son's height, y, on father's height, x, described above, is an example of *simple linear regression*. We refer to y as the *response, outcome* or *dependent* variable. There are many alternative terms for x most of which will be used in this book: *regressor*, *predictor* or *explanatory* variable or *covariate*. Sometimes x is called the *independent* variable, because a value of x is chosen and then used to predict y. In that sense y *depends* or is conditional on the independent variable but the terminology is confusing because the word *independent* has other meanings.

The equation of a straight line may be written $y = a + bx$, so fitting a line to a scatterplot amounts to choosing a and b; a is known as the intercept because it is the point at which the line cuts the y-axis; b is the slope of the line and is referred to as the *regression coefficient* of y on x. Either or both of a and b may be negative. The regression coefficient tells us the average amount by which y increases when x increases by a unit amount.

The line is fitted by least squares, that is the slope b and the intercept a are chosen to minimise the sum of squares of the vertical deviations about the line. This is a simple mathematical problem and its solution is as follows. Suppose that we have n pairs of values, $(x_1, y_1), (x_2, y_2) \ldots (x_n, y_n)$, the "errors", or residuals, i.e. the amounts by which the points deviate from the line, $y = a + bx$, are $e_i = y_i - a - bx_i$ for $i = 1, 2, \ldots, n$.

It can be shown mathematically that the sum of squares of these deviations, $\sum_{i=1}^{n} e_i^2$, will be minimised when

$$b = \frac{\text{cov}(x,y)}{\text{var}(x)} = \frac{\sum_{i=1}^{n}(x_i - \bar{x})(y_i - \bar{y})}{\sum_{i=1}^{n}(x_i - \bar{x})^2} \quad \text{and} \quad a = \bar{y} - b\bar{x}. \tag{6.1}$$

The fitted value of y for an observed value of x is therefore

$$\hat{y} = \bar{y} + b(x - \bar{x}), \tag{6.2}$$

which shows that the fitted line goes through the mean point, $(\bar{x}, \bar{y})$.

Note the use of the circumflex (hat), ˆ, to indicate a fitted value. Later we shall also use a circumflex to indicate that a population parameter, σ, say, is estimated from some data by $\hat{\sigma}$.

The regression coefficient, b, is closely related to r, Pearson's correlation coefficient,

$$r = \frac{\text{cov}(x, y)}{\sqrt{\text{var}(x)\text{var}(y)}} \quad \text{so that} \quad b = r\sqrt{\frac{\text{var}(y)}{\text{var}(x)}}.$$

The correlation coefficient and the regression coefficient are the same if and only if x and y have the same variance — in other words the regression coefficient depends on the scales of the variables whereas the correlation coefficient does not.

The form of the relationship between r and b alerts us to the fact that there could be *two* regression equations. The one we have determined is called the regression of y on x because it was constructed to predict y given x.

If we had approached the problem the other way round, we would have found the regression of x on y by minimising the sum of squares of the *horizontal* deviations. The necessary formulae are easily obtained by interchanging x and y in eq. (6.1). If we plot the two regression lines on the same diagram we will find that they are not the same and that the principal component lies between them, as shown in Figure 6.1 (d).

In multiple regression, to which we come shortly, where there are several xs, the symmetry between x and y is lost and this duality will not arise. One property of simple linear regression which becomes of central importance when we move on to multiple regression concerns the closeness of the points to the line, which is often described as the goodness of fit. This can be approached in several ways but here we do it in terms of the *analysis of variance* which links up with the partitioning of variance which occurs in PCA. The variance of y can be split into a part *explained* by the regression on x and a residual part. In terms of the sums of squares

$$\sum_{i=1}^{n}(y_i - \bar{y})^2 = \sum_{i=1}^{n}(\hat{y}_i - \bar{y})^2 + \sum_{i=1}^{n}(y_i - \hat{y}_i)^2, \tag{6.3}$$

where $\hat{y}_i$ is the value predicted by the regression line, eq. (6.2), for a given x_i. The first sum of squares on the right-hand side is described as the sum of squares *due* to the regression or, more simply, as the regression sum of squares and the second as the residual sum of squares or the sum of squares *about* the regression line. It is useful to express the sum of squares due to the regression as a proportion of the total sum of squares. This proportion is denoted by R^2 and is given by

$$R^2 = \frac{\sum_{i=1}^{n}(\hat{y}_i - \bar{y})^2}{\sum_{i=1}^{n}(y_i - \bar{y})^2}. \tag{6.4}$$

R is sometimes known as the multiple correlation coefficient because it can be shown to be the correlation between y and $\hat{y}$. The positive root of R^2 is taken because if the regression has any predictive value at all there must be a positive correlation between the two.

Example: GCSE scores, simple linear regression

The writing speeds of 110 girls at a large comprehensive school in Oxfordshire, England, were measured in the winter of 1996/7 by a standard method. The GCSE (General Certificate of Secondary Education) grades in English of these same girls in summer 1997 were coded from 1 (not graded) to 9 (grade A*). The data are reported in Barnett et al. (1999).

Some summary statistics for GCSE score (y) and writing speed in words per tenth of a minute (x) are $\bar{y} = 6.67$, $\text{var}(y) = 1.672$, $\bar{x} = 2.12$, $\text{var}(x) = 0.278$ and $\text{cov}(x, y) = 0.274$. Hence the slope and intercept of the fitted regression line are

$$b = \frac{0.274}{0.278} = 0.986 \quad \text{and} \quad a = 6.67 - 0.986 \times 2.12 = 4.58,$$

and the fitted values are given by $\hat{y}_i = 4.58 + 0.99x_i$ for $i = 1, 2, \ldots, 110$. Although the results are presented rounded to a few decimal places, the calculations were done carrying more figures for accuracy.

The analysis of variance table, Table 6.1, is obtained using eq. (6.3). DF stands for "degrees of freedom" and the Mean Square is the Sum of Squares divided by the degrees of freedom. The residual mean square measures the variance about the regression line. The total sample variance of y is 1.7 and the variance about the regression line is 1.4, so the residual standard deviation equals $\sqrt{1.4} = 1.2$. Also, using eq. (6.4), $R^2 = 29.5/182.2 = 0.16$. Writing speed appears to explain some (but not very much) of the variation in the girls' GCSE English grades.

TABLE 6.1: Analysis of variance for GCSE scores

Source of variation	DF	Sum of Squares	Mean Square
Regression on writing speed	1	$\sum (\hat{y}_i - \bar{y})^2 = 29.5$	29.5
Residual (about regression)	108	$\sum (y_i - \hat{y}_i)^2 = 152.7$	1.4
Total (about mean)	109	$\sum (y_i - \bar{y})^2 = 182.2$	1.7

Figure 6.2 shows the fitted line

$$\text{GCSE score} = 4.58 + 0.99 \times \text{Speed}$$

superimposed on a scatterplot of the data. The data in the scatterplot lie in six horizontal lines because GCSE grade is really an ordinal variable to which we have given integer scores. Although the relationship between the scores and writing speed shows some non-linearity there is a clear linear trend. For every extra word per tenth of a minute the GCSE score increases *on average* by almost one grade (but there is much variation about this average). When judged by eye the regression line drawn in Figure 6.2 might look too flat.

This is because it is chosen to predict GCSE score from writing speed, not to summarise the dataset. The reader may like to check our calculations that the slope is just under one. The first principal component has a higher slope, just under two and a half, and, were we to draw it in, would look more as if it went through the data. (Remember that for regression we minimise the sum of squared vertical distances from the line but for PCA we minimise the sum of squared perpendicular distances.)

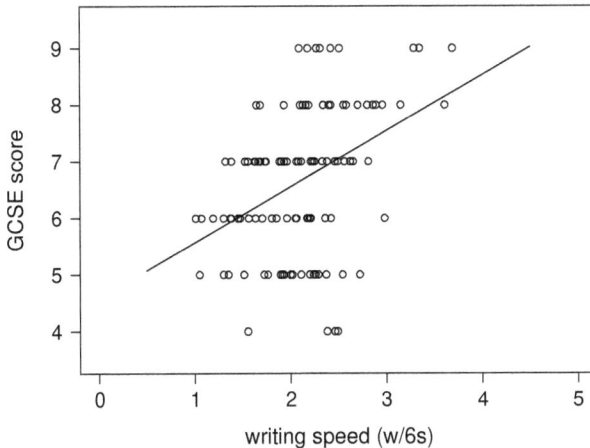

FIGURE 6.2: Scatterplot of GSCE English score against writing speed with the least squares regression line

6.3 A probability model for simple linear regression

We have now reached the turning point in the book when we pass from purely descriptive methods to model-based methods. Regression analysis as we have covered it so far describes the relationship between two variables in a particular way. The xs and ys were simply a set of observations and nothing has been assumed about how they arose. A probability model is a set of mathematical assumptions about how the data could have arisen. Because we are dealing with quantities which vary, we describe their variation by probability distributions. Our models are therefore *probability*, or *stochastic* models though we shall usually describe them simply, as "models". Once we have a model we can do very much more, though this freedom is bought at a price. All our conclusions depend on the model being correct — at least to an adequate approximation. There are sometimes good theoretical reasons for making assumptions and it may be possible to check them empirically. In any case, it can sometimes be shown that the conclusions we draw do not depend critically on the assumptions in which case we say that the method is *robust*.

Before we embark on the model for simple linear regression, which is the prototype for the many models which follow in later chapters, it may be helpful to give a broad overview of what we can expect to gain from the introduction of probability models.

If we knew the exact probability model by which the dataset or sample was generated we could work out what alternative samples *might* have been generated. In particular we could see how the calculated value of b using eq. (6.1) would vary from one possible sample to another. The distribution of the possible values of b is called the *sampling distribution* of b, its mean is referred to as the *expectation* of b, $E(b)$, and its standard deviation as the *standard error* of b, se(b). The idea of a sampling distribution is discussed more fully in many introductory text books such as Agresti and Finlay (2008) and can be illustrated by the use of computer simulation (generating a large number of samples from a given probability model).

Of course if we knew the exact probability model we would not need a sample, but in practice the model may not be fully specified and there may be unknown *parameters* that we wish to estimate. By treating our particular sample as one selected at random from a population of possible samples we can, for example, both calculate b and estimate the standard deviation of its sampling distribution, the *estimated standard error*, $\hat{\text{se}}(b)$. What the model enables us to do, therefore, is to make *inferences* from our particular sample to the unknown parameters of the model. We can use b to estimate its mean, $E(b)$, and $\hat{\text{se}}(b)$ as a measure of the accuracy of b. Our hope is that we will be able to generalise from our sample to some larger, real population, but this depends on the model being an adequate representation of that population.

In simple linear regression we think of y not as a single number but as a *random variable* which has a probability distribution — in particular, a normal distribution. We suppose that the distribution of y depends on x in a very simple way, namely that the mean is $\alpha + \beta x$. Here α is the intercept and β is the slope of the theoretical or "population" regression line of y on x, corresponding to a and b for the sample least squares regression line. The variance is assumed not to depend on x and may be unknown. The value of the regressor variable, x, is still treated as given. Formally, the model for a sample of n independent observations $y_1, \ldots, y_n$ for given $x_1, \ldots, x_n$ may be written:

$$y_i = \alpha + \beta x_i + e_i \text{ with } e_i \sim N(0, \sigma^2) \text{ independently for } i = 1, \ldots, n, \quad (6.5)$$

where $e_i \sim N(0, \sigma^2)$ is read as "e_i has a normal distribution with mean 0 and variance σ^2". This shows immediately that y is assumed to be made up of the linear regression on x, $\alpha + \beta x$, plus the residual or error term e; x is regarded as a fixed number whereas e is a random variable.

An alternative way of writing the same model is:

$$(y_i | x_1, \ldots, x_n) \sim N(\alpha + \beta x_i, \sigma^2) \text{ independently for } i = 1, \ldots, n. \quad (6.6)$$

This is read as "y_i, given $x_1, \ldots, x_n$, has a normal distribution with mean $\alpha + \beta x_i$ and variance σ^2, independently for i going from 1 to n".

The conditional expectation or mean of y given x is denoted by $E(y|x) = \alpha + \beta x$, which depends linearly on x and the conditional variance of y given x is denoted by $\text{Var}(y|x) = \sigma^2$, which does not depend on x. Once these assumptions are made we can go ahead and derive the sampling properties of the estimated regression.

6.4   Inference for the simple linear regression model

The essential assumptions in the model expressed by eq. (6.5) and (6.6) are, in decreasing order of importance:

1. that the n observations are independent in the sense that, once $x_1, \ldots,$ x_n are given, the values of $y_1, \ldots, y_n$ are mutually independent;
2. that y is linearly related to x in the sense that its mean is a linear function of x; that is, $E(y|x) = \alpha + \beta x$;
3. that the residual variance about the regression line is constant; that is, $\text{Var}(e) = \text{Var}(y|x) = \sigma^2 = \text{constant}$;
4. that the conditional distribution of y given x is normal.

It follows from assumptions 1 to 4 that the residuals $e_1, \ldots, e_n$ in eq. (6.5) each have mean 0 and variance σ^2 and are independent of each other and of $x_1, \ldots, x_n$.

The assumptions can be relaxed or modified to give different models. For example weighted least squares can be used when the residual variance is not constant and in Chapter 13 multilevel models for clustered observations relax assumption 1 to allow ys in the same group or cluster to be correlated. Assumption 4 justifies the use of the normal distribution for making inferences using the methods described below, but it can be shown that these methods are reasonably robust against moderate non-normality.

A new feature in the model-based approach is the unknown variance, σ^2, which measures the variability about the line. Thus there are three *parameters*, α, β and σ^2. If we use the method of maximum likelihood we find that the estimators of α and β are exactly the same as the least squares estimators, a and b, given in eq. (6.1). The maximum likelihood estimator of σ^2 is the sum of squares about the regression line divided by n, $\sum_{i=1}^{n}(y_i - \hat{y}_i)^2/n$; but this is biased because it does not take into account that α and β have been estimated. Instead we use the unbiased estimator

$$\hat{\sigma}^2 = \frac{1}{n-2}\sum_{i=1}^{n}(y_i - \hat{y}_i)^2 = \frac{1}{n-2}\sum_{i=1}^{n}(y_i - a - bx_i)^2, \tag{6.7}$$

where the divisor is the *degrees of freedom*, $(n-2)$, since there are n observations but each $\hat{y}_i$ is a function of two estimated parameters, the intercept, a, and the slope, b.

The sampling distribution of b can be derived mathematically from the model. It is

$$b \sim N\left(\beta, \frac{\sigma^2}{\sum_{i=1}^{n}(x_i - \bar{x})^2}\right). \tag{6.8}$$

Using this result it is possible to construct a confidence interval for the unknown β or to test a hypothesis about its value. The most obvious *null* hypothesis to test is that $\beta = 0$, that is, that the slope of the regression line is zero, in which case the distribution of y does not depend on x. If σ^2 were known the test statistic would be b divided by its standard error.

In practice, of course, σ^2 is rarely known so b is divided by its estimated standard error, which is

$$\hat{se}(b) = \frac{\hat{\sigma}}{\sqrt{\sum_{i=1}^{n}(x_i - \bar{x})^2}}.$$

To test the null hypothesis that $\beta = 0$, we would compare the absolute value of $[b/\hat{se}(b)]$ with t^*, where t^* is the upper 2.5 percentage point of Student's t-distribution with $(n-2)$ degrees of freedom (t_{n-2}), since $[(b - \beta)/\hat{se}(b)]$ has this distribution.

More usefully, a 95 percent confidence interval for β would be

$$\left(b - t^* \times \hat{se}(b), \; b + t^* \times \hat{se}(b)\right).$$

For large samples, when $(n-2)$ is greater than 30, t^* will be close to 2, so we commonly use $(b - 2 \times \hat{se}(b), \; b + 2 \times \hat{se}(b))$. If b was more than twice as large (positively or negatively) as its estimated standard error we might conclude that there was evidence that β was not zero and that the fitted regression on x might be useful in predicting or explaining y.

Users of statistics in both social and physical sciences sometimes confuse the estimator, b or $\hat{\beta}$, say, of a parameter with the parameter, β, say, itself. Where a model successfully reflects reality, the parameters will have a descriptive interpretation. For example we may refer to β as the *true* slope describing, for example, the average rate at which the son's height increased with the father's height in Victorian England. The estimated slope, b, in general, will not equal the true slope, although we hope it will be close and we use the estimated standard error to help us to judge its accuracy.

Statistical inference uses sample statistics to estimate and make judgements about the parameters of the probability model, but, to be useful, an extra stage of interpretation is needed. Galton (1890) estimated from a sample of men that the regression coefficient of the son's height on the father's height was about $\frac{1}{3}$, thus the fitted model predicted that if a father was 3 inches taller than average the son would be 1 inch taller than average. But how could this be used for the general population? If Galton's sample was representative of fathers and sons (in Victorian England), then the equation could be used to predict the adult height of a son born to a father of known height. Also it might be used as part of a *causal* explanation of the actual height of a man (that he shared genes with his father and maybe grew up in a similar environment to his father). Both uses require the sample to be representative of the population and the second use also depends on extra substantive assumptions about the determinants of growth.

Later, in discussing the interpretation of multiple regression and path analysis, we shall consider how taking different variables into consideration may show an apparent correlation to be spurious (explained by a third variable), or may reinforce it as indicative of a possible causal relationship.

6.5   Checking the assumptions

Initial scatterplots can be used to check that the relationship between x and y is approximately linear with constant variance. Residual plots of $\hat{e}_i = y_i - \hat{y}_i$, $i = 1, ..., n$, can be used to check the assumptions of normality, constant variance, linearity and independence. Information about the source of the data is also important in judging whether the assumption of independent errors is reasonable. In some cases one knows *a priori* that the distribution of y, or e cannot be normal. For example, if y is necessarily positive, as when it is an amount of money or a weight, its conditional distribution given x cannot be exactly normal because a normal distribution has a doubly infinite range (from minus infinity to plus infinity). Nevertheless it may still be close enough to normality for practical purposes. If it is not, it may be possible to make it more nearly normal by a transformation. For example, taking the logarithm of a positive y produces a variable with a suitable range and often induces an adequate approximation to normality. Taking the logarithm, or any other transformation, of y destroys the linearity as we can see by inspection of eq. (6.5). However, the linearity may not have been there in the first place and we may well get a better fitting model in the logarithmic form. If this is so we might have spotted the need for a transformation from the initial scatterplot if it exhibited something like an exponential growth. The preanalysis in any simple regression problem requires careful inspection of the data to see what transformation may bring it closest to the form required by the model, or whether the model might need to be adapted. Atkinson (1987) provides a comprehensive guide to diagnostic plots and analyses for checking assumptions and determining appropriate transformations.

When y is a categorical variable we cannot transform it in this way and it is better to use a different kind of model. For example, we describe logistic regression for binary y in Section 6.14.

Note, however, that there is nothing in the simple linear regression model (or in the multiple regression model that follows) that requires regressor variables to be continuous. They could be discrete or categorical. For example, Figure 6.3 shows the regression of birth weight, y, on a dummy binary variable, D. We introduce the dummy variable, D, so that $D = 0$ for a girl and $D = 1$ for a boy. The fitted line goes through the mean birth weight for 34 newborn girls (3.24 kg) and the mean birth weight for 31 newborn boys (3.43 kg) and the slope of the line (0.19 kg) is the difference in mean birth weight (boys minus girls since boys are coded 1). The reader may care to check these using the formulae given in eq. (6.1).

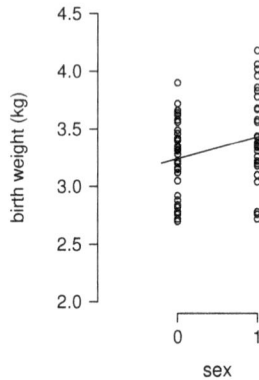

FIGURE 6.3: Plot of birth weight against sex illustrating regression on a binary variable

One might ask at this stage: What happens if the scatterplot does not suggest a linear relationship and if no suitable transformation suggests itself? The theory we have given may seem rather limited if it can only deal with this special kind of relationship. The answer is that a great many other kinds of relationship can be made linear by one means or another. This is a topic to which we shall return in Section 6.11.

6.6 Multiple regression

The reader who encounters regression in the social sciences will almost certainly become involved with multiple regression where there are several regressor (predictor or explanatory) variables. The model in this case may be written

$$y_i = \alpha + \beta_1 x_{i1} + \beta_2 x_{i2} + \ldots + \beta_k x_{ik} + e_i$$

with $e_i \sim N(0, \sigma^2)$ independently for $i = 1, \ldots, n$.   (6.9)

Exactly the same assumptions are made about the *e*s (residuals or error terms) as before, namely that they are mutually independent and normally distributed with mean 0 and variance σ^2.

The βs are still called regression coefficients or, more fully, *partial* regression coefficients, to distinguish them from the *marginal* regression coefficient in simple linear regression. The subscripts show to which *x* the partial regression coefficient belongs. Sometimes an even more complex notation is used to show which other variables are included in the model: thus $\beta_{y1.2}$ would be the partial regression coefficient of *y* on x_1 given, or conditional on, x_2 and would be conceptually and numerically different both from $\beta_{y1.23}$, the partial regression coefficient of *y* on x_1 conditional on x_2 and x_3, and from β_{y1}, the marginal or simple regression coefficient. Consider, for example, trying to predict a man's weight, *y*, from only the length of his left leg, x_1, or from both x_1 and his hip size, x_2, or from x_1, x_2 and his height, x_3. The marginal regression coefficient, β_{y1}, will be much larger than the partial regression coefficient, $\beta_{y1.2}$, and $\beta_{y1.23}$ could turn out to be zero or even negative! However, usually, the simpler notation is adequate and we only need to say whether a regression coefficient is marginal or partial for greater emphasis or for clarity when the context does not make it clear.

The aim of fitting the least squares regression is to estimate the βs so as to obtain the best prediction of *y*. In practice, however, regression equations are often not fitted primarily for predicting *y* but, as we shall see later, for investigating which predictor or explanatory variables are needed and what their relative importance might be. This use is sometimes suspect!

Before we continue, some examples of applications in the social sciences may help to focus on the main questions.

6.7 Examples of multiple regression

1. *Social deprivation.* Studies of social deprivation have attempted to relate this phenomenon to various demographic characteristics. Suppose we wish to relate an index of social deprivation for a large number of local authority areas to such demographic characteristics as: percentage of residents over 75 (x_1), average household income per annum (x_2), number of single parent families (x_3), infant mortality rate (x_4), unemployment rate (x_5), number of local benefit offices per ten thousand of population (x_6) and so on. In general we might be interested in which, if any, of these variables were important determinants of social deprivation. More specifically, whether there are any variables which it is within the power of local or central governments to influence, and if so, what reduction in the index of social deprivation might be achieved by policy changes. Two warnings:

(a) if the measure of social deprivation were itself a function of any of the xs, then regressing on the same xs would be circular and establish nothing,

(b) a relationship found using observational data may not be causal so that changing, say, the number of local benefit offices, x_6, would not necessarily have the effect of changing social deprivation. A variable useful for *prediction* may be useless for *manipulation* of the outcome variable.

2. *House prices.* Many economic examples attempt to relate prices to factors which are believed to influence them. Variables which might be used in such a study to determine the average house price in towns might include: per capita crime rate, average number of rooms per dwelling, proportion of owner occupied dwellings, pupil/teacher ratio in the local schools and the concentration of nitric oxides in the air in parts per 10 million. Multiple regression might help us to discover whether any or all of these variables are important.

3. *Hedonism.* In Section 6.17 and again in Chapter 13 on multilevel modelling, we shall use data from a survey in the European Union (Jowell 2003) to study the relationship of country of residence, gender, income and age to an index of hedonism. This will enable us to see what relationships exist and whether, for example, the relationship with age varies from one country to another.

4. *Recidivism.* Criminologists and those deciding on sentencing would like to be able to predict the probability of re-offending from information about an offender. (The results of a logistic multiple regression are presented in Section 6.17.)

5. *Health care expenditure.* Multiple regression of expenditure on a suite of socio-economic variables has been used (or misused) to determine the appropriate level of expenditure for a Primary Care Trust in England.

In some of the above examples there is a question of what might be the appropriate units of analysis (residents or local authorities, patients or Primary Care Trusts). The extension of regression to multilevel modelling in Chapter 13 provides for units of analysis at more than one level.

6.8   Estimation and inference about the parameters

The parameters of the model can be estimated by least squares or maximum likelihood in an analogous way to simple linear regression. The results can be written down very compactly using matrix notation but any standard software package will produce numerical values of the estimates, $a, b_1, \ldots, b_k$, together with their estimated standard errors as a matter of routine. The fitted or predicted value of y is

$$\hat{y} = a + b_1 x_1 + \ldots + b_k x_k. \tag{6.10}$$

A new feature when there are two or more regressor variables is that the correlations between the regressors give rise to correlations between the bs. The size of these correlations may be important when we come to interpretation.

Two important characteristics are almost the same for multiple as for simple regression, σ^2 and R^2. The residual variance, σ^2, can still be estimated from the sum of squares about the regression, the only difference being in the *degrees of freedom* used as the divisor. This is reduced because we have now fitted k regression coefficients and one intercept. This makes the degrees of freedom $(n - k - 1)$. We thus have

$$\hat{\sigma}^2 = \frac{1}{(n - k - 1)} \sum_{i=1}^{n} (y_i - \hat{y}_i)^2, \tag{6.11}$$

where $\hat{y}_i$ is given by eq. (6.10). Similarly, the squared multiple correlation coefficient, R^2, is still defined as in eq. (6.4) but with $\hat{y}_i$ given by eq. (6.10).

In multiple regression R^2, or preferably adjusted R^2 (adjR^2), can be used to compare the fits of various models. A large R^2 suggests that the model is a good fit. But adding a new predictor variable will always increase R^2, even if the new variable is irrelevant. Therefore, computer packages often print out adjR^2, where

$$\text{adj}R^2 = \frac{(n-1)R^2 - k}{n - 1 - k} = 1 - \frac{\hat{\sigma}^2}{\sum(y_i - \bar{y})^2 / (n-1)}. \tag{6.12}$$

Unlike R^2, adjR^2 will only increase when the addition of an extra explanatory variable reduces the estimated residual variance, $\hat{\sigma}^2$, which makes it a better indicator of whether the extra variable increases the accuracy of prediction.

It is common, in social science applications, to find that the predictor variables account for only a small part of the variation in y. Values as low as 0.3 (30%) or even less, are not uncommon in social research. Although these values may be both statistically highly significant and substantively important, it is clear that there must be many other unknown factors at work. The variation to which such unknown factors give rise will all be included in the residual term. This may have implications for the assumptions we make about its distribution and so it is important to check the distribution of the residuals empirically. As we add new predictor (regressor or explanatory) variables, we transfer variation from the residual to the regression. One can test the significance of individual regression coefficients to determine whether a particular predictor variable needs to be included. Some packages facilitate this by printing a t-value for each regression coefficient. This is the coefficient divided by its estimated standard error. As in the simple linear regression case, $[(b_j - \beta_j)/\hat{se}(b_j)]$ has a t-distribution. However, care must be taken in interpreting an individual b_j because its meaning will depend on which other explanatory variables are included in the model.

Example: GCSE scores, multiple regression

Continuing the example in Section 6.2 we can see whether using the results of a Cognitive Ability Test (Verbal Stanine), referred to as CATsVS, taken at about age 12 will improve the prediction of GCSE English scores for girls at the comprehensive school in Oxfordshire. We can also find out whether writing speed remains a useful predictor when CATsVS is taken into account.

Table 6.2 gives the results from fitting the model in eq. (6.9). The t-values for Speed and for CATsVS are both large (with small p-values) so it would appear that using CATsVS as well as writing speed increases the accuracy of prediction (R^2 has increased from 16% to 50%) and that writing speed is still a useful predictor even after the CATsVS test is taken into account. Also note the difference between the partial regression coefficient, $b_{y1.2} = 0.68$, and the marginal regression coefficient, $b_{y1} = 0.99$.

TABLE 6.2: Summary of regression of GCSE on Speed and CATsVS

	estimate	s.e.	t-value	p-value
Intercept	2.9412	0.4158	7.07	< 0.001
Speed	0.6780	0.1722	3.94	< 0.001
CATsVS	0.4476	0.0532	8.42	< 0.001
Residual standard deviation $\hat{\sigma} = 0.93$, $R^2 = 0.50$, adj$R^2 = 0.49$				

The fitted regression (with coefficients rounded to two decimal places) is

$$\widehat{\text{GCSE}} = \hat{y} = 2.94 + 0.68 \times \text{Speed} + 0.45 \times \text{CATsVS}. \qquad (6.13)$$

The analysis of variance table is given in Table 6.3. There are two degrees of freedom for the regression as there are two explanatory variables.

TABLE 6.3: Analysis of variance table for multiple regression of GCSE scores

Source of variation	DF	Sum of Squares	Mean Square
Regression	2	$\sum (\hat{y}_i - \bar{y})^2 = 90.4$	45.2
Residual (about regression)	107	$\sum (y_i - \hat{y}_i)^2 = 91.9$	0.9
Total (about mean)	109	$\sum (y_i - \bar{y})^2 = 182.2$	1.7

Since the sample is not representative of girls other than those at the single comprehensive school, and, since the response variable is ordinal, we cannot formally use the normal multiple regression model. This regression must be regarded as descriptive of the particular sample, rather than as an estimate of how GCSE English scores depend on CATsVS and writing speed in a wider population. To what extent we can generalise our conclusions beyond the girls at that school in that year is a matter for judgement and further investigation.

6.9 Interpretation of the regression coefficients

Interpretation lies at the heart of applied regression analysis in the social sciences and it is fraught with hazards. In the simple case we had only one predictor variable and its effect was directly measured by the size of the regression coefficient. In multiple regression there are several coefficients and their relative, as well as their absolute values become important. In the first place the importance of a regression coefficient cannot be judged from its magnitude alone. Suppose, for example, one of the predictor variables is an amount of money. This could be expressed in dollars or in cents. The coefficient will be much smaller in the second case because its effect on y is the product of the coefficient and the x, so if one is a larger number the other will be correspondingly smaller.

For this reason the regression function is sometimes expressed in terms of standardised explanatory variables to make them independent of the units in which they were measured, and, therefore, more directly comparable. If we standardise an x by dividing by its standard deviation we must multiply the corresponding b by the same factor to compensate. The fitted regression equation may then be written

$$\hat{y} = a + b_1^* \left(\frac{x_1}{s_1} \right) + b_2^* \left(\frac{x_2}{s_2} \right) + \ldots + b_k^* \left(\frac{x_k}{s_k} \right), \qquad (6.14)$$

where s_j is the estimated standard deviation of x_j and $b_j^* = b_j \times s_j$, $j = 1, 2, \ldots, k$. Thus b_j^*, the *standardised regression coefficient*, is the change in $\hat{y}$ when x_j changes by one standard deviation (provided the other xs do not change). The b^*s are sometimes called *beta coefficients* or *beta weights* in order to emphasise their difference from the bs calculated from unstandardised data (but do not be misled, they are *not* the theoretical parameters, denoted by βs, in the regression model).

The second problem of interpretation concerns the meaning of any particular b. In simple linear regression b was the estimated difference in average y for two individuals whose x-value differed by one unit. In the case of multiple regression a given b has the same interpretation but *only if all other explanatory variables are the same for both individuals*. For example, from eq. (6.13), if we consider only girls with the same CATsVS score, $x_2 = 2$, say, then the estimated difference in average GCSE score between girls with a writing speed of two words and those with a writing speed of three words per tenth of a minute is $b_{y1.2} = 0.68$ of a grade. Similarly if we take only girls with a writing speed of $x_1 = 2$ words per tenth of a minute, then the estimated difference in average GCSE score between those with CATsVS of 2 and those with CATsVS of 3 is $b_{y2.1} = 0.45$ of a grade.

But in practice the explanatory variables are likely to vary together! Girls who obtained higher CATsVS scores when 12 tend to be faster writers at age 16, which is why, in this case, the marginal regression coefficient, $b_{y1} = 0.99$, is larger than the partial regression coefficient, $b_{y1.2} = 0.68$.

When an increase in x_1 is associated not only with an increase in y but also with an increase in x_2 the effects of x_1 and x_2 on y are said to be *confounded* and it is not possible, without extra substantive assumptions, to disentangle the separate contributions of each explanatory variable. (Path analysis, see Section 6.15, and structural equation modelling, see Chapter 11, attempt to do just this by introducing extra structure.)

With many explanatory variables or predictors the inter-connections can be quite complicated. This can be particularly disconcerting if one hopes to effect policy changes by manipulating one of the predictors. For example, suppose that some measure of quality of life for a family (y) is regressed on, among other things, their annual expenditure on holidays (x_1) and on food (x_2). Suppose it turns out, as one would expect, that x_1 and x_2 both have positive regression coefficients. The obvious deduction is that an increase in food expenditure would result in an increase in quality of life. But the two expenditures may be related in such a way that when one is increased the other decreases. In that case, the increase in y, resulting from extra expenditure on food, would be more or less offset by the decrease, due to reduced holiday expenditure. A partial regression coefficient such as $b_{y1.2}$ is sometimes described as measuring the *effect* of x_1 on y adjusted for, or after allowing for, x_2.

Such problems would be avoided if the predictor variables were uncorrelated, or nearly so, in which case the partial regression coefficient would equal the marginal coefficient, or nearly so. Where the researcher has the opportunity to design a study and to choose the values of the predictor variables, they should be chosen, if possible, so that they are uncorrelated. For example an educational experiment on handwriting might divide a block of 100 boys into two groups of 50 at random, boys in one group would receive individual tuition in speed writing and in the other group not. Similarly a block of 100 girls would be divided into two groups of 50 at random, only one of which would be given individual tuition in speed writing. The two predictor variables, gender (the *blocking factor*) and speed-writing-tuition (the *intervention*), would be uncorrelated and their effects on, say, GCSE scores, would not be confounded. But most datasets in social science are observational and, even where the values of an intervention variable can be allocated to experimental units using random numbers, there are usually other covariables which cannot be controlled or chosen. Furthermore it is dangerous to infer from an intervention study what would happen if the predictor variables were allowed to vary and covary naturally (and conversely it is dangerous to infer from an observational study what would be the effect of an intervention).

Another way of obtaining uncorrelated regressor variables is to carry out PCA of the xs. The principal components are, by construction, uncorrelated so if we were to regress y on the most important components the interpretation would be more straightforward, provided the principal components were readily interpretable in their own right, which may often not be the case. But, if prediction is more important than interpretation, and, if the number of regressor variables is large, then a preliminary PCA to reduce their dimensionality might be useful even if the components did not have any clear meaning.

In practice it is more usual to use a selection procedure to decide which variables to use as predictors. As illustrated by the example of predicting height from the lengths of legs and arms, there is no point in adding extra predictor variables which are closely related to others which are already included. For example, if one variable was the percentage of people aged 70 or over, there would be little to be gained by adding the percentage aged 80 or over. In a stable population these percentages will be proportional but, more generally, will be highly correlated. It is intuitively clear that the second variable will add little that is not already implicit in the first but, if we overlook this fact, it will come to light in the computation. Where the regressor variables are highly inter-correlated it may not even be possible to fit the model numerically because the fitted regression coefficients will be unstable with large standard errors and high (positive or negative) correlations. This is a simple example of what is known as *multi-collinearity*. If it occurs, it is an indication that there are strong relationships among the regressor variables and thus that they can be reduced in number without serious loss of information about the response variable.

6.10 Selection of regressor variables

In most social science applications there will be many variables which could be used as regressor, predictor or explanatory variables. We need some means of selecting those which are best in some sense. Adding extra regressor variables will always increase R^2, but will not always increase the precision of the prediction. For this reason many computer packages also print adjusted R^2, given by eq. (6.12), which will decrease when the model is over-fitted. However, there may come a point before this when interpretability and parsimony dictate that the process should stop. There may be *a priori* reasons for including certain variables because they are of substantive interest, otherwise the aim is to select predictor variables that, together, provide a good prediction of the response variable.

Example: GCSE scores, selection of variables

In Sections 6.2 and 6.8 we regressed GCSE scores on writing speed and CATsVS, but further information is available on the website about each girl at the comprehensive school in Oxfordshire which might help in predicting their GCSE scores. We shall consider the regression of GCSE English score (the response variable) on up to four regressor variables: Speed; CATsVS; Joins (a binary variable coded 1 if there is a problem with joined-up writing and 0 if there is no problem); and CATsNVS, which is the score at age 12 on the Cognitive Ability Test (Non-Verbal Stanine).

TABLE 6.4: Summary of regressions of GCSE on 4 and 3 variables

	4 variables		3 variables	
	estimate (s.e.)	t-value	estimate (s.e.)	t-value
Intercept	4.05 (0.43)	9.34	4.10 (0.41)	9.94
Speed	0.41 (0.16)	2.62	0.42 (0.16)	2.69
CATsVS	0.39 (0.06)	6.93	0.40 (0.05)	8.52
CATsNVS	0.03 (0.06)	0.43		
Joins	−1.00 (0.17)	−5.86	−1.00 (0.17)	−5.89
	$\hat{\sigma} = 0.811$, adj$R^2 = 0.607$		$\hat{\sigma} = 0.808$, adj$R^2 = 0.610$	

The left-hand side of Table 6.4 gives results for the regression of GCSE score on all four variables. The residual standard deviation after fitting this model is $\hat{\sigma} = 0.811$ on 105 degrees of freedom. The estimated regression coefficient for CATsNVS is smaller than its standard error, giving a t-value of 0.43. Clearly if the other three variables are already included CATsNVS adds nothing to the predictive or explanatory usefulness of the regression equation.

The right-hand side of Table 6.4 gives corresponding results when CATsNVS is dropped. For the regression on the three variables, the coefficients of Speed, CATsVS and Joins are almost the same as before, as is the residual standard deviation, $\hat{\sigma} = 0.808$ on 106 degrees of freedom. The t-values are all greater than 2 in absolute value, so we conclude that these three variables are all useful in predicting and explaining GCSE English score. The fitted regression equation is

$$\widehat{\text{GCSE}} = 4.10 + 0.42 \times \text{Speed} + 0.40 \times \text{CATsVS} - 1.00 \times \text{Joins}. \quad (6.15)$$

Comparing Tables 6.4 and 6.2 shows that the effect of Speed adjusted for CATsVS and Joins ($b_{y1.24} = 0.41$) is reduced compared to the effect adjusted for CATsVS only ($b_{y1.2} = 0.68$). This is because girls who fail to join up their writing properly tend to be slower writers, so that Speed and Joins are confounded in their effects on GCSE score.

Faced with a large number of potential regressor variables the social scientist should first turn to theory, established or hypothetical, for guidance in the choice of regressor variables. This may still leave a large number of interrelated variables leading to multi-collinearity if they were all included.

Statistical packages provide automatic procedures for selecting a good subset of regressor variables that together contribute to explaining the variation in the response variable. But care should be exercised in interpreting the result. A regression analysis is not designed to uncover "social laws" connecting the dependent (response) variable to the explanatory variables. Confusion arises because it is commonly regarded in that light and regression coefficients are wrongly interpreted as measuring the strength of causal links. On its own, all that regression analysis tells us is that some combinations of variables have good predictive value.

Furthermore, where variables have been selected because they are good predictors for the particular sample available, there is a risk of *over-fitting* and they may not be such good predictors more generally even for the population from which the sample was obtained.

Consider the added problem when forecasting, for example, which criminal offenders will re-offend. Inevitably the model must be fitted to historical data but needs to be applied to the new population of current offenders.

6.11 Transformations and interactions

In the case of simple linear regression we noted that there was no need for the regressor variables to be continuous. This is equally true for multiple regression and this fact can be exploited to enlarge the scope of regression analysis.

We can change the functional form of the regression equation and incorporate other types of variable with virtually no modifications at all. For example if we make $x_2 = x_1^2$ we have a quadratic regression function. No change is required beyond writing the squares of the first variable in place of the second. In like manner any x can be replaced by any function of other variables or even of itself. For example in economic and biological applications it is common to work with the logarithms of variables. Further, if a new variable is formed as the product of two existing variables ($x_3 = x_1 \times x_2$), this introduces an *interaction* term into the model. Thus we might have

$$y_i = \alpha + \beta_1 x_{1i} + \beta_2 x_{2i} + \beta_3 x_{1i} x_{2i} + e_i, \tag{6.16}$$

where $e_i \sim N(0, \sigma^2)$ independently for $i = 1, \ldots, n$.

Example: GCSE scores, including an interaction

Continuing the example of Sections 6.2 and 6.8, the effect of writing speed on GCSE score might be different for girls with good joined-up writing (i.e. Joins = 0) than for girls whose writing exhibits problems (Joins = 1). Without an interaction term the model uses parallel partial regressions, implying that the slope of the regression of GCSE score on writing speed is the same for girls with Joins = 1 as for girls with Joins = 0, although the intercepts would differ. This would mean that the difference in mean GCSE score between girls with a given writing speed and Joins = 1, and girls with the same writing speed and Joins = 0, would be the same for any writing speed.

Table 6.5 give results from fitting the model in eq. (6.16). The coefficient of the interaction term, -1.26, has a significantly small p-value, suggesting that the relation between GCSE and Speed does depend on the value of Joins. The fitted model equation is formally

$$\widehat{\text{GCSE}} = 5.11 + 0.94 \times \text{Speed} + 1.26 \times \text{Joins} - 1.26 \times \text{Speed} \times \text{Joins}.$$

TABLE 6.5: Summary of regression of GCSE on Speed and Joins with interaction

	estimate	s.e.	t-value	p-value
Intercept	5.1078	0.5155	9.91	< 0.001
Speed	0.9395	0.2231	4.21	< 0.001
Joins	1.2645	0.9152	1.38	0.170
Speed×Joins	−1.2629	0.4477	−2.82	0.006
Residual standard deviation $\hat{\sigma} = 1.01$, $R^2 = 0.41$, adj$R^2 = 0.39$				

However, it is not straightforward to interpret the individual coefficients in this equation. For example, the coefficient of $+1.26$ for Joins looks counter-intuitive, but this does *not* imply that GCSE scores tend to be higher for girls with Joins = 1 because the interaction term also depends on Joins. We have effectively fitted a separate line relating GCSE to Speed for each value of Joins, the two lines having different intercepts *and* slopes. Substituting Joins = 0, we see that for girls with no problems the predicted GSCE score is $5.10 + 0.94 \times$Speed; and substituting Joins = 1, the predicted score for girls with problems is $6.37 - 0.32 \times$Speed. This latter equation looks strange as it seems to say that the predicted GCSE score for girls whose writing is not properly joined-up *decreases* as writing speed increases. However, it can be shown that the coefficient -0.32 has a large standard error and a nominal 95% confidence interval would go from -1.1 to $+0.5$. Speed does not appear to have any further effect on the GCSE grades of girls whose writing is not properly joined-up, so for them, rather than using the regression on Speed, we should just use their mean, 5.76, to predict their GCSE score.

Interpreting the fitted model without separating the two groups as we have done would not be easy. But even for more complicated models the meanings of the regression coefficients can be worked out logically.

The ability to transform and combine explanatory variables greatly increases the range of regression functions which can be used. Also categorical variables can be dealt with by a simple extension of the idea we used for binary explanatory variables in Section 6.5. For example, in a study of attitudes in *three* European countries, just *two* dummy variables can be used to code the three countries, Austria (A), Belgium (B) and the Czech Republic (C). For people in Austria let $D_1 = 1, D_2 = 0$, for people in Belgium let $D_1 = 0, D_2 = 1$ and for those in the Czech Republic, the *reference category*, let $D_1 = 0, D_2 = 0$. The fitted model would then be

$$\hat{y} = a + b_1 D_1 + b_2 D_2.$$

The predicted values of y for the three groups are: for A, $\hat{y} = a + b_1$; for B, $\hat{y} = a + b_2$; and for C, $\hat{y} = a$. C is the *reference* group and the estimated regression coefficients b_1 and b_2 are the estimated differences between A and C and between B and C. See Section 6.17 for a numerical example.

The problem we have been discussing is, of course, an example of a one-way analysis of variance but it is interesting to see that this can be included within the framework of the regression model.

When used in this broader way the model is often described as the *general linear model*. The adjective *linear* refers to the regression function being linear in the regression coefficients, not necessarily in the xs.

In Chapter 13 random effects or multilevel models will be developed. In these the group effects are treated as random variables rather than as fixed parameters.

6.12 Prediction, test error, cross-validation and regularised estimation

As mentioned previously, a multiple regression model is useful not only for inferring the relationship between variables but also for making predictions. We now elaborate further on the task of prediction.

Prediction and test error

Suppose that we are interested in predicting the values of the response variable for a set of observations. For these observations, we observe their regressors but not their responses. For example, in Galton's example about people's heights, these observations may correspond to study participants whose sons had not yet grown up at the time of the study (so that their adult heights were unknown). These observations are typically referred to as the test observations. We denote their regressors as $x_{i1}^{(test)}$, ..., $x_{ik}^{(test)}$, $i = 1, ..., n^{(test)}$. Further, suppose that we have collected a sample of data for which we observe both the regressors and responses. In Galton's example, this sample includes the study participants for whom both fathers' and sons' heights were observed. This sample is referred to as the training data because it will be used to train prediction models. We denote the training observations as $x_{i1}, ..., x_{ik}, y_i$, $i = 1, ..., n$. When using a multiple regression model as a prediction model, we first estimate the model parameters based on the training data, for example using methods mentioned in Section 6.8, and then use the estimated parameters to predict the response value for the test observations. More specifically, let a, b_1, ..., b_k be the estimated intercept and slope parameters, respectively. Then, for test observation i, we predict their response value by

$$\hat{y}_i^{(test)} = a + b_1 x_{i1}^{(test)} + \cdots + b_k x_{ik}^{(test)}.$$

When multiple models are available, for example models involving different subsets of regressors, we want to choose the one that yields the most accurate predictions. In that case, we hope to find a model whose predicted values are as close as possible to the corresponding actual values of the response variable. Imagine that we can observe the response variable for the test observations. We denote its values by $y_i^{(test)}$, $i = 1, ..., n^{(test)}$. Then, the prediction accuracy can be evaluated by the test Mean Squared Error (MSE), defined as

$$\text{MSE}^{(test)} = \frac{1}{n^{(test)}} \sum_{i=1}^{n^{(test)}} (y_i^{(test)} - \hat{y}_i^{(test)})^2,$$

which measures the average deviation between the predicted and observed responses. If the test MSEs could be computed for different models, we would choose the one with the smallest test MSE.

Cross-validation

Unfortunately, in most real-world scenarios, the response variable for the test observations cannot be observed, at least when prediction is needed (otherwise, the prediction task is pointless). Consequently, the test MSEs cannot be computed and, thus, cannot be used to choose the best prediction model. In that case, we may estimate the test MSEs for different models using only the training data and then choose the prediction model based on the estimated test MSEs. The question becomes: how do we sensibly estimate the test MSEs given only the training data?

A naive estimator for the test MSE of an estimated model is the training MSE, defined as

$$\text{MSE}^{(train)} = \frac{1}{n} \sum_{i=1}^{n} (y_i - \hat{y}_i)^2,$$

where $\hat{y}_i$ is the predicted value for the ith observation in the training set. However, $\text{MSE}^{(train)}$ is not a good estimator for $\text{MSE}^{(test)}$. A major issue is that the value of $\text{MSE}^{(train)}$ always decreases as more regressors are included in the regression model, in which case more parameters are estimated. Consequently, the largest model, i.e. the model with the most parameters, will be chosen. This model is typically not the best prediction model due to the issue of over-fitting. That is, the more parameters a model has, the more information it tends to capture about the independent noises e_i in the training data, which leads to deterioration in its prediction performance when applied to the test data.

The issue with $\text{MSE}^{(train)}$ is due to its "double dipping" of the training data. That is, the training dataset is used twice when calculating the training MSE – first to estimate the model parameters and second to calculate the squared differences $(y_i - \hat{y}_i)^2$. The cross-validation method is a popular method for estimating the test MSE. It addresses the double-dipping problem by data splitting. In what follows, we describe the steps of performing a p-fold cross-validation for a given multiple regression model. In practice, p is typically chosen as 5 or 10.

1. Randomly split the training set into p folds (i.e. subsets). The sample sizes are approximately the same across the folds, and every observation in the training set belongs to one and only one fold. Figure 6.4 illustrates the data splitting for a five-fold cross-validation.

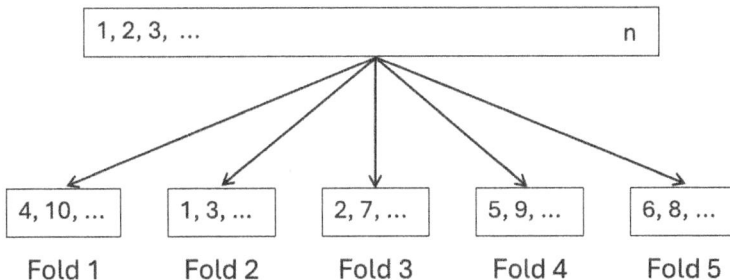

FIGURE 6.4: A visual illustration of the data splitting step of a five-fold cross-validation

2. Fit the multiple regression model p times, each time using $p-1$ folds of data for parameter estimation and one fold of data for computing a validation MSE defined below. More specifically, the following p steps are performed.

   (1) Use all the training observations except for those in the first fold to estimate the multiple regression model, and use the estimated parameters to predict the responses of observations in the first fold. Denote these observations as $\hat{y}_i^{(1)}$ for all the observations i in the first fold. Compute the validation MSE for the first fold, denoted as MSE_1, by averaging the squared differences $(y_i - \hat{y}_i^{(1)})^2$ for all i in the first fold.

   $\vdots$

   (p) Use all the training observations except for those in the pth fold to estimate the multiple regression model, and use the estimated parameters to predict the responses of observations in the pth fold. Denote these observations as $\hat{y}_i^{(p)}$ for all the observations i in the pth fold. Compute the validation MSE for the pth fold, denoted as MSE_p, by averaging the squared differences $(y_i - \hat{y}_i^{(p)})^2$ for all i in the pth fold.

3. Compute the cross-validation MSE as

$$MSE_{CV} = \frac{1}{p}\sum_{i=1}^{p} MSE_i,$$

which is used as the final estimate of the test MSE.

We note that the calculation of each MSE_i does not suffer from data double dipping, as data in the ith fold are used to compute the MSE but not to estimate the model parameters. In other words, the ith fold of data mimics test data when we calculate MSE_i. Therefore, each MSE_i gives a sensible estimate of the test MSE, though it can suffer from high variability if the sample size in each fold is small. The cross-validation method uses the average of the validation MSEs to reduce the variability of the individual MSE_is. We illustrate the cross-validation procedure using the example of GSCE scores. A five-fold cross-validation is performed to compare the five models considered earlier in this chapter that involve different regressors or an interaction term. As the sample size is an exact multiple of five, each fold includes 22 observations.

The models and their cross-validation MSEs are given in Table 6.6. As we can see, the third model with Speed, CATsVS, and Joins as regressors has the smallest cross-validation error. In other words, based on the five-fold cross-validation, we conclude that this model is estimated to have the best prediction performance among the five models. We note that although the fourth model has an additional regressor compared to the third model, and thus a smaller training MSE, its cross-validation MSE is slightly larger. This is likely due to the additional variability from estimating the additional regression coefficient.

TABLE 6.6: Result of five-fold cross-validation, GCSE data

Model	MSE_{CV}
1. Speed	1.41
2. Speed + CATsVS	0.87
3. Speed + CATsVS + Joins	0.66
4. Speed + CATsVS + CATsNVS + Joins	0.68
5. Speed + Joins + Speed × Joins	1.09

Further topic: Regularised estimation

We provide a brief introduction to the method of regularised estimation, a popular way to avoid the overfitting issue for a linear regression model. Instead of minimising the training MSE as in the least squares estimator, a regularised estimator introduces a modified loss function, which adds a penalty term to the training MSE, and finds the parameter estimates by minimising this modified loss function. The most widely used penalty functions are the Ridge and LASSO penalties, taking the forms of $\lambda(\beta_1^2 + \cdots + \beta_k^2)$ and $\lambda(|\beta_1| + \cdots + |\beta_k|)$, respectively. Here, λ is a positive tuning parameter that determines the relative weight of the penalty term to the training MSE. The larger the value of λ, the more the parameters are penalised. This tuning parameter is typically chosen by a cross-validation procedure.

As mentioned previously, the least squares estimator tends to overfit when the model contains many parameters. A penalty term essentially imposes a constraint on the magnitude of the model parameters, reducing the possible values they can take. The reduced flexibility tends to discourage the estimated model from capturing information about the independent noises e_i in the training data, and thus reduces the risk of overfitting.

Regularised estimation is most commonly used to find good prediction models. However, certain regularised estimators, such as the one based on the LASSO penalty, tend to shrink many parameters to be exactly zero and thus automatically perform variable selection. Therefore, these estimators tend to find models that simultaneously possess good predictive power and interpretability, which may be particularly useful in social science research.

We refer readers to Chapter 6.2 of James et al. (2021) for more details about regularised estimators based on the Ridge and LASSO penalties.

6.13 Principal component regression

Principal component regression is a regression technique that combines PCA (introduced in Chapter 5) and multiple regression to find good prediction models. This method is useful in regression problems that suffer from the overfitting issue due to the presence of many regressor variables.

Principal component regression involves two steps. The first step can be seen as a representation learning step, which performs PCA on the regressor variables to extract the top PCs as an informative representation of the original regressor variables. Recall that these PCs explain the most variation in the original regressors. In the second step, multiple linear regression is performed, using the extracted top PCs as the regressor variables. By replacing the original regressor variables with the top PCs, the number of regression coefficients is substantially reduced, and so is the risk of overfitting. More specifically, in the PCA step, the score of individual i for the jth PC is computed as

$$z_{ij} = a_{1j}x_{i1} + a_{2j}x_{i2} + \cdots + a_{kj}x_{ik}, \qquad (6.17)$$

$j = 1, ..., q$ for some value of q smaller than k. In this calculation, a_{1j}, ..., a_{kj} are the component weights computed by PCA as described in Section 5.4 of Chapter 5. Then, the multiple regression step fits the following regression model:

$$y_i = \tilde{\alpha} + \tilde{\beta}_1 z_{i1} + \cdots + \tilde{\beta}_q z_{iq} + \tilde{e}_i, \qquad (6.18)$$

where we add the "$\sim$" (tilde) notation above the parameters and the error term to distinguish them from those defined in the regression model for the original regressor variables given in Section 6.6. The parameters $\tilde{\alpha}$, $\tilde{\beta}_1$, ..., $\tilde{\beta}_q$ are still estimated by the least squares estimator.

The number of PCs in principal component regression is chosen so that the resulting model predicts as accurately as possible. This is typically done by cross-validation, which differs from the methods described in Section 5.4 of Chapter 5 for choosing the number of PCs. That is, the cross-validation errors for models with different values of q are calculated, and then the one with the smallest cross-validation error is chosen.

We provide several remarks about this approach. First, the PCs, except for the very top ones which explain the most variation in the xs, are often hard to interpret. Therefore, unless the value of q is very small, interpreting a principal component regression model in the form of eq. (6.18) can be challenging. Second, the model remains linear in the original regressor variables. That is, by substituting the z_{ij}s in eq. (6.18) with their expression in eq. (6.17), we have

$$y_i = \tilde{\alpha} + (\tilde{\beta}_1 a_{11} + \tilde{\beta}_2 a_{12} + \cdots + \tilde{\beta}_q a_{1q})x_{i1} + \cdots + (\tilde{\beta}_1 a_{k1} + \tilde{\beta}_2 a_{k2} + \cdots + \tilde{\beta}_q a_{kq})x_{ik} + \tilde{e}_i, \qquad (6.19)$$

where the regression coefficient for each variable x_j is $\beta_j = \sum_{l=1}^{q} \tilde{\beta}_l a_{jl}$. However, unlike some variable selection techniques, principal component regression does not yield a parsimonious model where many β_js are estimated to be zero. Consequently, when there are many predictors, interpreting a principal component regression model based on the form of eq. (6.19) can be difficult. In other words, principal component regression tends to be good for prediction but not interpretation. This is in contrast to some variable selection techniques that tend to perform well in both prediction and interpretation.

Example: Predicting student achievement

We illustrate principal component regression using a student achievement dataset from two Portuguese schools studied in Cortez and Silva (2008). The goal is to predict students' final year grade for the subject of the Portuguese language using their demographic, social, and school-related information. If students' final grades could be predicted accurately, then early interventions may be applied to help students who are predicted to fall behind.

In this analysis, the response variable is the students' final year grade, which takes values from 0 to 20. The regressor variables include 30 continuous, dichotomous, and categorical variables, including age, sex, parents' educational levels, quality of family relationships, and extra-curricular activities. The dummy coding of the dichotomous and categorical variables leads to $k = 39$ variables for principal component regression. The dataset contains 649 observations. To evaluate the prediction performance of principal component regression, we randomly split the dataset into a training dataset with 500 observations and a test dataset with 149 observations.

We apply principal component regression to the training data. A ten-fold cross-validation is performed to select q, the number of PCs, which can take a value from 0 to 39. Figure 6.5 shows the cross-validation MSE as a function of the number of PCs. It decays quickly as the number of PCs increases, stablises when the number of PCs is between 22 and 32, and increases slowly when the number of PCs goes beyond 32. The smallest cross-validation error is achieved by choosing $q = 31$, marked by a cross symbol in the plot. We note that the overfitting issue is mild for this example, as even if we use all the PCs, which is equivalent to running a standard multiple linear regression with all the original regressor variables, the cross-validation MSE is only slightly larger than the selected model when $q = 31$. The PCs are hardly interpretable, resulting in the low interpretability of the selected model.

The models' prediction performance on the test data is consistent with the cross-validation result. The selected model achieves a test MSE of 7.31 on the test dataset, while the model with all the PCs has a test MSE of 7.71. This result shows that the dimension reduction step in principal component regression leads to better prediction performance, though the improvement is relatively small in the current example.

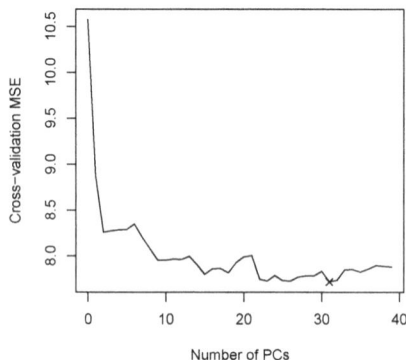

FIGURE 6.5: Plot of the cross-validation MSE as a function of the number of PCs in principal component regression for student achievement data

6.14 Logistic regression

If y is categorical the position is a little more complicated. Here we consider the simplest case of a binary response coded 0 or 1. To obtain a clue about how to construct a suitable model we return to the thinking behind eq. (6.9). We chose a suitable distribution for y (normal) and then made that distribution depend on the regressor variables in a convenient way (by making its mean linear in the xs). Now we have a binary response variable taking only two values, 0 and 1. Let us denote the conditional probability that $y = 1$ for given values of $x_1, \ldots, x_k$ by $\pi(x_1, \ldots, x_k)$. Next we have to find an appropriate way to express the dependence of π on the regressor variables. We cannot simply write

$$\pi(x_1, \ldots, x_k) = \alpha + \beta_1 x_1 + \beta_2 x_2 + \ldots + \beta_k x_k$$

because the right-hand side of this expression is not, in general, contained in the interval $[0, 1]$. Probabilities cannot be negative or greater than 1. However, there is merit in trying to retain some linearity and this can be done by using the logistic regression model with the *systematic component*

$$\pi(x_1, \ldots, x_k) = \frac{\exp L}{1 + \exp L}, \tag{6.20}$$

where

$$L = \alpha + \beta_1 x_1 + \beta_2 x_2 + \ldots + \beta_k x_k,$$

and the *random component*

$$y|(x_1, \ldots, x_k) \sim \text{Bernoulli}(\pi(x_1, \ldots, x_k)). \qquad (6.21)$$

The systematic component shows how π depends on the xs and the random component shows how y varies about π. (In the linear regression model the random component specifies the distribution of the residual, e.)

The Bernoulli distribution is a special case of the Binomial distribution and simply specifies that a single random variable, y, say, will take the value 1 with probability π and the value 0 with probability $(1-\pi)$. Thus the model in eq. (6.20) and (6.21) specifies that the probability that $y = 1$ is the given function of the x values.

The exponential function used in eq. (6.20), $\exp(L) = M$, say, gives the exponential transformation of L. Its inverse is the logarithmic or log function, $\log(M) = L$. The log odds that $y = 1$ is called $\text{logit}(\pi) = \log[\pi/(1-\pi)]$ and the systematic component can be written as

$$\text{logit}(\pi) = \log\left(\frac{\pi}{1-\pi}\right) = L = \alpha + \beta_1 x_1 + \beta_2 x_2 + \ldots + \beta_k x_k. \qquad (6.22)$$

To fit the model we need a sample of independent observations, that is a set $\{y_i, x_{i1}, \ldots, x_{ik}, i = 1, \ldots, n\}$ such that, given all the xs, the ys are mutually independent. Then the logistic regression or *logit-linear* model can be fitted by the method of maximum likelihood. Most standard statistical computer packages include logistic regression. The estimates of the α and βs obtained in this way are interpreted in a way analogous to that for the standard linear regression model, but they relate to the log odds, $\text{logit}(\pi)$, rather than to π. It is common to instead interpret e^{β_k}, which is the multiplicative effect of a 1-unit change in x_k on the odds $\pi/(1-\pi)$. We can also calculate fitted probabilities $\hat{\pi}$ for different values of the xs. Both approaches to interpretation are illustrated in the example below. We shall meet the use of logits again in Chapter 8.

Example: GCSE scores, logistic regression

Suppose instead of scoring GCSE English grades from 1 to 9 and fitting a linear regression as in Section 6.8, we merely recorded the binary variable, $y = 1$ for grades A and A* and $y = 0$ for grades B and below. Then the fitted logistic regression on Speed and CATsVS would be

$$\text{logit}(\hat{\pi}) = \hat{L} = -13.26 + 3.32 \times \text{Speed} + 0.88 \times \text{CATsVS}, \qquad (6.23)$$

where π denotes the probability of scoring grade A or A*. The fitted logistic regression coefficient for Speed conditional on CATsVS is 3.32 which can be interpreted as meaning that every extra word per tenth of a minute adds 3.32 to $\hat{L}$, the estimated log odds of obtaining an A or A*, which means that the estimated *odds* are *multiplied* by $\exp(3.32) = 27.7$.

The left panel of Figure 6.6 gives the parallel lines, $\hat{L}_2$ and $\hat{L}_5$, showing how the fitted value of L increases with writing speed when CATsVS equals 2 ($\hat{L}_2$) and when CATsVS equals 5 ($\hat{L}_5$). The right panel gives the equivalent logistic curves showing how $\hat{\pi}$ increases with writing speed when CATsVS equals 2 ($\hat{\pi}_2$) and when CATsVS equals 5 ($\hat{\pi}_5$). The lines in the left panel of Figure 6.6 are parallel because we have not included an interaction term in the model.

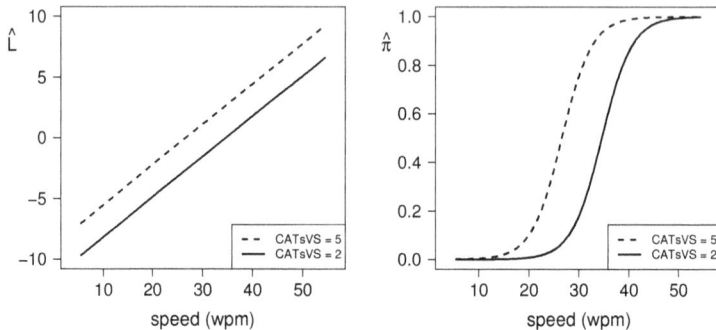

FIGURE 6.6: Fitted curves $\hat{L} = \text{logit}(\hat{\pi})$ (left panel) and $\hat{\pi}$ (right panel) against writing speed in words per minute using eq. (6.23) for CATsVS = 2 (lower curve) and CATsVS = 5 (upper curve)

Clearly, interpreting logistic regression coefficients is harder than interpreting normal multiple regression coefficients. As well as plotting the fitted curves, a table of estimated or fitted probabilities can help. Fitted probabilities, such as those in Table 6.7, are obtained from $\hat{L}$ using the inverse logit transformation

$$\hat{\pi} = \frac{\exp(\hat{L})}{1 + \exp(\hat{L})}.$$

For example, substituting Speed $= 2$ words per tenth of a minute (i.e. 20 words per minute) and CATsVS=6 gives $\hat{L} = -13.26 + 3.32 \times 2 + 6 \times 6 = -1.34$; $\exp(-1.34) = 0.26$; and therefore

$$\hat{\pi} = \frac{0.26}{1 + 0.26} = 0.21,$$

as in Table 6.7.

TABLE 6.7: Fitted probabilities of an A or A*

		CATsVS score			
		2	4	6	8
writing speed	10	0.00	0.00	0.00	0.05
(words per minute)	20	0.00	0.04	0.21	0.61
	30	0.18	0.41	0.88	0.98

Since the dataset analysed comes from a single school in a single year it would be unwise to generalise to all girls at all English schools without further investigation. But for this school it does appear that girls who wrote faster had a higher chance of getting a top grade (A or A*) even after taking their CATsVS score at age 12 into account.

Note that in dichotomising the GCSE score we lose some detailed information but avoid the technically invalid treatment of the original ordinal response variable as if it were continuous and normal (compare with the analysis in Section 6.8). A better analysis would use *ordinal logistic regression* an extension of logistic regression to cope with ordered categorical response variables.

A brief account of ordinal logistic regression is given in Agresti and Finlay (2008) and fuller accounts can be found in Agresti (2007) or in Hosmer and Lemeshow (2000). In Chapter 8 the idea of logistic regression is applied in an extension of factor analysis to binary variables. Similarly Chapter 9 uses the idea of ordinal logistic regression to extend factor analysis to ordered categorical variables.

6.15   Path analysis

Path analysis uses regression methods to try to study the patterns of causation in networks. Here we shall only give a brief outline of the main idea and thereby establish a link with structural equation modelling to be discussed in Chapter 11. Causation implies a temporal ordering of variables. If one variable exerts a causal influence on another it is necessary that the value of the first is determined first. This idea is already implicit in some applications of multiple regression where the explanatory or predictor variables are thought of as determined before the response or dependent variable which they are used to predict. Path analysis begins with a network of variables which specifies the assumed paths of causation.

We shall illustrate the idea using a simplified version of a path analysis (Ridge 1974, Chapter 2) of social mobility using data collected in 1949. We discuss PCA and FA of these data in Sections 5.9 and 7.12. Occupations were categorised according to their prestige or social and financial desirability. Following Ridge, we shall treat Occupational Category (OC) as if it were a continuous variable, although, like much social science data, it is ordinal. Lack of social mobility was measured by the extent to which a man's occupational category (OC) depended on his father's OC.

Suppose it is thought that a man's OC, x_1, might depend both on the level of education he attained, x_2, and on his father's OC, x_3. The subscripts are in reverse temporal order because x_3 occurs before x_2 which occurs before x_1. Part of the association with the father's OC might be *direct* through the amount of financial help given, influence, encouragement to apply for posts and such like. Part might be *indirect*, being mediated through education. The *path diagram* (Figure 6.7) indicates the possible causal relationships by arrows in the direction of causation.

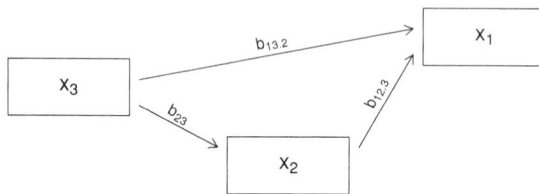

FIGURE 6.7: Path diagram for three variables

Path analysis helps to disentangle the direct and indirect influences. Suppose we fit the regression of x_1 on x_2 and x_3 obtaining the equation

$$\hat{x}_1 = a_1 + b_{12.3}x_2 + b_{13.2}x_3. \tag{6.24}$$

The coefficient $b_{13.2}$ measures the direct effect of father's OC, x_3, on the son's OC, x_1, for a fixed education level, x_2. Suppose next we regress x_2 on x_3 and obtain the equation

$$\hat{x}_2 = a_2 + b_{23}x_3. \tag{6.25}$$

The coefficient b_{23} measures the effect of father's OC on education. Finally, if we substitute $\hat{x}_2$ from eq. (6.25) for x_2 in eq. (6.24) we obtain

$$\hat{x}_1 = a_1 + b_{12.3}(a_2 + b_{23}x_3) + b_{13.2}x_3. \tag{6.26}$$

In this version x_3 appears twice. The last term on the right gives the direct effect but in the middle term x_3 has the coefficient $b_{12.3}b_{23}$. This represents the indirect effect of father's OC via education. The bs are called *path coefficients* and indirect effects are found by multiplying together the coefficients of the paths along which the influence is transmitted. Although we have only illustrated this for two consecutive paths the result applies quite generally. It can be shown algebraically that eq. (6.26) is identical to the equation obtained by regressing x_1 simply on x_3

$$\hat{x}_1 = a'_1 + b_{13}x_3. \tag{6.27}$$

Comparing eq. (6.26) and (6.27), the intercept a'_1 equals $(a_1 + b_{12.3}a_2)$ and the regression coefficient b_{13} equals $b_{12.3}b_{23} + b_{13.2}$. The direct and indirect effects add up to the total effect. If it were found that the direct effect was negligible, this would suggest that there was no direct causal relationship between the father's and the son's OCs after taking into account the indirect effect via the level of education.

Ridge (1974) used a more complicated path analysis and concluded that the son's educational level was strongly associated with the father's OC and, in its turn, influenced the son's OC, but that there was also a (smaller) direct effect from the father's OC to the son's OC, over and above that mediated by educational level.

An alternative use of the same path diagram, Figure 6.7, could be for a child's writing ability (x_1), the child's footsize (x_2), and the child's age (x_3). The simple regression of x_1 on x_2 would give a positive regression coefficient, $b_{12} > 0$, but (if our understanding of the world is correct) when x_3 is added the partial regression coefficient, $b_{12.3}$, of writing ability on footsize *given* age would be close to zero (not significantly different from zero). Thus path analysis can be used to demonstrate that an observed correlation is spurious.

Note that any *causal* interpretation of the results of a path analysis depends on the prior assumption that the variables are causally related as shown by the path diagram. But, even when causal explanations are not appropriate path analysis can help to clarify the difference between the marginal regression coefficient when the response variable is regressed on a single regressor and the partial regression coefficient when one or more extra regressors are included.

Example: GCSE score, path analysis

Returning to the example of GCSE English scores in Section 6.8, since CATsVS is taken at age 12 it cannot be causally dependent on either writing speed or GCSE score at age 16. On the other hand a high CATsVS score may be indicative of general ability, which could contribute to both writing speed and GCSE score. The path diagram (Figure 6.8) therefore allows for both direct and indirect effects of CATsVS score on GCSE score. The path coefficients written above the lines are the simple regression coefficients of writing speed on CATsVS score, and the partial regression coefficients of GCSE score on writing speed and CATsVS score. Because the analysis has been carried out without standardising the variables, the path coefficients show the estimated change in the response for an increase of one grade in CATsVS or for an increase of one word per tenth of a minute in Speed.

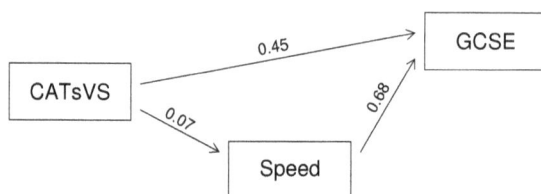

FIGURE 6.8: Path diagram for GCSE English score

The path coefficient from CATsVS to GCSE (0.45) is the estimated direct effect of CATsVS on GCSE score. By multiplying together the path coefficients from CATsVS to Speed (0.07) and from Speed to GCSE (0.68) we obtain 0.04, which is the estimated indirect effect of CATsVS on GCSE English score via writing speed. The direct effect is clearly more important. The strong link between the Cognitive Ability Test (Verbal Stanine) at age 12 and the GCSE English grade at age 16 is, at most, only slightly mediated by writing speed. The reader is reminded that this study was made of girls at one school in one year and may not apply more widely.

6.16   Missing data

Most statistical methods, including the regression models discussed in this chapter, require data to be complete. If there are very few missing values, or if they are located in one particular part of the data matrix, it may be possible to proceed without risk of serious bias. For example, if most of the missing values are in one column of the data matrix (i.e. they relate to only one variable), that column may be deleted. Something will be lost by excluding what may be a key variable, but it should still be possible to discover what the other variables are telling us. Similarly, if the missing values are concentrated in relatively few rows of the data matrix (meaning that some respondents have given few responses), those rows can be omitted. This is sometimes known as listwise deletion or complete case analysis. This, too, may introduce small biases, but the broad picture should not be seriously affected.

If there are many missing entries spread out over the data matrix, the list-wise deletion method becomes statistically inefficient and may lead to severe bias. To deal with such data using statistical methods, we need to understand the three mechanisms of data missingness – "missing completely at random", "missing at random", and "missing not at random" (Little and Rubin 2019).

If the probability that the data are missing is independent of the observed and unobserved data, then we say the data are missing completely at random. In that case, there is no systematic difference between observation units with missing data and those with complete data. For example, consider a study on blood pressure levels in which the researchers used a blood pressure monitor to collect data from a sample of participants. Due to a sudden battery failure in the monitor, the blood pressure readings for some participants were lost. These observations are missing completely at random. The listwise deletion method is unbiased under this missingness mechanism but may be statistically inefficient due to the reduction in sample size. This missingness mechanism is often unrealistic for real-world data, especially for those in social science research.

If the probability that the data are missing is fully explained by the observed data or, in other words, if it is conditionally independent of the unobserved data given the observed data then the data are missing at random. Missing completely at random is a special case of this mechanism. Consider, for example, a survey examining people's attitude towards a certain social topic. If the probability of completing the survey is related only to the participants' sex (which is fully observed), then the data are missing at random but not completely at random. If the distribution of the attitude variable also depends on sex, then listwise deletion will lead to a biased estimate of the average attitude of the population due to the difference in missing rates between male and female participants. Missing at random may still be a relatively strong assumption for many real-world applications. Nevertheless, modern missing data methods generally rely on this assumption. A popular method for dealing with missing data is to impose a joint model for all the variables. Most models introduced in the subsequent chapters, such as factor models, graphical models, and multilevel models, are joint models for multivariate data. Under these models, and if the data are missing at random, then statistical inference and predictions can be made based on the distribution of the observed data entries derived from the joint model. Another popular method for handling missing data is multiple *imputation*. This method creates multiple complete datasets by imputing the missing data entries multiple times in some way judged optimal. We can then perform statistical inference and prediction tasks by applying a standard method to each of the complete datasets and then aggregating their results. The performance of an analysis based on multiple imputations relies on the quality of the imputation model/algorithm, the method used to analyse each complete dataset, and the way the results from multiple complete datasets are aggregated.

Finally, missing not at random occurs when the probability that the data are missing depends on the unobserved data; in other words, it is not conditionally independent of the unobserved data given the observed ones. Continuing the previous example where the response of interest is people's attitude towards a certain social topic, if the probability of completing the survey is not only related to the participants' sex but also their attitude towards the topic, then the data are missing not at random. Missing not at random is the most complex but realistic missing data mechanism that covers a wide range of data generation processes. If domain knowledge is available on the relationship between the data and its missingness (e.g. those with weaker attitude respond less), then we can run a sensitivity analysis to find out how robust the results obtained by a standard missing data method (which typically assumes data to be missing at random) are under various scenarios suggested by the domain knowledge. This analysis is typically performed by simulating data under a statistical model that reflects the domain knowledge. This book will not go into depth on methods to deal with non-ignorable missingness, except for providing some discussion and references. Further discussions about missing data can be found in Chapters 7 and 14.

6.17   Additional examples and further work

GCSE scores for boys

As well as the 110 girls whose GCSE scores have been analysed in this chapter, data on 88 boys at the same school in Oxfordshire are given on the website. Fitting the regression of GCSE scores on Speed (words per tenth of a minute), CATsVS score and Joins (1 indicates a problem with joined up writing and 0 indicates no problem) for boys gives

$$\widehat{\text{GCSE}} = 3.5 + 0.3 \times \text{Speed} + 0.4 \times \text{CATsVS} - 0.5 \times \text{Joins}.$$

Compare this with eq. (6.15). What further information would you need to judge whether the differences in coefficients could have arisen from chance rather than from differences between the genders? Try fitting a single model to all 198 pupils with Gender, Speed, CATsVs and Joins as the explanatory variables. Do you need to include any interaction terms?

Hedonism in different countries

We use the model given in Section 6.11 to analyse scores on hedonism or attitude to pleasure seeking. The data come from the 2002 European Social Surveys (Jowell 2003) and more details and fuller analyses for all 20 countries are given in Chapter 13. Here we use only the data for Austria, Belgium and the Czech Republic. Before carrying out an analysis it is wise to examine the data. In this case the histograms in Figure 6.9 are sufficiently informative for no further analysis to be necessary, except that we shall later suggest the addition of extra explanatory variables.

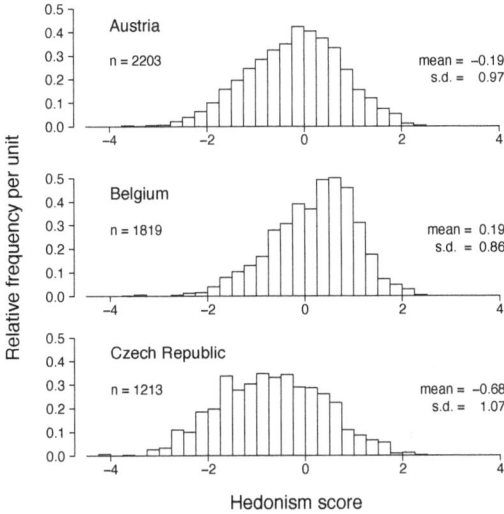

FIGURE 6.9: Histograms of hedonism score for three countries

The histograms show that there are shifts between the three countries (Czech residents tend to have lower hedonism scores), but a wide spread within countries. The distributions are clearly not normal, but the departure from normality is not so great as to invalidate an analysis assuming normality.

We fit the model

$$y_i = \alpha + \beta_1 D_1 + \beta_2 D_2 + e_i, \tag{6.28}$$

where $e_i \sim N(0, \sigma^2)$ for $i = 1, 2, \ldots, n$. Here y_i is the hedonism score; $D_1 = 1$ for a person in Austria and 0 otherwise; $D_2 = 1$ for a person in Belgium and 0 otherwise; so that for a person in the Czech Republic (the reference or baseline country) both D_1 and D_2 will be 0.

The least squares estimates of α, β_1 and β_2 are $a = \bar{y}_C = -0.68$, the sample mean for the Czech Republic, $b_1 = (\bar{y}_A - \bar{y}_C) = 0.49$, the difference in average score between an Austrian and a Czech resident, and $b_2 = (\bar{y}_B - \bar{y}_C) = 0.87$, the difference in average between a Belgium and a Czech resident.

We suggest that you extend the analysis by adding extra explanatory variables such as the age and/or the household income of each respondent. You can add a new variable to the model eq. (6.28) in different ways. For example, if you regress hedonism on x (household income), D_1 and D_2 (the dummy country variables) you will obtain the *analysis of covariance* model which allows comparison of the country means *adjusted* for income. However, if you subtract the country mean income, say, from each observation, giving $x^* = x - \bar{x}_A$, for Austria, $x^* = x - \bar{x}_B$, for Belgium and $x^* = x - \bar{x}_C$ for the Czech Republic, and regress hedonism score on x^*, D_1 and D_2, you will obtain the within-group regression model which assumes that the slope of the regression of hedonism on income is the same for each country but that the intercept might vary. This describes how hedonism changes with income *within* a country but keeps the same *between*-country effects as in eq. (6.28).

There are many missing values in this dataset. As discussed in Section 6.16, computer packages typically apply listwise deletion for variables included in the model, which may introduce bias.

Offender Group Reconviction Scale

Copas and Marshall (1998) used logistic regression to provide the probation service in England and Wales with the Offender Group Reconviction Scale (OGRS) as a guide in assessing the risk of reoffending. The reader is recommended to refer to this article, which is a case study in the use of logistic multiple regression explicitly for prediction. The article gives details of the datasets used, of how the authors chose, scored and transformed the predictor variables, and discusses the use and interpretation of the prediction formula they produced.

Their final formula for OGRS, representing the approximate probability of reconviction, is

$$
\begin{aligned}
\text{OGRS} \;=\; & 31 - (\text{age in years}) + (\text{ 3 if the offender is male}) \\
& - (\text{number of youth custody sentences}) + (\text{severity of offence score}) \\
& + \sqrt{\text{rate of offending}}.
\end{aligned} \tag{6.29}
$$

This was obtained from the logistic regression of a binary variable taking the value 1 if an offender was reconvicted and 0 otherwise, giving

$$
\text{logit}(\hat{\pi}) = \hat{L},
$$

where π is the probability of reconviction and $\hat{L}$ is a linear function of the predictor variables in the above formula. $\hat{L}$ was then rescaled and the coefficients rounded to integer values to give OGRS as in eq. (6.29).

This example has several interesting features.

i) It is designed for use in assessing the risk of reoffending (although it actually measures the risk of reconviction).

ii) The formula has been scaled and the regression coefficients have been rounded to integers to make it easier to use. Copas and Marshall explain that OGRS is approximately equal to the estimated percentage probability of reconviction.

iii) The coefficients are partial regression coefficients and should only be interpreted jointly, not individually, so the counter-intuitive negative coefficient does not imply that an increase in the number of youth custody offences reduces the chance of reconviction. A strong positive marginal association has been reversed by conditioning on the other predictor variables. An offender who has served a youth custodial sentence is also likely to be male and to have had a number of court appearances. This example reinforces our comments in Section 6.9 on the difficulty of interpreting regression coefficients.

iv) In accordance with good practice, the variables were selected and the form of the model chosen from an earlier (1987) cohort of offenders and then fitted to a later (1990) cohort. This was done to avoid combining the model selection effects with the model fitting effects both of which can result in a formula working well for the data used to develop it but not for new data. But after examining the 1990 cohort, they decided that it would improve the formula to use the *square root* of the rate of offending rather than the actual rate. This illustrates that the application of statistical methods is typically a compromise between textbook rules and judgement.

The real test would have been how well the formula predicted later cohorts of offenders; unfortunately this assessment was not done.

Algebra

Those who enjoy algebra can check that in the identity

$$\sum_{i=1}^{n}(y_i - \bar{y})^2 = \sum_{i=1}^{n}(y_i - \hat{y}_i)^2 + 2\sum_{i=1}^{n}(y_i - \bar{y})(\hat{y}_i - \bar{y}) + \sum_{i=1}^{n}(\hat{y}_i - \bar{y})^2,$$

the cross-product term sums to zero, leading to eq. (6.3). You might also check algebraically that in simple linear regression with a binary regressor variable the least squares line goes through the two group means (see Section 6.5 and Figure 6.3).

Software

All the methods described in this chapter can be implemented using standard statistical computer packages. We have mainly used Stata, SPSS and R (Stata and SPSS are more accessible for the non-specialist, but we prefer R for data visualisation).

6.18   Further reading

There are many introductory textbooks that both cover the basic statistics required for this chapter and have a chapter on regression. We recommend:

> Agresti, A. and Finlay, B. (2008). *Statistical Methods for the Social Sciences* (4th ed.). Englewood Cliffs, NJ: Prentice Hall.

Regression is also included in many books on multivariate analysis, for example:

> Everitt, B.S. and Dunn, G. (2001). *Applied Multivariate Data Analysis* (2nd ed.). London: Arnold.

More discussions about regression models from a prediction perspective can be found in Chapters 2, 3, 5, and 6 of the book:

> James, G., Witten, D., Hastie, T., and Tibshirani, R. (2021). *An Introduction to Statistical Learning with Applications in* R (2nd ed.). New York: Springer.

Finally we draw the reader's attention to the on-line multilevel modelling course developed by the Centre for Multilevel Modelling at the University of Bristol (www.bristol.ac.uk/cmm/) which includes a module on regression with examples to work through.

7

Factor Analysis

7.1 Introduction to latent variable models

Factor analysis belongs to a family of methods which involve what are called *latent variables*. Often, particularly in social science research, we cannot directly measure the variables of major interest. Examples of such concepts are intelligence, political ideology (left-wing, moderate, or right-wing), and socio-economic status. Although we use these concepts in social science discourse as if they were just like any other variable, they differ from other variables in that they cannot be observed — which is why they are called latent. In some cases, a concept may be represented by a single latent variable, but often, they are multidimensional in nature and so involve more than one latent variable. Suppose that there are q latent variables, denoted by $f_1, f_2, \ldots, f_q$. These latent variables are commonly called *factors* which is why we denote them by f. Latent variable methods, of which factor analysis is the oldest and most widely used, form the subject of this and the next four chapters.

There is a close link between factor analysis and principal components analysis. It is common to regard PCA as a method of factor analysis. Some books (e.g. Basilevsky 1994 and the SPSS computer package) treat both methods within the same framework. We shall explain the justification for this at the end of the chapter, but we prefer to keep them quite distinct at this stage for two reasons. We have introduced PCA as a descriptive method concerned with summarising a data matrix in a manner which expresses its structure in a small number of dimensions. Factor analysis, on the other hand, is a model-based technique. That is, it involves assumptions about the joint distributions over some relevant population of the variables involved. This enables us to make inferences about the population using the notions of goodness-of-fit, statistical significance and the precision of estimation. We link the observable to the unobservable variables by a probability model, as we shall see later.

The second reason for emphasising the difference between PCA and factor analysis is that we want to emphasise the strong link between factor analysis and the other latent variable methods treated in the following four chapters. Traditionally, latent trait (which is factor analysis for categorical data) and latent class analysis have been treated quite separately from factor analysis. Their essential unity has been obscured by the use of different notation and the practices of the rather different scientific cultures in which they have been used. They differ in the level of measurement used for the variables involved but share a common basis of interpretation, which we shall seek to emphasise.

To measure the latent variable of interest, we often collect observable variables which we feel are likely to be indicators of the latent variable(s). Suppose that we collect p observed variables, denoted by $y_1, y_2, \ldots, y_p$. The ys are also called *indicators, items* or *manifest variables*.

Examples of problems involving latent and manifest variables

i) There is great interest in measuring intelligence. This is considered an important characteristic of individuals who possess it to a greater or lesser extent. However, it is not like weight or age, for which there are some ready-made measuring instruments. Intelligence is a construct; it is a concept we find useful and meaningful, but it does not exist in the same sense that more tangible properties, such as weight, do. We can, however, introduce it into a mathematical model and treat it just like any other variable. Intelligence is a good example of a latent variable. In this case, the indicator variables are quantities presumed to be influenced by the latent variable. In this case, these are usually the scores obtained in a series of test items chosen because it is believed that more intelligent people will perform better. Some items might be verbal or arithmetical. Others might involve spatial exercises designed to test the ability to see patterns. If the items all require the same sort of basic mental ability, we would expect the scores on the items to be positively correlated. The problem is to see whether a single latent variable can account for that correlation and, if so, how we can determine where to place individuals on the latent scale.

ii) The measurement of political attitude is very similar to the case of intelligence. We describe individuals as left- or right-wing, and some, for example, as more right-wing than others. Implicit in this kind of language is the idea that there is a scale on which individuals can be located, extending from extreme left at one end to extreme right at the other. This is a latent scale, and if we are to construct such a scale, we require appropriate indicators. These could be provided by a social survey in which respondents are asked about their attitudes on a range of political issues, such as private healthcare, private education, and trade unions.

iii) In order to measure a latent variable such as the socio-economic status of a household, we might similarly collect information about household income, as well as the occupations and education levels of household members.

In each of these examples, we have used our intuitive understanding of the latent variable in which we are interested to identify some manifest variables which we expect will reveal something about that underlying latent variable. In effect, we started with a latent variable and looked for manifest variables, which would serve as indicators because we already had an idea of what the key latent variables would be. Sometimes, we proceed in the opposite direction. If, for example, we have a large general-purpose survey, we might wonder whether the large number of manifest dimensions represented by, perhaps, 50 questions could be reduced to a small number of dimensions without significant loss of information. This second approach is essentially the one we follow when using PCA. In practice, the true situation usually lies somewhere between these two extremes. The enquiry may have been motivated by the desire to investigate the existence of some latent variables, but we wish to carry out the investigation in an open enough fashion for unexpected features to be picked up.

Latent variable models

Latent variable models are closely related to the standard regression model. It may, therefore, be helpful to describe the central idea of factor analysis in terms of regression analysis. A regression model, as explained in Chapter 6, expresses the relationship between a dependent (response or criterion) variable and one or more explanatory (predictor or regressor) variables. In factor analysis, the regression relationship is between a manifest variable and the latent variables. In both cases, we add distributional assumptions about the residual or error terms, which enable us to make inferences. The essence of the problem that factor analysis or other latent variable analysis has to solve is that of inverting the regression relationships to tell us about the latent variables when the manifest variables are given. Since we can never observe the latent variables, we can only ever learn about this relationship indirectly.

The assumption is that several manifest or observed variables depend on the same latent variable or variables, and this dependence induces a correlation between them. Indeed, the existence of a correlation between two indicators may be taken as evidence of a common source of influence. As long as any correlation remains, we may, therefore, suspect the existence of a further common source of influence. The aim of a latent variable analysis is to determine whether the dependencies among the observed variables may be explained by a small number of latent variables. As we observed above, latent variable models may be used either in an exploratory way (as in this chapter) to identify the latent variables underlying a set of items, or in a confirmatory way (as in Chapter 11), to test whether a set of items designed to measure particular concepts is indeed consistent with the assumed structure.

There are various types of latent variable models. These models are distinguished by the level of measurement of the observable variables and the assumptions made about the level of measurement of the latent variables. Table 7.1 shows a classification of latent variable models.

TABLE 7.1: Classification of latent variable models

	Observed variables (y)	
	Continuous	Categorical
Latent variables (f)	(interval/ratio)	(nominal/ordinal)
Continuous (interval/ratio)	Factor analysis	Latent trait analysis
Categorical (nominal/ordinal)	Latent profile analysis	Latent class analysis

This table does not exhaust the possibilities because, for example, the manifest variables may be a mixture of continuous and categorical variables. However, this classification is sufficient for the purposes of this book.

We begin, in this chapter, with a discussion of *factor analysis*, which is an appropriate technique when all observed variables are measured on a continuous (interval or ratio) scale. The factor model assumes that the latent variables are also continuous.

7.2 The linear single-factor model

The simplest factor model involves only one factor. Charles Spearman, who invented factor analysis (Spearman 1904), introduced this model in the study of human intelligence. For rather special reasons connected with that particular application, he referred to it as a two-factor model, but that usage has long been abandoned.

We introduce the model using a practical example, which thus forms a bridge between what is familiar and what is novel. This will serve to show that a factor analysis model is simply a set of regression models in which some of the variables (the latent variables) are unobserved. The central ideas should be made clearer by repeating the argument of the last section in relation to a special case.

Factor analysis aims to explain the correlations among the set of manifest variables. Such correlations are often spurious because no direct causal link exists between the variables concerned. They sometimes arise because the variables in question have a common dependence on one or more other variables.

For example, the fact that the size of a child's feet is positively correlated with his or her writing ability does not mean that large feet help the child to write better. The correlation is, rather, an incidental consequence of the fact that both are correlated with age — the older the child, the larger the feet and the better the handwriting. When one finds correlations among variables like this, it is important to investigate whether a common dependence on some other variables can explain them.

In some circumstances, there may be an obvious candidate for the role of an "other variable". Suppose, for example, that we look at weekly family expenditure for a large sample of families on a variety of things: food, travel, entertainment, clothes, and so on. Suppose we also find that the correlations (between pairs of purchases) are positive. It would not be credible to claim that high spending on clothes, say, *causes* high spending on travel. It seems more plausible to suppose that high spending on these things results from a high income. To investigate this hypothesis, we might obtain further data on the incomes of each family. This would enable us to see whether the size of each item of expenditure was related to total income and, if so, whether that relationship wholly explained the correlations between expenditures.

How might we investigate this empirically? One way would be to specify how each expenditure might be related to income. To get a preliminary idea of how to do this, we might plot expenditure on food against income. Suppose it turned out to be roughly linear and that a similar result was obtained for each item. We are then in the familiar territory of regression analysis, described in Chapter 6, and we could fit simple regressions of the form:

$$C_i = \alpha_i + \beta_i I + e_i \qquad (i = 1, 2, \ldots), \tag{7.1}$$

where C_i is consumption or expenditure on the ith item, I is the income of the family, α_i and β_i the intercept and slope, respectively, of the regression, and e_i a random component or residual, specific to C_i with mean zero, independent of I, which explains the residual variation about the line. If we found that this model was a good fit for all items of expenditure and that the residual e_is were uncorrelated with each other, then we would have shown that income was the only detectable determinant of expenditure. For fixed income, expenditure on item i would behave like a random quantity with mean $\alpha_i + \beta_i I$ and standard deviation given by the standard deviation of e_i and because the residuals are independent, all correlation between the observed variables would have been removed. If all this proved to be the case (and there are many ifs), we would be satisfied that the mutual correlations among the original expenditures were explained by their common dependence on income. Furthermore, the regression coefficients, β_i, would tell us how strongly each item of expenditure depended on income.

In most practical problems, there is no ready-made variable, like the income of this example, to invoke as an explanation. (Even if there were, it might be impractical to collect it because, for example, the question was held to be too intrusive.) In the absence of any such observable variable, we have to ask whether there could be *any* such latent variable (or variables) which could play the same role.

Whether or not the latent variable is a real variable, which we are unable to observe, or a construct, we are faced with the same fundamental question: is there any way of estimating the regression models of eq. (7.1) without knowing the values of I? This is the technical problem that factor analysis seeks to solve. We shall see below that, rather surprisingly, the set of correlations does contain enough information to enable us to estimate the regression relationships and hence to infer that there could be some common factor.

Suppose that our p manifest variables, $y_1, y_2, \ldots, y_p$ are believed to depend on a single factor or latent variable, f. The simplest way to express the regression of each y on f is by means of the linear model,

$$y_i = \alpha_i + \beta_i f + e_i \qquad (i = 1, 2, \ldots, p), \qquad (7.2)$$

where f may be called the *common factor* since it is common to all y_is. The e_is were sometimes called *specific* or *unique factors*, since they are unique to a particular y_i. (It was because Spearman thought of the e_i as factors that he called his model a two-factor model. Modern terminology counts only the number of common factors.) In the one-factor model, we make the standard regression assumption that e_i is independent of f and has a normal distribution with a mean of zero and a standard deviation of σ_i. We also assume that $e_1, e_2, \ldots, e_p$ are independent so that $y_1, y_2, \ldots, y_p$ are conditionally independent, given f. Then, we can make some deductions about the distribution of the ys, particularly their covariances and correlations. We can choose the scale and origin of f as we please, because this does not affect the form of the regression equation, so we choose to make f have zero mean and unit standard deviation. It turns out that, under this model, the theoretical covariance coefficients have a very simple form, namely

$$\mathrm{Cov}(y_i, y_k) = \beta_i \beta_k \qquad (i, k = 1, \ldots, p; \ i \neq k).$$

The important thing about this formula is that the covariance is a product of two numbers, one depending only on i and the other only on k. From these equations, it is possible to deduce something about the regression coefficients in the model. For example,

$$\mathrm{Cov}(y_1, y_2)\mathrm{Cov}(y_2, y_3)/\mathrm{Cov}(y_1, y_3) = \beta_2^2,$$

which serves to determine β_2 from the covariances. However, we can construct other expressions like this, which should also give β_2: for example, if we replace the subscripts 1 and 3 by any other pair in the range 1 to p, the right-hand side will be the same. If the model is correct and if we knew the true covariances, $\mathrm{Cov}(y_i, y_k)$, then the different equations would all give exactly the same value of the regression coefficient, β_2.

Since, in analysing real data, we would only have estimated or fitted covariances (denoted by $\mathrm{cov}(y_i, y_k)$ with a lower case "c"), we would not get identical estimated values $\hat{\beta}_i$ of β_i even if the model were correct. However, if all the "estimates" of β_i were similar, that would suggest the model is a good fit. In the early days, factor models were fitted by a method very similar to this, tedious to apply and not easily extended to the case of several factors, but exploiting the basic result behind fitting all factor models, which is that we can determine the parameters of the model from the covariances between the manifest variables without knowing the values of the factors themselves.

The one-factor model can be easily extended to allow for an arbitrary number of factors. We replace the simple linear regression equation with a multiple regression equation. In doing so, we shall introduce a more flexible notation and terminology, which will also be useful for the models of the following three chapters.

7.3 The general linear factor model

The general linear factor model for p observed variables and q factors or latent variables take the form:

$$y_i = \alpha_{i0} + \alpha_{i1}f_1 + \alpha_{i2}f_2 + \cdots + \alpha_{iq}f_q + e_i \qquad (i = 1, \ldots, p), \qquad (7.3)$$

where $f_1, f_2, \ldots, f_q$ are the common factors or latent variables, e_i are residuals, and α_{i1}, α_{i2}, and α_{iq} are called the *factor loadings*. The constant term α_{i0} plays no role in fitting or interpreting the model; it can be dispensed with if we assume the ys are measured about their means. The other αs play a key role in interpreting the factors. For this purpose, it is useful to know that the factor loadings turn out to be the covariances between the latent variables and the ys (or correlations if the ys are standardised). As in the simple model, we scale and locate the fs so that they have mean zero and standard deviation one.

Here, and throughout this chapter, linear equations have been written out in full. They can be expressed more concisely in matrix notation, and readers may wish to have some practice in doing this by making the translation.

The linear factor model is based on the idea of multiple linear regression. Still, it is more complicated in that instead of having just one response or criterion variable, it has p, which, conditionally, are mutually uncorrelated given the explanatory variables, and the explanatory variables are unobserved.

We list the assumptions of the model as follows:

i) $f_1, f_2, \ldots, f_q$ are uncorrelated with each other (though we relax this assumption later — see Section 7.6)

ii) $f_1, f_2, \ldots, f_q$ each has a mean of zero and variance of one

iii) $e_1, e_2, \ldots, e_p$ are uncorrelated with each other

iv) each e_i has a mean of zero, but they may have different variances, $\text{Var}(e_i) = \sigma_i^2$, $(i = 1, \ldots, p)$

v) the fs are uncorrelated with the es.

Sometimes, and for some purposes, we make the following additional assumptions:

vi) $f_1, f_2, \ldots, f_q$ follow a multivariate normal distribution

vii) $e_1, e_2, \ldots, e_p$ follow a multivariate normal distribution.

Assumptions vi) and vii) imply that $y_1, y_2, \ldots, y_p$ also have a multivariate normal distribution. These assumptions lead to the *normal linear factor model*. Assumptions (iii) and (v) imply that the correlations among the ys are wholly explained by the factors.

Properties of the linear factor model

An alternative notation for writing the general linear factor model, given by eq. (7.3) and assumptions (i) through (v) above is:

$$
\begin{aligned}
\text{E}(y_i \mid \mathbf{f}) &= \alpha_{i0} + \alpha_{i1} f_1 + \cdots + \alpha_{iq} f_q && (i = 1, 2, \ldots, p), \\
\text{SD}(y_i \mid \mathbf{f}) &= \sigma_i && (i = 1, 2, \ldots, p), \\
\text{Cov}(y_i, y_k \mid \mathbf{f}) &= 0 && (i, k = 1, 2, \ldots, p;\ i \neq k),
\end{aligned}
$$

where $\text{E}(y_i \mid \mathbf{f})$ is read as the conditional expectation (or mean value) of y_i for fixed $\mathbf{f}$ (where $\mathbf{f}$ is the column vector with elements $f_1, f_2, \ldots, f_q$). Similarly, $\text{SD}(y_i \mid \mathbf{f})$ is the conditional standard deviation of y_i given $\mathbf{f}$, which is, of course, just the standard deviation of e_i. The final statement states that the conditional covariance is zero. This implies that the conditional correlation is zero.

Consider the earlier example of children's writing: if y_1 is foot size, y_2 is writing ability and f is the single variable, age, then y_1 and y_2 are positively correlated, but *conditional* on f they are uncorrelated:

$$
\begin{aligned}
\text{Corr}(y_1, y_2) &> 0, \\
\text{Corr}(y_1, y_2 \mid f) &= 0.
\end{aligned}
$$

Differences in age fully account for the apparent correlation between foot size and writing ability.

If the conditional distributions of $y_1, \cdots, y_p$ given $\mathbf{f}$ are normal, then zero conditional correlation implies conditional independence. The normal linear factor model is a conditional or local independence model.

Returning to the general linear factor model, we deduce that the unconditional or marginal mean of y_i is:

$$E(y_i) = \alpha_{i0} \qquad (i = 1, 2, \ldots, p), \qquad (7.4)$$

that the unconditional or marginal variance is:

$$\text{Var}(y_i) = \alpha_{i1}^2 + \alpha_{i2}^2 + \cdots + \alpha_{iq}^2 + \sigma_i^2 \qquad (i = 1, 2, \ldots, p), \qquad (7.5)$$

and that the unconditional covariance between two observed variables, y_i and y_k, takes the form:

$$\text{Cov}(y_i, y_k) = \alpha_{i1}\alpha_{k1} + \alpha_{i2}\alpha_{k2} + \cdots + \alpha_{iq}\alpha_{kq}, \qquad (7.6)$$

where $(i, k = 1, 2, \ldots, p; \ i \neq k)$.

Therefore, the variance is composed of two parts: $\alpha_{i1}^2 + \alpha_{i2}^2 + \cdots + \alpha_{iq}^2$, the part of the variance of y_i explained by the common factors (also called the *communality*), and σ_i^2, the residual or specific variance. The covariances between the ys depend solely on the regression coefficients linking them with the common factors. If the common factors are held constant, there will be no remaining source of covariance among the ys.

From the above expressions for the variances and covariances of the ys, we obtain the form of the covariance matrix assumed under the factor model. For example, under the one-factor model the covariance matrix of the ys is:

$$\begin{pmatrix} \alpha_{11}^2 + \sigma_1^2 & \alpha_{11}\alpha_{21} & \cdots & \alpha_{11}\alpha_{p1} \\ \alpha_{21}\alpha_{11} & \alpha_{21}^2 + \sigma_2^2 & \cdots & \alpha_{21}\alpha_{p1} \\ \vdots & & & \vdots \\ \alpha_{p1}\alpha_{11} & \alpha_{p1}\alpha_{21} & \cdots & \alpha_{p1}^2 + \sigma_p^2 \end{pmatrix}.$$

In the general case, the elements in this matrix are replaced by the expressions given in eq. (7.5) and eq. (7.6) above.

Whereas in PCA the choice of the scale of the variables changes the components — variables with large variances tending to dominate the first few components — in FA, because it is model-based, the factors are the same whatever the scale of measurement used for the observable variables. However, it is common practice to scale or standardise the variables to zero mean and unit variance. As we have already chosen to standardise the latent variables to unit variance, this will give factor loadings on the scale of correlations. The factor loading between an observable variable and a factor will be their correlation. Hence, the communality between an observable variable and the set of factors will also be the squared multiple correlation coefficient, R^2. Thus, using standardised variables (and therefore analysing the correlation matrix rather than the covariance matrix) makes the interpretation of the results easier.

Fitting the model

The usual starting point for a factor analysis is the correlation matrix of the ys. The correlations should first be examined. If the correlations between the ys are low, then factor analysis is unlikely to be useful since the ys are unlikely to share common factors. Inspection may also reveal interesting patterns or undesirable anomalies of various kinds. For example, if two highly correlated ys have been included inadvertently, meaning that one adds very little information to the other, the correlation close to one will be immediately apparent. The problem is that factors common to the other ys would not explain this particularly high correlation, and we would not wish to fit an extra factor just to explain one correlation.

Fitting the factor model itself involves finding the values of the parameters which make the observed correlation matrix as close as possible to that predicted by the model. In the case of the one-factor model, we saw (Section 7.2) that this could be done by a rather *ad hoc* procedure. What we need is a numerical routine which can be programmed to fit any model. Inspection of any text or computer software package will reveal a bewildering array of methods with names like ordinary least squares, generalised least squares and maximum likelihood. All of these methods start by constructing a measure of the distance between the observed and predicted correlation matrices; they differ in the measure they choose. Ordinary least squares, as its name suggests, simply sums the squares of the differences between the corresponding elements of the two matrices. Maximum likelihood uses a distance which arises naturally when we make the normality assumptions (vi) and (vii) above, but it can still be used when this is not the case. In practice, one usually finds that all methods give rather similar results, and it is instructive to try several methods since all are very fast on personal computers. There is some theoretical advantage in using either maximum likelihood or weighted least squares.

Fitting the model does not, of course, guarantee that the fit will be acceptable. We shall describe methods of judging the suitability of a model below.

7.4   Interpretation

The factor loadings

The factor loadings have a similar interpretation to the component loadings in PCA. If the correlation matrix is analysed, and if the factors are constrained to be uncorrelated (assumption (i), Section 7.3), the factor loading $\hat{\alpha}_{ij}$ is the correlation between the observed variable y_i and the latent variable f_j. A factor may be interpreted or labelled, by examining the pattern of the loadings on that factor across the observed variables. To illustrate the interpretation of factor loadings, we re-analyse two datasets that were previously analysed using PCA.

The linear factor model with two factors was fitted to the subject marks data described in Section 5.5. The estimated loadings obtained by the maximum likelihood method are shown in Table 7.2. Since the correlation matrix of marks was analysed, the loadings may be interpreted as correlations between the mark in a subject and a factor. For example, the correlation between Gaelic and the first factor is estimated as 0.56. In attempting to interpret the factor, we have to ask ourselves what it is that is correlated positively and fairly strongly with each of the subject marks. The position is very similar to the one we faced when interpreting the component loadings obtained from PCA for this set of data. Therefore, we may interpret the first two factors similarly to the first two components. The first factor measures overall ability in the six subjects, while the second contrasts humanities and mathematics subjects.

TABLE 7.2: Estimated factor loadings from a two-factor model of the subject marks data

Subject	$\hat{\alpha}_{i1}$	$\hat{\alpha}_{i2}$
Gaelic	0.56	0.43
English	0.57	0.29
History	0.39	0.45
Arithmetic	0.74	−0.28
Algebra	0.72	−0.21
Geometry	0.60	−0.13

A two-factor model was also fitted to the children's personality trait data described in Section 5.5. The loadings are shown in Table 7.3. Again, the interpretation of the factors is essentially the same as that of the principal components. The first factor represents some overall personality measure, while the second contrasts indicators, such as sociability, of how a child relates to other people with those who are internal to the individual, like guilt.

Communalities

The communality of a standardised observable variable is the squared multiple correlation coefficient or the proportion of the variance that is explained by the common factors. The estimated communalities from the factor analysis of the subject marks data are shown in Table 7.4. These show, for example, that 49% of the variance in Gaelic scores is explained by the two common factors. Recall also from Section 7.3 that the communality of a variable is calculated as the sum of the squared loadings for that variable. For example, the communality for Gaelic scores is calculated as $0.56^2 + 0.43^2 = 0.49$. The larger the communality, the better the variable serves as an indicator of the associated factors. Or, put another way, a variable, y_i, with a large communality is a "purer" indicator of the common factors, $\mathbf{f}$, with less contamination from the specific component, e_i. The sum of the communalities is the variance explained by the factor model. From Table 7.4, this is 2.81 or 47% of 6, which is the total variance for the subject marks data.

TABLE 7.3: Estimated factor loadings from a two-factor model of the children's personality trait data

Variable (personality trait)	$\hat{\alpha}_{i1}$	$\hat{\alpha}_{i2}$
Mannerliness	0.65	0.57
Approval seeking	0.54	0.54
Initiative	0.61	−0.45
Guilt	0.63	−0.54
Sociability	0.56	0.54
Creativity	0.72	−0.59
Adult role	0.67	−0.45
Cooperativeness	0.64	0.60

TABLE 7.4: Communalities from fitting a linear two-factor model to the subject marks data

	Communalities
Gaelic	0.49
English	0.41
History	0.36
Arithmetic	0.62
Algebra	0.56
Geometry	0.37

7.5 Adequacy of the model and choice of the number of factors

A primary goal of factor analysis (FA) is to reduce the dimensionality of the multivariate dataset while retaining sufficient dimensions to provide a good approximate representation of the original data. There are several ways in which the adequacy of a factor model may be assessed.

i) *Percentage of variance explained by the factors*

Although factor analysis aims to explain the covariances or, equivalently, the correlations between the observed variables rather than their variances, the proportion of variance explained by the common factors should be reasonably high. The two common factors fitted to the subject marks data together explain approximately 47% of the total variance, which is roughly the same as the first principal component. Also, the communalities can be used to check that the factors adequately explain the individual observable variables. Table 7.4 shows that the Arithmetic marks are better explained than the History marks.

ii) *Reproduced correlation matrix*

A good way of assessing the fit of a model is to compare the fitted (reproduced) correlation matrix of the ys with the correlation matrix computed from the sample data. Table 7.5 shows the reproduced correlation matrix obtained from fitting a two-factor model to the subject marks data. The diagonal entries of the upper section of the table are the communalities (also given in Table 7.4). The off-diagonal entries of this section of the table are the reproduced correlations. For example, the correlation between Gaelic and English marks is estimated from the model as

$$\text{corr}(y_2, y_1) = \hat{a}_{21}\hat{a}_{11} + \hat{a}_{22}\hat{a}_{12} = (0.57 \times 0.56) + (0.29 \times 0.43) = 0.44.$$

TABLE 7.5: Reproduced correlations and communalities (top section) for a linear two-factor model fitted to the subject marks data, and discrepancies between observed and reproduced correlations (bottom section), subject marks data

Correlation	Gaelic	English	History	Arithmetic	Algebra	Geometry
Gaelic	0.49	0.44	0.41	0.29	0.31	0.28
English		0.41	0.35	0.34	0.35	0.30
History			0.36	0.16	0.19	0.17
Arithmetic				0.62	0.59	0.48
Algebra					0.56	0.46
Geometry						0.37
Discrepancy						
Gaelic		0.00	0.00	0.00	0.02	−0.03
English			0.00	0.01	−0.03	0.03
History				0.00	0.00	0.01
Arithmetic					0.00	−0.01
Algebra						0.00
Geometry						

iii) *Goodness-of-fit test*

The reproduced correlations should be compared with the sample correlation matrix given in Table 5.2. The lower section of the table shows the discrepancies or differences between the observed sample correlations and the reproduced correlations. Here, the differences are small, suggesting that the two-factor model is a good fit.

If the normal factor analysis model is assumed, we can carry out a likelihood ratio test or goodness-of-fit test to test the null hypothesis that the covariance matrix of the ys has the form specified by the factor model. Failure to reject this null hypothesis would suggest that the covariance matrix is compatible with the factor model (an adequate fit). The test statistic, denoted by W, has a chi-squared distribution under the null hypothesis with $\{(p-q)^2 - (p+q)\}/2$ degrees of freedom.

The test statistic for the two-factor model fitted to the subject marks data was 2.18 on 4 degrees of freedom, suggesting that the model is a very good fit.

If a model with a given number of factors is deemed to be a poor fit, more factors may be added until a good fit is achieved. However, as always, we have to bear in mind the balance between interpretability and goodness-of-fit. A well-fitting model with a large number of factors may not be interpretable, while a poorer-fitting model may still reveal some interesting features of the data. For large sample sizes, the test becomes sensitive to small departures from the model, which may not be of practical relevance.

Also, a statistically significant result may be due to departures from multivariate normality rather than from the need for an extra factor.

iv) *Standard errors of factor loadings*

Traditionally, standard errors of factor loadings have not been routinely quoted, and some packages still do not provide them. However, in interpreting the factors and deciding how many factors are needed, it would be useful to examine the standard errors. For example, if the absolute values of the estimated loadings for a factor were all less than twice as large as their standard errors, then the imprecision of that factor would render it useless. As with the goodness-of-fit test statistic, it is necessary to know the sample size to calculate standard errors. There is a theoretical aspect that needs to be taken into account when standard errors are computed. Typically, software for structural equation models provides standard errors for the factor loadings of the one-factor model but not for the model with more than one factor unless a confirmatory factor analysis model is fitted. The reason is that no unique solution exists when the number of factors is greater than one (see Section 7.6). A unique solution can be obtained by fixing some of the factor loadings to a pre-specified value. The number of factor loadings to be fixed depends on the number of factors fitted. For example, in the two-factor model, one loading needs to be fixed to obtain a unique solution.

Choosing the number of factors

The number of factors, q, must be small enough for $\{(p-q)^2 - (p+q)\}/2$, the degrees of freedom, to be greater than or equal to zero. So when $p = 3$ or $p = 4$, q cannot be greater than one, but when $p = 20$, q could be as large as 14. When choosing the number of factors to fit, a useful first step is to conduct a principal components analysis, as the number of components required is often a good guide to the number of factors. The number of principal components required is judged according to the criteria described in Section 5.4. A factor model with the same number of factors can then be fitted. The rationale for this procedure is given in Section 7.10, where we examine the relationship between PCA and factor analysis in more detail. To assess the adequacy of a model with a given number of factors, we use the percentage of variance explained, the communalities, the discrepancies between the observed and reproduced correlations, the goodness-of-fit test and the standard errors of the estimated factor loadings, as described above.

There are also other formal methods of choosing the optimum number of factors based on what are called information or model selection criteria (Akaike, Bayesian etc.). A discussion of those criteria can be found in Sclove (1987). In particular, under suitable conditions, the Bayesian information criterion theoretically guarantees the correct selection of the number of factors with high probability when the sample size is large; see Claeskens (2016) for further discussion.

7.6 Rotation

When we posed the problem of fitting a factor model, we tacitly assumed that there was just one set of parameter values which would minimise the chosen measure of distance between the observed and predicted correlation matrices. This is true for the one-factor model, but with two or more factors, infinitely many sets of values give the same minimum distance. At first sight, this seems to be a serious drawback, but it allows us to introduce other criteria for choosing solutions. However, this must not be seen as granting a license to pick and choose among the solutions until one finds one which suits one's preconceptions. A criticism often levelled against factor analysts is that subjectivity seems to play too big a role. This is a misreading of the situation in two respects. First, one cannot obtain *any* solution one wants. Secondly, the situation is more accurately described as one in which the *same* solution can be expressed in different ways. Certain features, such as the communalities, remain the same in all versions of the solution.

Rotation is the name given to the process by which we move from one solution to another. The name comes from the geometrical representation of the procedure.

Once a factor model is fitted, the factors may be transformed to give a new set of factors, say $f_1^*, f_2^*, \ldots, f_q^*$. In the process, the estimated factor loadings are also transformed to give $\hat{\alpha}_{i1}^*, \hat{\alpha}_{i2}^*, \ldots, \hat{\alpha}_{iq}^*$. Our model specifies (assumption (i), Section 7.3) that $f_1, \ldots, f_q$ are mutually uncorrelated. Following an *orthogonal* transformation, the transformed $f_1^*, \ldots, f_q^*$ will also be mutually uncorrelated. Geometrically, the axes will have been rotated while being kept at right angles.

However, sometimes an *oblique* rotation might yield transformed factors that are easier to interpret. In such a case, assumption (i) in Section 7.3 would need to be relaxed to allow the transformed factors to be correlated. Note that the factor variances are still set at 1.

Factor rotation is used to clarify the underlying structure of the factors. Rotation is usually performed by optimising an objective function; see Browne (2001). The usual motivation is to find a pattern of loadings which is relatively easy to interpret. One of the most useful patterns is described as *simple structure*. The loadings are said to have simple structure if each variable has a large contribution from only one factor, with close to zero contributions from the other factors. An illustration of simple structure for a three-factor model fitted to eight observed variables is shown in Table 7.6. The observed variables have been partitioned into three groups, each associated with one of the latent variables. For example, the third factor has large positive loadings on y_4, y_5 and y_6. This factor may be interpreted by considering what these three variables have in common, exactly as if we had analysed these variables alone. In effect, we are reducing the interpretation problem to the one we faced with a single factor.

There is no guarantee, of course, that it will be possible to find a solution with something close to a simple structure. But if we can find one, it will make the interpretation much easier. The rotation routines provided by the various software packages are designed to search for that solution among the solution set, which is as close as possible to a simple structure.

TABLE 7.6: Illustration of factor loadings with simple structure from a three-factor model (+ indicates a large positive loading and - indicates a small, close to zero loading)

	α_{i1}	α_{i2}	α_{i3}
y_1	+	-	-
y_2	-	+	-
y_3	+	-	-
y_4	-	-	+
y_5	-	-	+
y_6	-	-	+
y_7	+	-	-
y_8	-	+	-

Illustration of orthogonal rotation in a two-factor model

The unrotated factor loadings from fitting a two-factor model to a fictional set of seven observed variables are given in the first two columns of Table 7.7. These loadings are plotted in Figure 7.1. An orthogonal rotation is applied to the axes in Figure 7.1 to achieve a new set of loadings with a simple structure. In geometrical terms, this means that we are looking to rotate the axes so that the points lie close to one or the other of the axes. The dashed lines in Figure 7.1 represent rotated axes having this property. In this case, an almost perfect simple structure has been achieved through rotation. The coordinates of the points with respect to these rotated axes are the new factor loadings given in the last two columns of Table 7.7. The rotated factor f_1^* contributes to variables y_5, y_6 and y_7 but makes virtually no contribution to the first four variables. The rotated factor f_2^* contributes to the first four variables, but makes almost no contribution to y_5, y_6, and y_7.

TABLE 7.7: Unrotated and rotated factor loadings from a two-factor model

	Unrotated		Rotated	
	$\hat{\alpha}_{i1}$	$\hat{\alpha}_{i2}$	$\hat{\alpha}_{i1}^*$	$\hat{\alpha}_{i2}^*$
y_1	0.2	0.3	0.0	0.4
y_2	0.4	0.5	0.0	0.6
y_3	0.6	0.7	0.0	0.9
y_4	0.7	0.7	0.0	1.0
y_5	0.5	−0.5	0.7	0.0
y_6	0.7	−0.6	0.9	0.0
y_7	0.3	−0.2	0.4	0.0

It is important to note that rotation does not alter the fit of the model. Rotation does not change the reproduced correlation matrix or the goodness-of-fit test statistic. The communalities also remain unchanged. This is because rotation has not changed the relative positions of the loadings. In the plot of the loadings, loadings that appear close together before rotation also appear close together after rotation. However, since rotation alters the loadings, the interpretation of the new factors will be different. Also, although the overall percentage of variance explained by the common factors remains the same after rotation, the percentage of variance explained by each factor will change. Rotation redistributes the explained variance across the factors.

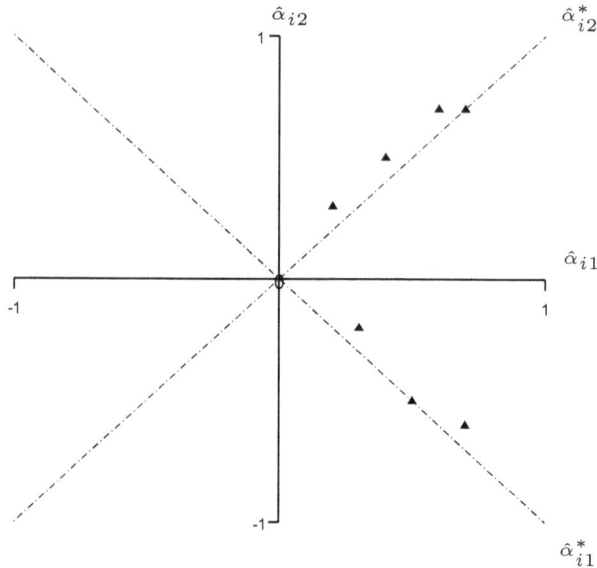

FIGURE 7.1: Plot of rotated factor loadings from a two-factor model

Procedures for orthogonal rotation

In the example above, it is possible to examine a plot of the unrotated loadings to find a suitable rotation that will lead to a simple structure. However, it is not always clear from the loadings plot which rotation should be carried out, particularly if there are more than two factors in the model. Several procedures have been developed to search for a suitable rotation automatically. For example, the VARIMAX procedure attempts to find an orthogonal rotation that is close to a simple structure by finding factors with few large loadings and as many near-zero loadings as possible.

Non-orthogonal (oblique) rotations

Sometimes, simple structure can be achieved only using a non-orthogonal (oblique) rotation. This type of rotation requires us to relax the original assumption of the linear factor model that the latent variables are uncorrelated. An oblique rotation leads to correlated factors. Although this complicates the interpretation of the factors, it is often reasonable to expect the latent variables to be correlated. For example, one might expect a child's mathematical ability to correlate positively with their verbal ability. In that case, a factor analysis that assumes the latent variables to be uncorrelated may not uncover latent variables representing mathematical ability and verbal ability. Figure 7.2 shows how an oblique rotation for the subject marks data can produce new oblique axes, one going through the cluster of History, Gaelic, and English, and the other through the cluster Geometry, Algebra, and Arithmetic. The correlation between these transformed factors is 0.512.

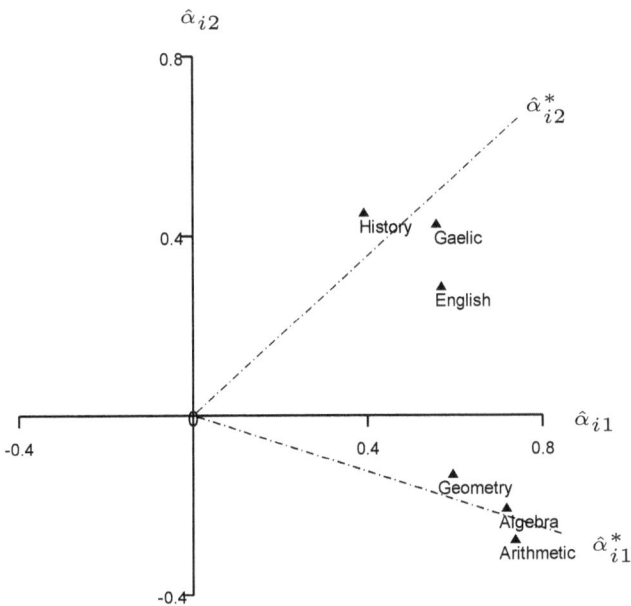

FIGURE 7.2: Plot of unrotated and rotated factor loadings for the subject marks data

7.7   Factor scores

Sometimes, we want to calculate an individual's score on the latent variable(s), perhaps for use in subsequent analysis. Such scores can be determined exactly in PCA by the simple expedient of substituting the values of an individual's manifest variables into the expressions for the principal components (see Section 5.6). It is not so straightforward in factor analysis because the factors are random variables with a probability distribution. There are various methods for obtaining predicted factor scores. All of them lead to expressions of the form:

$$\hat{f}_1 = c_{11}y_1 + c_{12}y_2 + \cdots + c_{1p}y_p,$$
$$\hat{f}_2 = c_{21}y_1 + c_{22}y_2 + \cdots + c_{2p}y_p,$$
$$\vdots$$
$$\hat{f}_q = c_{q1}y_1 + c_{q2}y_2 + \cdots + c_{qp}y_p.$$

The cs are called estimated factor score coefficients (the different methods giving rise to different cs). We shall use the Thomson or regression estimates which are estimates of the conditional expectations or mean values of the factor given the values of the manifest variables. Estimated factor score coefficients for computing scores for the two factors for the subject marks data are given in Table 7.8. To obtain the scores for an individual boy, we would need to know his marks, $y_1, \ldots, y_6$.

TABLE 7.8: Coefficients for calculating factor scores (regression method) for the subject marks data

Subject	c_{1i}	c_{2i}
Gaelic	0.20	0.39
English	0.17	0.23
History	0.11	0.33
Arithmetic	0.35	−0.35
Algebra	0.29	−0.22
Geometry	0.17	−0.10

Unlike principal component score coefficients, factor score coefficients are not simple multiples of the loadings, so their interpretation differs from that of the factor loadings. The factor score coefficients allow the scores to be calculated from the original (standardised) variables. Examination of the coefficients is most relevant when we are concerned with individuals and wish to understand what their scores mean. The common practice of examining the factor loadings is more relevant when our interest is in the population structure and the interrelationships between the variables. Sometimes, one factor score coefficient is dominant, so the scores are virtually proportional to that variable. In such a case, anyone using the factor scores should realise that they are calculated almost entirely from a single variable.

7.8 A worked example: the test anxiety inventory

The following example illustrates a typical factor analysis, including the use of oblique rotation. This example is based on data from the test anxiety inventory, which assesses overall anxiety associated with taking tests. The inventory has been used in many countries with similar results. The data analysed here are from a study by Gierl and Rogers (1996) who applied a test with 20 items to 335 male and 389 female grade 12 students in British Columbia. A factor analysis of these data is also discussed in Bartholomew et al. (2011). The analysis of the male sample is presented here.

Students were asked to report how frequently they experienced various symptoms of anxiety in taking tests. A brief description of the items is given below.

1. Lack of confidence during tests
2. Uneasy, upset feeling
3. Thinking about grades
4. Freeze up
5. Thinking about getting through school
6. The harder I work, the more confused I get
7. Thoughts interfere with concentration
8. Jittery when taking tests
9. Even when prepared, get nervous
10. Uneasy before getting the test back
11. Tense during test
12. Exams bother me
13. Tense/ stomach upset
14. Defeat myself during tests
15. Panicky during tests
16. Worry before important tests
17. Think about failing
18. Heart beating fast during tests
19. Can't stop worrying
20. Nervous during test, forget facts

The question of how, precisely, students were scored on these items would be relevant for judging the suitability of making the normality assumption, but not for our more limited object here.

The correlation matrix is given in Table 7.9, showing that pairs of responses positively correlate to a modest degree. This suggests that there is at least one common factor underlying the scores, which is what one would expect when the items have been specifically constructed to reflect test anxiety.

TABLE 7.9: Pairwise correlations between test anxiety inventory items

	y_1	y_2	y_3	y_4	y_5	y_6	y_7	y_8	y_9	y_{10}
y_1	1.00	.51	.24	.42	.31	.29	.39	.43	.43	.26
y_2		1.00	.30	.41	.23	.34	.39	.48	.44	.28
y_3			1.00	.37	.40	.27	.54	.38	.34	.26
y_4	.			1.00	.35	.34	.42	.41	.42	.30
y_5					1.00	.34	.49	.30	.22	.19
y_6						1.00	.39	.26	.27	.25
y_7							1.00	.45	.41	.32
y_8								1.00	.54	.33
y_9									1.00	.36
y_{10}										1.00

	y_{11}	y_{12}	y_{13}	y_{14}	y_{15}	y_{16}	y_{17}	y_{18}	y_{19}	y_{20}
y_1	.50	.39	.40	.41	.52	.47	.40	.36	.39	.38
y_2	.47	.41	.43	.37	.49	.48	.30	.46	.34	.33
y_3	.33	.28	.29	.35	.40	.38	.43	.35	.38	.29
y_4	.39	.42	.35	.49	.60	.44	.46	.41	.36	.53
y_5	.34	.26	.31	.43	.34	.27	.55	.30	.26	.30
y_6	.35	.33	.28	.42	.36	.31	.37	.27	.23	.40
y_7	.45	.43	.38	.48	.49	.41	.51	.41	.44	.45
y_8	.59	.48	.43	.34	.56	.53	.36	.48	.45	.44
y_9	.53	.42	.37	.36	.52	.55	.35	.48	.43	.41
y_{10}	.39	.33	.29	.27	.39	.43	.32	.35	.42	.30
y_{11}	1.00	.53	.47	.45	.55	.60	.38	.54	.45	.45
y_{12}		1.00	.32	.41	.52	.53	.41	.43	.38	.51
y_{13}			1.00	.41	.44	.40	.31	.50	.46	.35
y_{14}				1.00	.52	.44	.40	.39	.38	.50
y_{15}					1.00	.65	.52	.52	.55	.56
y_{16}						1.00	.45	.51	.54	.49
y_{17}							1.00	.35	.46	.40
y_{18}								1.00	.48	.43
y_{19}									1.00	.46
y_{20}										1.00

However, it is not clear whether one factor is sufficient to account for the observed correlations, and this is where we might hope that factor analysis will reveal something further. To get some idea of the number of factors likely to be necessary, we first carry out a PCA. The result, given in Table 7.10, shows that there is one dominant eigenvalue and a second that is greater than one. These two together account for about half of the variance. Beyond the second eigenvalue, the rate of decrease is very slow. This suggests that at least two factors might be needed.

TABLE 7.10: Variance explained by each principal component, test anxiety inventory items

Component	Variance	Variance explained %	Cumulative %
1	8.78	43.90	43.90
2	1.35	6.75	50.65
3	0.97	4.86	55.51
4	0.89	4.44	59.95
5	0.77	3.87	63.82
6	0.74	3.71	67.53
7	0.71	3.53	71.06
8	0.66	3.27	74.33
9	0.57	2.85	77.18
10	0.54	2.72	79.90
11	0.54	2.71	82.61
12	0.51	2.53	85.14
13	0.47	2.36	87.50
14	0.44	2.18	89.68
15	0.42	2.10	91.78
16	0.38	1.89	93.67
17	0.35	1.74	95.41
18	0.33	1.66	97.07
19	0.31	1.57	98.64
20	0.28	1.38	100.00

The unrotated factor loadings from a two-factor model are given in Table 7.11: y_1 corresponds to the first item, y_2 to the second item, and so on. A plot of these loadings is given in Figure 7.3. The first factor is positively correlated with each item and may be interpreted as a measure of overall anxiety during exams. However, the interpretation of the second factor is not immediately clear. It is obvious from Figure 7.3 that no orthogonal rotation will result in a simple structure.

In an attempt to clarify the interpretation of the factors, an oblique rotation was carried out, and the new oblique axes (transformed factors) have been added to Figure 7.3. The rotated factor loadings (the *pattern matrix*) are shown in Table 7.11. While the loadings do not show a simple structure, they do have an interpretable pattern. Factor 1, f_1^*, has high positive loadings on variables $y_1, y_2, y_4, y_8, y_9, y_{10}, y_{11}, y_{12}, y_{13}, y_{15}, y_{16}, y_{18}, y_{19}$ and y_{20}. These items are largely measures of physiological reactions of the nervous system. Factor 2, f_2^*, has high positive loadings on items $y_3, y_5, y_6, y_7, y_{14}$ and y_{17}, which could be measures of psychological anxiety (several involve thinking). Gierl and Rogers (1996) identified $y_2, y_8, y_9, y_{10}, y_{15}, y_{16}$ and y_{18} as indicators of what they called "emotionality" and $y_3, y_4, y_5, y_6, y_7, y_{14}, y_{17}$ and y_{20} as indicators of "worry". Although the items plotted in Figure 7.3 do not lie along one or other axis, they do fall roughly into two groups.

TABLE 7.11: Pattern matrices giving the loadings of the unrotated factors and of the OBLIMIN rotated factors from a two-factor model for the test anxiety inventory items, males. (Rotated loadings larger than 0.4 are printed in bold.)

	Unrotated		Rotated	
	$\hat{\alpha}_{i1}$	$\hat{\alpha}_{i2}$	$\hat{\alpha}_{i1}^{*}$	$\hat{\alpha}_{i2}^{*}$
y_1	0.62	−0.07	**0.56**	0.09
y_2	0.62	−0.16	**0.67**	−0.04
y_3	0.54	0.25	0.14	**0.49**
y_4	0.65	0.09	**0.41**	0.31
y_5	0.51	0.50	−0.16	**0.82**
y_6	0.49	0.20	0.16	**0.41**
y_7	0.67	0.29	0.21	**0.58**
y_8	0.69	−0.18	**0.75**	−0.04
y_9	0.66	−0.21	**0.76**	−0.10
y_{10}	0.50	−0.09	**0.49**	0.02
y_{11}	0.73	−0.18	**0.78**	−0.03
y_{12}	0.65	−0.10	**0.62**	0.05
y_{13}	0.59	−0.06	**0.53**	0.09
y_{14}	0.64	0.18	0.30	**0.42**
y_{15}	0.80	−0.07	**0.71**	0.13
y_{16}	0.75	−0.21	**0.83**	−0.07
y_{17}	0.64	0.34	0.12	**0.64**
y_{18}	0.67	−0.14	**0.68**	0.01
y_{19}	0.65	−0.06	**0.59**	0.10
y_{20}	0.66	0.02	**0.50**	0.21

After an oblique rotation, the factors will be correlated. The correlation between the two factors from the analysis of the test anxiety inventory items is estimated as 0.68, indicating that the two factors are quite closely related. Because the factors are themselves correlated, the rotated factor loadings are now given in the pattern matrix, which are no longer the correlations between the observed variables and the factors. These correlations are given separately in the *structure matrix*. The structure matrix from the oblique rotation of the two-factor model of the test anxiety inventory data is given in Table 7.12. All the variables are positively correlated with both factors, which means that each item is, to some extent, a measure of each factor.

7.9   How rotation helps interpretation

Now that we have seen the process of rotation in action, it is worth looking again at the issues raised by having an infinite number of solutions available when there are two or more factors. All estimation routines are designed to yield what we might call a "standard" solution, which is similar to the PCA solution. Often, this solution lends itself to direct interpretation, as in the case

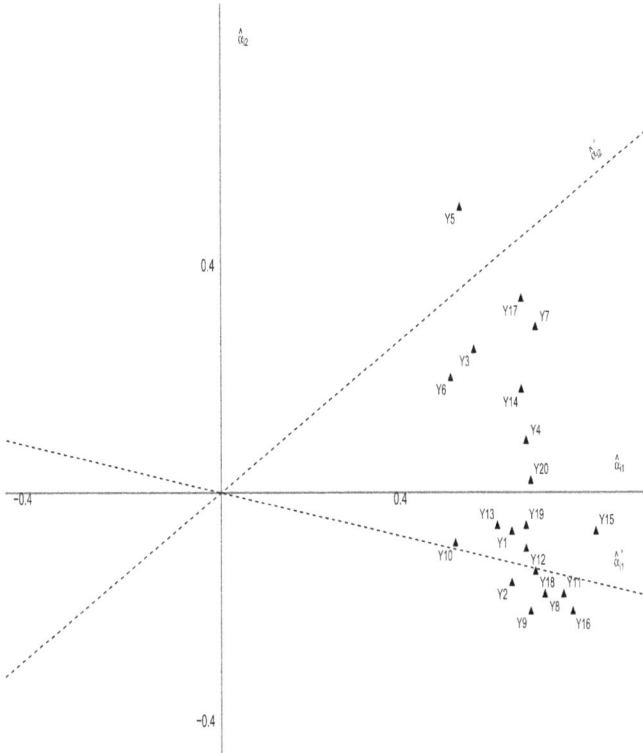

FIGURE 7.3: Plot of unrotated loadings, with rotated oblique axes shown for a two-factor model of the test anxiety inventory items, males

of the example of the subject marks considered earlier, and other examples in Section 7.12 below. In other cases, rotation may lead to a more straightforward interpretation. We argued earlier that rotation gives us alternative ways of describing the same solution rather than providing different solutions. In this example, the fundamental thing about the solution is that it takes two dimensions to describe the latent variation between individuals. Or, put negatively, one dimension does not appear to be sufficient to describe the variation among individuals in their responses to these test items as, one supposes, the original designers of the test had intended.

The standard solution, like PCA, produced a dominant factor with fairly large and positive loadings on all variables. This indicated that there is something common to all test items that we may equate with the "test anxiety" that the test was designed to measure. But it also showed that, even among those at the same point on this scale, there was some further variation which we could not readily interpret. Rotation enabled us to look at the variation from another angle. It proved possible to describe the solution in terms which are roughly the same as what earlier researchers, using a slightly different

TABLE 7.12: Structure matrix giving correlations between test anxiety inventory items and the two rotated factors after an OBLIMIN rotation, males. (Correlations larger than 0.7 are printed in bold.)

	Factor 1	Factor 2
y_1	0.62	0.47
y_2	0.64	0.41
y_3	0.48	0.59
y_4	0.62	0.59
y_5	0.40	**0.71**
y_6	0.44	0.52
y_7	0.60	**0.72**
y_8	**0.72**	0.47
y_9	0.69	0.42
y_{10}	0.50	0.35
y_{11}	**0.75**	0.50
y_{12}	0.66	0.50
y_{13}	0.59	0.45
y_{14}	0.59	0.63
y_{15}	**0.80**	0.61
y_{16}	**0.78**	0.50
y_{17}	0.56	**0.72**
y_{18}	0.68	0.47
y_{19}	0.65	0.50
y_{20}	0.64	0.55

analysis, have called "emotionality" and "worry". Although these two dimensions are highly correlated, they appear to be distinct aspects of what we usually call "anxiety". The analysis shows us that anxiety could be compounded of something which has a physiological origin in the nervous system and something which seems to be more directly psychological. In retrospect, this may not seem particularly surprising, but the distinction does not appear to have been recognised until it was suggested by factor analysis (see Gierl and Rogers (1996) and references therein). To check this interpretation, we would need to collect further data and then, perhaps, carry out a confirmatory factor analysis as described in Chapter 11.

7.10 A comparison of factor analysis and principal components analysis

PCA and factor analysis share some aims. Both methods attempt to reduce the dimensionality of a set of correlated variables, $y_1, \ldots, y_p$, by obtaining a small set of components, $z_1, \ldots, z_q$, or of factors, $f_1, \ldots, f_q$. PCA finds components that account for as much as possible of the total variance, $\sum_i \text{var}(y_i)$, whereas FA tries to match the reconstructed correlations to the observed sample correlations.

Neither method is useful if the ys have low correlations; in PCA, the components will be close to the original variables, and there will be little correlation to explain in FA. Although the methods have common aims, the procedures by which they achieve these aims are rather different.

i) *Descriptive versus model-based.* PCA is a descriptive technique which does not assume an underlying statistical model. The principal components are transformed versions of the ys. It makes no prior assumptions about how many components are being looked for or what they might represent. The most that is hoped for is that a few zs will provide a good summary of the observed ys for the given sample. Factor analysis assumes a statistical model which incorporates a fixed number of factors, and there may be some prior notion of what they represent. This means that factor analysis may be used to make inferences about the population from which the sample was drawn. For example, in confirmatory factor analysis, we might test whether the factor loadings follow a pre-specified pattern.

Confirmatory factor analysis and structural equation modelling, which we discuss in Chapter 11 are powerful tools for social science research, building on the ideas of regression, path analysis and factor analysis. Unfortunately, they are open to misuse, and correctly interpreting the results is not always easy. We hope that by developing a basic understanding of FA, the reader will be better prepared to learn how and when to use them.

ii) *Rotation.* The first principal component is always the same, no matter how many further components are extracted. However, the single factor for a one-factor model will not be the same as the first factor in a two-factor model. Thus, if we choose to rotate principal components, for example, in searching for a simple structure, we lose some information. With FA, all solutions for the q-factor model are equivalent, defining the same q-dimensional subspace in which the correlation structure can best be represented. Rotating a solution for FA neither adds nor loses information, but it may help interpretation.

iii) *Score coefficients.* Often, the component loadings for PCA and FA will be very similar when both analyses are performed on a dataset with strong correlation structure. However, in converting to score coefficients, differences may arise. PCA aims to explain the variation in all the variables, and the score coefficients are the loadings divided by the eigenvalue:

$$\tilde{a}_{ij} = a_{ij}^*/\lambda_j \quad (i = 1, \ldots, p; \; j = 1, \ldots, q).$$

FA, on the other hand, only aims to explain the correlations, so variables with small loadings (correlations) with a factor will contribute even less to the factor scores. The larger loadings are multiplied by a larger amount, and the differences between the factor loadings are accentuated. As already mentioned in Section 7.7, in the extreme case, the scores for a factor may be almost a linear function of a single variable with factor score coefficients for other variables close to zero.

A formal comparison of PCA and factor analysis

Despite these important differences, there are circumstances under which PCA is a good approximation to factor analysis, and it is this fact that provides the basis for using PCA as a guide to choosing the number of factors. As outlined in Section 5.8, in PCA, a set of p uncorrelated components, each with unit variance, can be derived from the original p (standardised) variables:

$$\tilde{z}_j = \tilde{a}_{j1}y_1 + \tilde{a}_{j2}y_2 + \cdots + \tilde{a}_{jp}y_p \qquad (j = 1, \ldots, p).$$

We can invert these equations to obtain a set of p equations with the ys on the left-hand side, and the zs on the right-hand side giving:

$$y_i = a_{i1}^{*}\tilde{z}_1 + a_{i2}^{*}\tilde{z}_2 + \cdots + a_{ip}^{*}\tilde{z}_p \qquad (i = 1, \ldots, p),$$

where the a^{*}s are the loadings (or correlations) with the principal components. Suppose we retain only the first q components, then we obtain:

$$y_i = a_{i1}^{*}\tilde{z}_1 + a_{i2}^{*}\tilde{z}_2 + \cdots + a_{iq}^{*}\tilde{z}_q + u_i \qquad (i = 1, \ldots, p).$$

This equation now has the same form as the factor analysis model. That is, each y is expressed as a linear combination of q uncorrelated variables, each with variance one. But for this to be close to a factor analysis model, the residual terms, u_i, would need to behave like the uncorrelated residuals or specific factors, e_i, in the factor model. In general, the u_i will not be uncorrelated. However, the two analyses will often give similar results, as when, for example, in the factor model, the variances, σ_i^2, of the e_i, are roughly equal or are all small. The reader is recommended to compare the results of PCA and FA carried out on the same datasets.

7.11   Missing data

When data are collected from social or educational surveys, respondents may refuse or be unable to respond to some items, resulting in missing values in the multivariate data. Factor analysis and principal component analysis can still be conducted in the presence of missing values. As discussed in Section 6.16, the treatment of missing values depends on the mechanism generating the missing data. Simple and ad hoc methods for handling missing values are list-wise, pairwise deletion and single imputation. Listwise deletion excludes from the analysis any case with missing data, resulting in a reduced sample size, and produces biased results unless the missing data mechanism is completely at random (MCAR), which is rare in real applications. Pairwise deletion uses all available data for each pair of variables, resulting in variable sample sizes for estimating the correlations among the variables, which can lead to inconsistent and non-positive definite matrices. Single imputation replaces the missing values with the mean or median of the variable, computed from the available data. In contrast, regression imputation predicts the missing values based on a regression model fitted to the available data. Single imputation, although easy to implement, can lead to biased results and does not account for the variability in missing data prediction or estimation through methods such as the mean, median, or regression estimation.

To avoid the shortcomings of single imputation, multiple imputation (MI) has been proposed (see, e.g. Rubin 1987; Schafer 1997; Schafer and Olsen 1998) which works well under a missing-at-random (MAR) data mechanism. MI creates multiple datasets (e.g. m in number) by replacing the missing values with values drawn from some predictive distribution, and each imputed dataset is analysed separately using the same statistical model. The results from each analysis on the m imputed datasets are combined using Rubin's rules (Rubin 1987), which average the estimates across the separate analyses and provide estimates for both within-imputation and between-imputation variance. The variability across the m imputations measures the uncertainty associated with predicting the missing values. MI is available in standard software, and readers are advised to use it as a method for handling missing data in a multivariate setting. However, it is more computationally intensive than the other methods. The predictive distribution, used to draw the imputed values, is a distribution of the missing data given the observed data and specific values of the model parameters. More on the predictive distribution can be found in the references provided.

Finally, maximum likelihood estimation utilises all available data to estimate the model parameters (factor loadings and error variances), providing unbiased results under the MAR assumption. In the presence of missing values, practitioners are advised to use packages that offer maximum likelihood estimation as a method of estimation. In PCA, it is recommended to use pairwise deletion if the amount of missing data is small; otherwise, multiple imputation is used.

In the case of a non-ignorable missing data mechanism, none of the above methods are applicable unless the amount of missingness is low. We briefly discuss the case of non-ignorable nonresponse. The non-ignorable case is the most challenging to handle, as it requires incorporating a model for the missing data mechanism into the model of interest. In the literature, models have been proposed for handling non-ignorable nonresponse in latent variable models.

The approach adds a latent dimension to the substantive latent dimensions, response propensity, which represents a general tendency to respond, varying across individuals. Suppose that we have p observed items to analyse and there is a proportion of missing data in each item. We create p response indicators as follows: when an individual gives a response to an item, the response indicator for that item takes the value 1, and 0 otherwise. Those p additional response indicators provide information about an individual's response propensity, and they are modelled together with the p attitudinal items. Different model specifications offer different insights to the non-ignorability issue. Therefore, we advise the reader to consult the available literature (O'Muircheartaigh and Moustaki 1999; Moustaki and Knott 2000; Glas and Pimentel 2008; Rose, von Davier, and Nagengast 2015; Kuha, Katsikatsou, and Moustaki 2018) for different model specifications that incorporate the model for the missing data mechanism. The choice of the model depends on the focus of the analysis. All models discussed in the above references can be used to test for non-ignorability. Some of the available models, especially those that allow for co-variates, can provide information on the non-respondent's socio-demographic profile as well as test for non-ignorability at the item level. The addition of the response propensity dimension results in a multidimensional latent trait model, which can be fitted using all available software for latent variable models and the R packages discussed in Chapters 8, 9, and 11.

The methods discussed here apply to all models for multivariate data. Therefore, we do not repeat them in the remaining chapters unless there are noted differences or additional material to discuss.

7.12 Additional examples and further work

Psychomotor tests

A sample of 197 airmen was subjected to a range of tests (Fleishman and Hempel 1954). A subset of these tests has been selected for analysis, and brief descriptions of those selected are given in Table 7.13. Further details can be found in Fleishman and Hempel (1954). The first of these tests was a criterion practice task in which the airmen's performance was assessed in repeated trials over two days. The practice period was divided into eight time segments, and the scores in each were obtained to give variables y_1 to y_8. The remaining tests were of two types: written tests, which aimed to assess speed in performing verbal, spatial, and arithmetic tasks (y_9, y_{10}, and y_{11}), and practical tests to assess speed and accuracy in operating apparatus and reaction times to various stimuli (y_{12}, y_{13}, and y_{14}).

TABLE 7.13: Descriptions of psychomotor test items

Type of test	Test	Variable
Criterion task: complex co-ordination	Stage 1	y_1
Criterion task: complex co-ordination	Stage 2	y_2
Criterion task: complex co-ordination	Stage 3	y_3
Criterion task: complex co-ordination	Stage 4	y_4
Criterion task: complex co-ordination	Stage 5	y_5
Criterion task: complex co-ordination	Stage 6	y_6
Criterion task: complex co-ordination	Stage 7	y_7
Criterion task: complex co-ordination	Stage 8	y_8
Printed tests of comprehension and speed	Numerical operations	y_9
Printed tests of comprehension and speed	Dial and table reading	y_{10}
Printed tests of comprehension and speed	Mechanical principles	y_{11}
Apparatus tests	Plane control	y_{12}
Apparatus tests	Reaction time	y_{13}
Apparatus tests	Rate of movement	y_{14}

The pairwise correlations between the 14 test scores are given in Table 7.14. As would be expected, you can see some very high correlations between the criterion test scores over the eight time segments (y_1 to y_8), particularly between scores taken for segments that are close together in time. Among the printed tests, there are quite high correlations between y_9 and y_{10}, and between y_{10} and y_{11}, while the correlations among the apparatus tests are low to moderate.

TABLE 7.14: Pairwise correlations ($\times$ 100) between psychomotor test items

	y_1	y_2	y_3	y_4	y_5	y_6	y_7	y_8	y_9	y_{10}	y_{11}	y_{12}	y_{13}	y_{14}
y_1	100	75	73	66	64	57	63	59	28	51	49	40	8	25
y_2		100	85	85	84	79	77	79	30	46	40	45	22	32
y_3			100	85	83	79	81	79	30	45	39	44	27	31
y_4				100	90	88	86	85	26	40	36	44	30	28
y_5					100	90	87	86	22	37	36	42	30	34
y_6						100	85	86	23	34	29	39	27	37
y_7							100	90	23	36	33	39	33	30
y_8								100	24	34	30	36	27	32
y_9									100	63	32	8	9	12
y_{10}										100	54	22	5	24
y_{11}											100	22	-5	12
y_{12}												100	20	20
y_{13}													100	30
y_{14}														100

Before conducting a factor analysis of these data, you should conduct a principal components analysis. The scree plot from a PCA is shown in Figure 7.4. You can see that the first component is highly dominant. From a PCA, you will find that the first component explains 54% of the total variance, while the second and third explain 12% and 8%, respectively. Only the first three eigenvalues are greater than one. The results from the PCA suggest that a two- or three-factor model should be fitted. Here, we examine only the two-factor solution, but you should also consider the three-factor model.

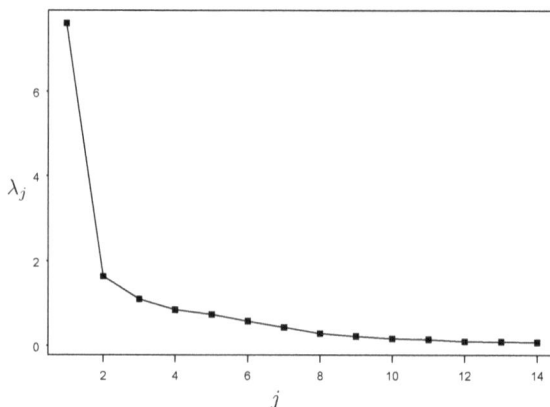

FIGURE 7.4: Scree plot of eigenvalue versus number of components from principal components analysis of the psychomotor test data

Table 7.15 shows the communalities obtained from the two-factor model. These are very high for seven of the criterion task variables (y_2 to y_8) and high for y_1 and y_{10}. In contrast, the variances of the apparatus test scores are not well explained by two common factors. You might like to see whether adding a third factor leads to higher communalities for these variables. The likelihood ratio test statistic for overall goodness-of-fit also indicates that the two-factor model is not a good fit ($W = 184.2$, degrees of freedom=64, $p < 0.001$). To further check the fit, you should examine the reproduced correlation matrix.

TABLE 7.15: Communalities from fitting a two-factor model to the psychomotor test data

y_1	y_2	y_3	y_4	y_5	y_6	y_7	y_8	y_9	y_{10}	y_{11}	y_{12}	y_{13}	y_{14}
0.64	0.82	0.82	0.89	0.90	0.87	0.85	0.84	0.35	0.64	0.40	0.21	0.12	0.12

Although the two-factor model does not appear to be a good fit, it may still reveal key patterns in the data. Figure 7.5 shows a plot of the factor loadings from the two-factor model. From this plot, you can see that the tests are roughly separated into three groups. The printed tests of comprehension and speed (y_9, y_{10}, and y_{11}) all have moderate, positive loadings on the first factor and high, positive loadings on the second factor. The apparatus tests (y_{12}, y_{13}, and y_{14}) form another cluster with positive, moderate loadings on the first factor and close-to-zero loadings on the second factor. Seven of the criterion variables (y_2 to y_7) constitute a third cluster with high loadings on factor 1 and small contributions from factor 2. The first criterion variable, y_1, appears between the cluster of the remaining criterion variables and the cluster of the printed tests. You can also, see that the criterion variables are approximately ordered on the second factor, with y_1 having the highest loading and the scores at the end of the practice period (y_6, y_7, and y_8) the lowest loadings. This might reflect a learning effect on the criterion task; in the first segment, when the task is new, the airmen's performance (y_1) is correlated with their performance on (y_9, y_{10}, y_{11}), but with practice, their performance on the criterion task becomes less correlated with their scores on the printed tests. To investigate further any possible learning effects, we would need to revisit the original data and examine the mean values as well as the correlations.

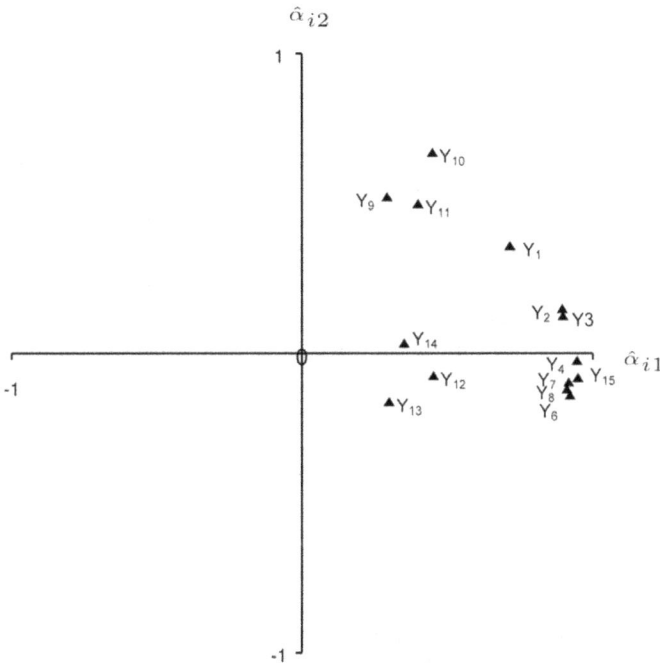

FIGURE 7.5: Plot of loadings from a two-factor model of the psychomotor test data

Social mobility

In Section 5.9, a principal components analysis was carried out of a correlation matrix of variables relating to the occupational and educational status of three generations of family members (Table 5.14). A description of the ten variables used in the analysis was given in Table 5.13. The correlation matrix may also be analysed using factor analysis.

A PCA of these data found that three, possibly four, components were needed, and the first four components were all interpretable. This suggests that we should begin with a three-factor model. The likelihood ratio test indicates that the three-factor solution is not a good fit to the correlation matrix ($W = 143.8$, degrees of freedom=18, $p < 0.001$). Since the fourth component from PCA was interpretable, the next step might be to fit a four-factor model. If you do this, you will find that the four-factor model is a good fit ($W = 16.6$, degrees of freedom= 11, $p=0.12$). However, there is a problem with this model. Most software packages will inform you that the estimates of one or more communalities exceeded a value of 1 during the fitting process. As the communality is a proportion, a value greater than 1 is not permissible, and therefore, a solution where such values have occurred should be treated with caution. This phenomenon is known as a *Heywood case*. If you examine the communalities from the four-factor solution, you will find that the value for y_6 is extremely close to 1 — an indication of a Heywood case. One solution to the problem might be to omit y_6 from the analysis. Still, if you try this, you will find that the problem is merely shifted elsewhere as the communality of another variable is estimated very close to 1. In such cases, it is recommended that a factor be dropped. Therefore, we return to the three-factor solution even though it is a much poorer fit than the four-factor one.

The factor loadings obtained from the three-factor model are given in Table 7.16. If you examine the pattern of loadings for each factor, you will find that the interpretation of the factors is the same as the interpretation of the first three principal components. To aid interpretation, relatively large loadings on the second and third factors are printed in bold. You can see that the first factor is positively correlated with all ten variables; the second factor contrasts occupational status of family members (y_1, y_2, y_5) with qualifications (y_4, y_7, y_9); while the third factor contrasts variables relating to the firstborn son (y_8, y_9, y_{10}) with the educational status of his mother (y_6, y_7).

TABLE 7.16: Loading matrix giving the unrotated loadings from a three-factor model of the social mobility data

		$\hat{\alpha}_{i1}$	$\hat{\alpha}_{i2}$	$\hat{\alpha}_{i3}$
y_1	HF/O	0.426	**0.403**	0.053
y_2	WF/O	0.404	**0.343**	0.008
y_3	H/FE	0.592	−0.026	0.116
y_4	H/Q	0.558	**−0.240**	0.118
y_5	H/O	0.575	**0.481**	0.031
y_6	W/FE	0.451	−0.126	**0.369**
y_7	W/Q	0.477	**−0.296**	**0.462**
y_8	FB/FE	0.615	−0.191	**−0.289**
y_9	FB/Q	0.519	**−0.358**	**−0.381**
y_{10}	FB/O	0.602	0.168	**−0.219**

Although the unrotated factors are interpretable, rotations can be carried out to determine whether a simple structure can be achieved. The factor loadings obtained from an orthogonal (VARIMAX) rotation and an oblique (OBLIMIN) rotation of the three-factor solution are shown in Table 7.17. Again, relatively large loadings are printed in bold. A VARIMAX rotation has not led to a simple structure since several variables have moderate loadings on more than one factor. However, the interpretation of the factors has been made clearer. The first factor has large loadings on the occupation variables and might be labelled "occupational status", while the second and third factors might be labelled "first born son" and "parents' education", respectively.

TABLE 7.17: Loading matrices giving the VARIMAX and OBLIMIN rotated loadings from a three-factor model of the social mobility data

		VARIMAX			OBLIMIN		
		$\hat{\alpha}_{i1}^*$	$\hat{\alpha}_{i2}^*$	$\hat{\alpha}_{i3}^*$	$\hat{\alpha}_{i1}^*$	$\hat{\alpha}_{i2}^*$	$\hat{\alpha}_{i3}^*$
y_1	HF/O	**0.576**	0.042	0.111	−0.064	**0.599**	0.025
y_2	WF/O	**0.516**	0.086	0.090	−0.003	**0.530**	0.002
y_3	H/FE	0.329	0.288	**0.416**	0.183	0.246	**0.353**
y_4	H/Q	0.135	0.360	**0.485**	0.279	0.015	**0.445**
y_5	H/O	**0.728**	0.113	0.144	−0.016	**0.747**	0.025
y_6	W/FE	0.163	0.078	**0.568**	−0.051	0.074	**0.585**
y_7	W/Q	0.042	0.106	**0.718**	−0.032	−0.085	**0.765**
y_8	FB/FE	0.209	**0.645**	0.194	**0.637**	0.101	0.058
y_9	FB/Q	0.018	**0.723**	0.140	**0.762**	−0.109	0.014
y_{10}	FB/O	**0.491**	**0.434**	0.098	**0.381**	**0.452**	−0.052

A pattern of loadings that is close to a simple structure is achieved by allowing the factors to correlate via an OBLIMIN rotation. An oblique rotation seems reasonable in this case since, for example, you would expect a son's occupational and educational status to be positively correlated with that of his parents. If you look at the correlations between the OBLIMIN rotated factors, you will find that they are all positive, ranging from 0.35 to 0.40. However, an oblique rotation does not change the overall interpretation of the solution.

7.13 Software

Factor analysis is included in all main statistical packages, as well as the R programming language. The reader is warned that some packages do not distinguish between principal component analysis and factor analysis.

7.14   Further reading

Bartholomew, D. J., Knott, M. and Moustaki, I. (2011). *Latent Variable Models and Factor Analysis: A Unified Approach* (3rd ed.). John Wiley & Sons.

Basilevsky, A. (1994). *Statistical Factor Analysis and Related Methods.* New York: Wiley.

Mulaik, S. A, (2010). *Foundations of Factor Analysis* (2nd ed.). Boca Raton, FL: Chapman and Hall/CRC.

Skrondal, A. and Rabe-Hesketh, S. (2004). *Generalized Latent Variable Modelling: Multilevel, Longitudinal, and Structural Equation Models.* Boca Raton, FL: Chapman and Hall/CRC.

8

Factor Analysis for Binary Data

8.1 Latent trait models

In this chapter, we move to the top right-hand cell of Table 7.1 and discuss methods based on models where the manifest variables are categorical. We start with the case where they are all binary — that is, where they are based on responses of the kind yes/no or right/wrong. Some methods appropriate when there are more than two categories will be given in Chapters 9 and 11. The word "trait" in the name of these models is often used because it arises from one of the principal applications for which they were devised, namely the measurement of psychological traits. In this book, they are used in a much broader context, and so it seemed appropriate to make this clear in the title of the chapter. Nevertheless, we have also retained the original terminology to keep the link with a very important field of application.

Conceptually, there is no difference between the problems treated here and those in the previous chapter on factor analysis. We start with a probability model linking the observed variables to a set of latent variables. We then discuss how to fit the models, judge their goodness-of-fit, interpret their parameters, and so forth. The difference lies in the special problems posed by having to deal with a data matrix consisting of binary items. The basic objectives are the same, namely:

   i) to explore the interrelationships between the observed responses,

  ii) to determine whether the interrelationships can be explained by a small number of latent variables,

 iii) to assign a score to each individual for each latent variable on the basis of the responses.

The binary data matrix

We have already met a data matrix for categorical data in the discussions of cluster analysis, multidimensional scaling and, in passing, correspondence

DOI: 10.1201/9781003483342-8

analysis. If the responses are binary, the ys record whether the response was positive or negative. A convenient convention, also used in earlier chapters, is to use 1 to indicate a "success" or a positive response, that is "correct" or "yes" as the case may be, and 0 for the "failure" or negative response. This convention has the advantage in the present context that if we sum the responses in any row of the data matrix, we get the total number of positive responses. This is a useful summary measure in its own right, and we shall use it in the subsequent analyses. The response coded 1 is sometimes referred to as the *keyed response*. A typical row of the data matrix might then be as follows:

$$00101110011.$$

The methods about to be described all start from a data matrix consisting of a set of rows like that above, one for each individual or object. However, the restriction to binary data sometimes makes it possible to express the matrix in a more compact and informative way.

Any row of the data matrix is referred to as a *score pattern* or a *response pattern*. If there are p variables, there are 2^p possible response patterns. When $p = 3$, for example, they are

$$000, 001, 010, 011, 100, 101, 110, 111.$$

If the sample size is much larger than 2^p, many of the response patterns will be repeated. It is, therefore, much more economical to present the matrix as a list of the possible response patterns together with their associated frequencies as follows:

000	175
001	64
010	17
100	12
011	9
101	3
110	33
111	98

The second column records how many times each response pattern occurs in the sample. This grouped form of the data matrix is used whenever the sample size is large. However, when the number of variables p is large, many response patterns may not occur at all, in which case they are omitted from the table to save space.

Latent trait methods were introduced in educational testing where most of their development has taken place; this is now a highly specialised field with a substantial literature of its own. Our emphasis in this chapter will be mainly on their general use as tools for social research in the factor analysis tradition.

An example

To illustrate the various steps in the analysis, we shall use a dataset with only four variables extracted from the 1986 British Social Attitudes Survey

(McGrath and Waterton, 1986). The data are the responses given by 410 individuals to four out of seven items concerning attitude to abortion. A small proportion of nonresponse occurs for each item, the proportions being (0.03, 0.03, 0.05, 0.04). To avoid the distraction of having to deal with missing values, we have slightly adjusted the data to eliminate missing values. An analysis that includes all respondents and uses a factor analysis (FA) model for binary items that takes account of missing values was carried out by Knott, Albanese, and Galbraith (1990). The results were not substantially different from those reported here. After eliminating the missing values, we are left with 379 respondents. For each item, respondents were asked if the law should allow abortion under the circumstances presented under each item. The four items used in the analysis are given below:

1. The woman decides on her own that she does not [WomanDecide]
2. The couple agree that they do not wish to have the child [CoupleDecide]
3. The woman is not married and does not wish to marry the man [NotMarried]
4. The couple cannot afford any more children [CannotAfford]

The frequency of each response pattern is given in Table 8.1.

TABLE 8.1: Frequencies of response patterns, attitude towards abortion

Response patterns	Frequency
1111	141
0000	103
0111	44
0011	21
0001	13
1110	12
0010	10
0100	9
0110	7
1011	6
0101	6
1101	3
1100	3
1000	1
1010	0
1001	0
Total	379

We find that the percentage of individuals agreeing that abortion should be legal under the circumstances described by items 1 to 4 are 43.8, 59.4, 63.6, and 61.7%, respectively. If we were doing a factor analysis, we would next compute the correlations between pairs of variables and inspect the result, looking for evidence of positive correlations which suggests that there might be one or more common underlying factors. In the case of binary data, the corresponding things to look at are the pairwise associations between variables. We can do this by constructing 2×2 contingency tables. For example, Table 8.2 cross-tabulates the first two items that show a strong association.

TABLE 8.2: Cross tabulation of items 1 and 2, attitude towards abortion

		Item 2	
		Yes	No
Item 1	Yes	159	7
	No	66	147

A similar analysis for other pairs of variables produces similar results. This suggests that it would be worth asking whether these associations can be attributed to one or more common factors. This is what a latent trait model enables us to do. If we can identify common factors, we may then wish to go on to compute scores for individuals on the latent dimensions.

8.2 Why is the factor analysis model for continuous variables invalid for binary responses?

Since the approach for binary and continuous variables has been so similar up to this point, it is natural to think of treating the binary data as if they were continuous. What is to prevent us from computing the product-moment correlations and doing a factor analysis in the usual way? There is no practical bar to doing just that, and one sometimes finds such factor analyses in the research literature. However, such an analysis is inappropriate because it is based on a model which assumes that the observed or manifest variables $(y_1, \ldots, y_p)$ are continuous rather than binary. To see why this is so, we briefly return to the factor analysis model. The model was written as:

$$y_i = \alpha_{i0} + \alpha_{i1}f_1 + \cdots + \alpha_{iq}f_q + e_i \qquad (i = 1, \ldots, p), \qquad (8.1)$$

where p denotes the total number of observed items, y_i denotes the ith continuous observed item, $\mathbf{f} = (f_1, \ldots, f_q)$ denotes the vector of latent variables and e_i denotes the residual. We assume that the residual follows a normal distribution with mean 0 and variance σ_i^2, the latent variables are assumed to be independent with standard normal distributions $f_j \sim N(0,1)$ for all j. Since $\mathbf{f}$ and e_i can take any value and are independent of each other, y_i can also take any value. Therefore, the linear factor model is invalid for categorical variables in general and for binary variables in particular.

We need a different model to relate the latent variables $\mathbf{f}$ to the manifest variables. Two approaches have been adopted to meet this need. The oldest is to try to retain as much as possible of the factor analysis method. This is done by imagining a fictitious (underlying) variable for each i, which is partially revealed to us by y_i. This enables us to retain the factor model for the (unobserved) underlying variable. In section 5.10, we introduced measures of association for binary and ordinal variables called tetrachoric and polychoric correlations, respectively, based on those underlying variables. This method is still widely used and we shall describe it in Section 8.7.

A better approach is to start, as we did in factor analysis, with the idea of a regression model. We want an appropriate model for the regression of each y_i on the latent variables. The usual regression method used for an observable binary response on a set of observable explanatory variables is known as *logistic* regression. It takes its name from the logistic function used in the regression equation (see also Section 6.14).

To motivate the choice of this function, we first remind ourselves that the regression of y_i on the latent variables is the expected value of y_i given the fs. Since y_i is binary, the expected value of y_i given the fs is the same as $\Pr(y_i = 1 \mid \mathbf{f}) = \pi_i(\mathbf{f})$ where $\pi_i(\mathbf{f})$ is the conditional probability that binary variable, y_i, equals one given the values of the q latent variables $f_1, \ldots, f_q$. We, therefore, have to specify the form of the probability $\pi_i(\mathbf{f})$ as a function of $f_1, \ldots, f_q$. The function chosen is known as the *link* function.

An identical linear link function would be the simplest, giving:

$$\pi_i(\mathbf{f}) = \alpha_{i0} + \alpha_{i1} f_1 + \cdots + \alpha_{iq} f_q \qquad (i = 1, \ldots, p). \qquad (8.2)$$

But such a linear relationship between the probability of a correct response and the latent variables has two flaws.

i) The left-hand side of eq. (8.2) is a probability that takes values between 0 and 1, and the right-hand side is not restricted in any way and can take any real value.

ii) We might expect that the rate of change in the probability of a correct/positive response will not be the same for the whole range of $\mathbf{f} = (f_1, \ldots, f_q)$. In that case, a curvilinear relationship might be more appropriate.

To take into account both those points, we need to introduce a different link function between the probability and the latent variables. That link should map the range $[0, 1]$ onto the range $(-\infty, +\infty)$. It should also be a monotonic function of each f. Possible links are the logit and the probit. We shall use the logit link mainly because it possesses theoretical and practical advantages (see Section 8.3). The logit model for binary data presented in Section 8.3 is one of the many item response models developed within the Item Response Theory (IRT) approach. IRT was developed mainly in connection with educational measurement. We shall also, in Section 8.7, briefly discuss the use of the probit link (also known as the normit) as an alternative when we consider the underlying variable (UV) approach for analysing binary variables with factor models.

8.3 Factor model for binary data using the Item Response Theory approach

The logit model for binary data is defined as:

$$\text{logit}[\pi_i(\mathbf{f})] = \log_e \frac{\pi_i(\mathbf{f})}{1 - \pi_i(\mathbf{f})} = \alpha_{i0} + \sum_{j=1}^{q} \alpha_{ij} f_j. \tag{8.3}$$

By transforming $\pi_i(\mathbf{f})$ using the logit transformation, we have been able to write the model as linear in the latent variables, which will greatly facilitate the interpretation. The probability $\pi_i(\mathbf{f})$ denotes the probability of "success" and the ratio $\pi_i(\mathbf{f})/(1-\pi_i(\mathbf{f}))$ is also known as the odds of "success". Eq. (8.3) is introduced in Section 6.14 with observed explanatory variables rather than latent variables. We can rearrange eq. (8.3) to get an expression for $\pi_i(\mathbf{f})$:

$$\pi_i(\mathbf{f}) = \frac{\exp(\alpha_{i0} + \sum_{j=1}^{q} \alpha_{ij} f_j)}{1 + \exp(\alpha_{i0} + \sum_{j=1}^{q} \alpha_{ij} f_j)}. \tag{8.4}$$

It may easily be checked that this expression behaves in the right way, namely that it lies between 0 and 1 and is monotonic in each f.

An important special case is obtained by setting $q = 1$. It is this case with which *item response analysis* is mainly concerned. Thus, we have the unidimensional latent trait model:

$$\pi_i(f_1) = \frac{\exp(\alpha_{i0} + \alpha_{i1} f_1)}{1 + \exp(\alpha_{i0} + \alpha_{i1} f_1)}.$$

The unidimensional latent trait model is also known as the two-parameter model. In the psychometric literature, $\pi_i(f_1)$ is referred to as the item characteristic curve or item response function (IRF). It shows how the probability of a correct response increases with ability, say.

The logit model with one latent variable is plotted in Figure 8.1 for $\alpha_{i0} = 0.5$ and for different positive values of the parameter α_{i1} and in Figure 8.2 for different values of α_{i0} and for $\alpha_{i1} = 0.5$.

It is clear that the parameter α_{i1} determines the steepness of the curve over the middle of the range. This means that a given change in the value of f_1 will produce a larger change in the probability of a positive response when this parameter is large than when it is small. For this reason, it is known in educational testing as the *discrimination* parameter. Increasing the parameter α_{i0} increases the probability for all values of f_1 and so it is referred to as the *difficulty* parameter.

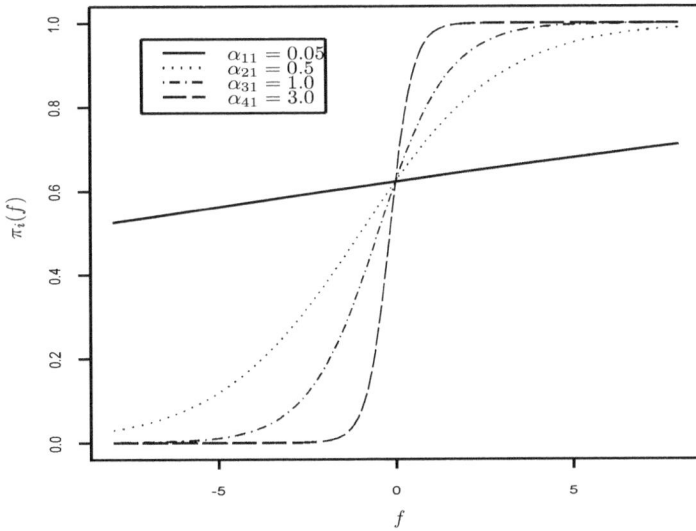

FIGURE 8.1: Item characteristic curves for different values of the discrimination coefficient α_{i1} and $\alpha_{i0} = 0.5$

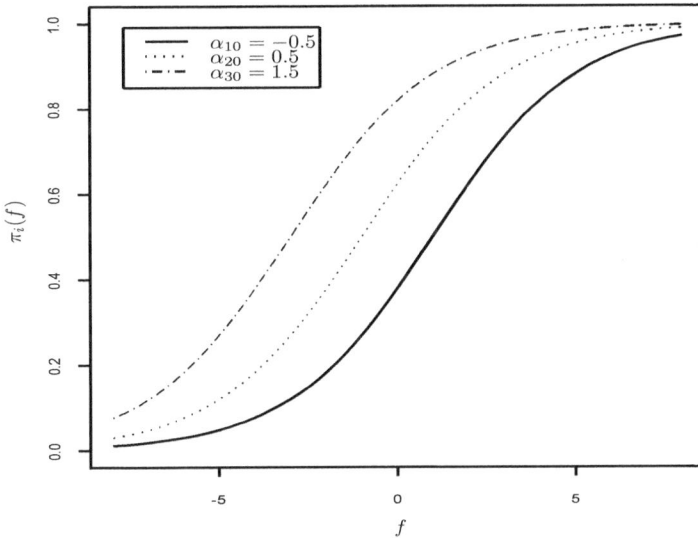

FIGURE 8.2: Item characteristic curves for different values of the "difficult" parameter α_{i0} and $\alpha_{i1} = 0.5$

The parameters of the model are the intercepts or difficulty parameters α_{i0}, $i = 1, 2, \cdots, p$, and the factor loadings or discrimination parameters α_{ij}, $i = 1, 2, \ldots, p, j = 1, 2, \ldots, q$. Since for $q > 1$, the matrix with the factor loadings $A = (\alpha_{ij})$ of dimension $p \times q$ is determined only up to an orthogonal transformation of order $q \times q$, (see also Section 7.6) there are only $pq - q(q-1)/2$ independent factor loadings. Hence, the total number of independent parameters of the model is $p + pq - q(q-1)/2$. All factor models are not identified unless q^2 constraints are imposed on the model parameters. Some of those constraints are taken care of by setting the variances of the latent variables to one and the covariances to zero $(q(q+1)/2)$, leaving us with $(q(q-1)/2)$ yet to be set. Those constraints can be applied directly to the factor loadings by fixing some of them to zero. In the case of the one-factor model, no further constraints are necessary. In the two-factor model, we still need an additional constraint (e.g. the R package mirt sets one loading to zero). The solution obtained is still subject to rotation as with the factor analysis model in Chapter 7. The package mirt provides orthogonal and oblique rotation options, whereas ltm does not.

We summarise and complete the specification of the factor model for binary data by listing the assumptions on which it depends as follows:

i) Conditional independence: the responses to the p observed items are independent conditional on the latent variables. In other words, the latent variables (factors) account for all the associations among the observed items. Since the latent variables are unobserved, the assumption of conditional independence can only be tested indirectly by checking whether the model fits the data. A latent variable model is accepted as a good fit when the latent variables account for most of the association among the observed responses.

ii) The link function: $\text{logit}[\pi_i(\mathbf{f})] = \alpha_{i0} + \sum_{j=1}^{q} \alpha_{ij} f_j$, where $\Pr(y_i = 1 \mid \mathbf{f}) = \pi_i(\mathbf{f}); (i = 1, \ldots, p)$. A possible alternative would be the probit link, see Section 8.7, which gives very similar results in practice.

iii) The latent variables or factors $f_1, \ldots, f_q$ are independent with standard normal distributions. That is $f_j \sim N(0, 1)$ for $(j = 1, \ldots, q)$. The choice of the normal distribution for the latent variables has rotational advantages as we will see later but other distributions could be used.

The Rasch model

A special case of the unidimensional model is obtained when all the discrimination parameters are equal $(\alpha_{11} = \alpha_{21} = \cdots = \alpha_{p1})$. This model was first discussed by Rasch (Rasch 1960) and it is usually written as:

$$\Pr(y_{ik} = 1 \mid \alpha_{i0}, \beta_k) = \pi_{ik} = \frac{\exp(\alpha_{i0} + \beta_k)}{1 + \exp(\alpha_{i0} + \beta_k)}.$$

Since the α_{i1} are all equal, $\alpha_{i1} f_k$ has been replaced by β_k, where f_k is the value of f for individual k, $(k = 1, \ldots, n)$. This formulation is useful in educational testing where the ability of each individual in the sample is of interest. The Rasch model is still quite popular in educational testing because of its simplicity and its attractive theoretical properties. In particular:

i) The total score $\sum_{i=1}^{p} y_{ik}$ is sufficient for β_k — that is, it contains all the information in the data about the β_k if the model is true.

ii) The total number of positive/correct responses for item i, $\sum_{k=1}^{n} y_{ik}$, is sufficient for α_{i0}.

In other applications, the interest is generally in the population from which the sample has been drawn, and f is treated as a variable with a probability distribution.

Fitting the logit model

Recall that in factor analysis, we fitted the model by choosing the parameter values to make the covariance matrix predicted by the model as close as possible to the observed matrix. For that model, the joint distribution was completely determined by the covariances, so, in effect, we were making the observed and predicted distributions as close as possible. We do essentially the same thing when fitting the latent trait model. We choose the parameter values which make the frequency distribution across responses predicted by the model as close as possible to the observed one. As in factor analysis, there are various ways in which this distance can be measured, but the one for which software is currently available is based on the likelihood function — the maximum likelihood method.

Interpretation of model parameters

In the latent trait model for each observed item i, we have $q + 1$ parameters to estimate, the intercept α_{i0} and the factor loadings $\alpha_{i1}, \ldots, \alpha_{iq}$. We have already noted that α_{i0} is called the difficulty parameter in educational testing because of its effect on the probability of a positive response. This effect can be seen more clearly if we consider the position when $\mathbf{f} = \mathbf{0}$. Since the fs are assumed to have standard normal distributions, an individual at this point in the latent space may be described as the "median" individual because, on each dimension, half the population lie on either side. In those circumstances we find,

$$\Pr(y_i = 1 \mid \mathbf{f} = \mathbf{0}) = \pi_i(\mathbf{0}) = \frac{\exp(\alpha_{i0})}{1 + \exp(\alpha_{i0})}.$$

This is the probability that the median individual will respond correctly or positively to item i. For example all four curves in Figure 8.1 have the same $\alpha_{i0} = 0.5$ and hence the same value $\pi_i(0) = 0.62$. For practical purposes, $\pi_i(0)$ is more directly interpretable than α_{i0}.

The α_{ij}s $(j = 1, \ldots, q)$ are factor loadings, but we have already noted that they are known in educational testing as discrimination parameters. The larger the value of α_{ij}, the greater is the effect of factor j on the probability of a positive response to item i; equivalently, the higher the value of α_{ij} for an item, the greater the difference in the probabilities of getting a correct/positive response between two individuals who are located at some distance apart on the latent dimensions. As a result, it will be easier to discriminate between those two individuals on the evidence of their responses to that item. The factor loadings α_{ij} are not bounded in any way, and for some items, they may take very large values, indicating a very steep slope for the item response curve. This phenomenon is known as a "threshold effect", and we shall meet it again in Chapter 10. Large estimates of the discrimination parameters often have large standard errors, which means that their values are poorly determined. The maximum likelihood estimates for the attitude to abortion data are given in Table 8.3 along with their asymptotic (i.e. estimated using large sample theory) standard error for a one-factor model.

TABLE 8.3: Parameter estimates and standard errors in brackets and standardised loadings for the one-factor model, attitude to abortion

Item	$\hat{\alpha}_{i0}$	s.e.	$\hat{\alpha}_{i1}$	s.e.	st$\hat{\alpha}_{i1}$	$\hat{\pi}_i(0)$
WomanDecide	−0.72	(0.31)	4.16	(0.81)	0.97	0.33
CoupleDecide	1.11	(0.34)	4.50	(0.82)	0.98	0.75
NotMarried	2.18	(0.67)	6.21	(1.69)	0.99	0.90
CannotAfford	1.15	(0.29)	3.49	(0.53)	0.96	0.76

The last column of the table gives the estimated probabilities that the median individual will respond positively to items 1-4. Item 1 stands out from the other items by being much less likely to be answered positively by the median individual. The loadings in the $\hat{\alpha}_{i1}$ column are all positive and very large, suggesting an underlying factor which is common to all items. In this context, one might identify this with a pro/anti-abortion attitude. It should be noted that the standard errors are all fairly large in relation to the differences in the estimates. This should caution us against placing undue weight on small inequalities among the loadings. In the present case, the broad conclusion we have drawn about a common factor seems unlikely to be sensitive to the effects of sampling variation. Taking the loadings at their face value, it appears that "the couple being unmarried" is the best discriminator between a pro- and an anti-abortion attitude and "inability to afford the baby" the worst discriminator.

The column headed stâ$_{i1}$ requires some further explanation. In factor analysis, when the correlation matrix is analysed, the factor loading α_{ij} is the correlation between the observed item y_i and the latent variable f_j. This was very convenient as an aid to interpretation. In the latent trait case, the loadings cannot be interpreted as correlation coefficients; indeed, as we have seen, the loadings are not bounded by 0 and 1 as a correlation would be. However, it is possible to transform to standardised loadings that can be interpreted as correlation coefficients in the same way as in factor analysis. This transformation arises naturally out of the alternative way of analysing binary items, which we shall consider in Section 8.7. We shall defer consideration of this point to later, but here we merely observe that all the standardised loadings are close to one, indicating a close link between each item and the common factor.

8.4   Goodness-of-fit

The goodness-of-fit of the model can be checked in several different ways.

i) *Global goodness-of-fit test*
One way is to use a standard goodness-of-fit test to compare the observed and expected frequencies across the response patterns. Strictly, we compare observed frequencies and estimates of the expected frequencies under the model being tested — but conventionally, these estimates are referred to as "expected frequencies" when carrying out likelihood ratio or Pearson chi-squared goodness-of-fit tests as below. Since we fit the models by choosing the parameter values so that these distributions are as close as possible, the minimum closeness would be an obvious measure to use for goodness-of-fit. A test based on such a measure is the likelihood ratio test. The likelihood ratio test statistic, G^2, is defined as:

$$G^2 = 2 \sum_{r=1}^{2^p} O(r) \log_e \frac{O(r)}{E(r)}, \tag{8.5}$$

where r represents a response pattern, and $O(r)$ and $E(r)$ represent the observed and expected frequencies, respectively, of response pattern r. An alternative is to use the Pearson chi-squared goodness-of-fit test statistic, X^2, given by:

$$X^2 = \sum_{r=1}^{2^p} \frac{(O(r) - E(r))^2}{E(r)}. \tag{8.6}$$

If the model holds, both statistics are distributed approximately as χ^2 with degrees of freedom equal to the number of different response patterns minus the number of independent parameters minus one $(2^p - p(q+1) - q(q-1)/2 - 1)$. If the sample size n is much bigger than the total number of distinct responses given by 2^p, then the observed and expected frequencies will be reasonably large, and the approximation on which the test is based will be valid. However, when the number of binary variables is large, many response patterns will have expected frequencies which are very small. It is usually recommended that all expected frequencies should be at least five for either test to be valid. If, for example, $p = 20$, there are $2^p = 1048576$ possible response patterns, and even with a sample size of several thousand, there will be many expected frequencies which are exceedingly small. In those cases, the chi-squared test and the likelihood ratio test will not follow a chi-squared distribution, and so from the practical point of view these tests cannot be used. The problem can be overcome to some extent by pooling response patterns with expected frequencies less than 5, but that might quickly lead to a situation where no degrees of freedom are left to perform the test. In such cases, we need another approach.

For the attitude to abortion dataset in Table 8.6, the chi-squared statistic is $X^2 = 18.84$, on seven degrees of freedom, indicating a bad fit at the 1% and 5% significance level (the 1% and 5% critical values for chi-squared with seven degrees of freedom are 18.475 and 14.067 respectively). The X^2 test statistic is computed using all sixteen possible response patterns (including those that did not occur in the sample) and without pooling patterns with small expected frequencies. The likelihood ratio test statistic G^2 cannot be computed when there is a response pattern with zero observed frequency, as the log function is involved. The X^2 and G^2 test statistics can also be computed after pooling the response patterns when the expected frequencies are less than five. In general, we have no trust in the asymptotic distribution of both statistics, whether there are patterns with expected frequencies less than five or when patterns are being pooled; therefore, we do not provide the pooled versions of these statistics here.

We also propose inspecting the discrepancies between observed and expected score patterns. These are given in the first two columns of Table 8.6. There are no large deviations, except perhaps for three response patterns. In a sparse table with many more response patterns, it would be much more difficult to judge this, and alternative approaches would be needed.

ii) *Goodness-of-fit for margins*

Rather than look at the whole set of response patterns, we can look at the two-way margins. That is, we can construct the 2×2 contingency tables obtained by taking the variables two at a time. We have already done this at the beginning of the chapter when we looked at the pairwise associations among variables. The reason for doing that was to bring out the parallel with factor analysis. The two-way tables provided the same sort of information for binary variables as the correlations do for factor analysis. The two-way margins are the cell frequencies in these two-way tables. Comparing the observed and expected two-way margins is therefore analogous to comparing the observed and expected correlations when judging the fit of a factor model. The comparison is made using what we call *chi-squared residuals*. These are the contributions to the chi-squared statistic for the 2×2 table which would arise from the cell. Thus, if O is the observed frequency and E the expected frequency, then the residual is $(O-E)^2/E$. Tables 8.4 and 8.5 give the observed and expected frequencies for the two-way and some of the three-way margins, respectively, for the attitude to abortion data when the one-factor model is fitted. The last column of the tables gives the chi-squared residual as a measure of the discrepancy between the observed and the predicted frequency. From Table 8.4, we see that 147 respondents responded negatively to items 1 and 2; the model predicted 143.74 responses for that cell, giving a residual equal to 0.07. The same calculations are done for all the pairs of items. The residuals computed for each cell are not independent, and therefore, they cannot be summed to give an overall test distributed as chi-squared.

TABLE 8.4: Chi-squared residuals for the second-order margins for the one-factor model, attitude towards abortion

Response	Item i	Item j	Observed frequency (O)	Expected frequency (E)	$(O-E)^2/E$
(0,0)	2	1	147	143.74	0.07
	3	1	131	133.17	0.04
	3	2	117	119.69	0.06
	4	1	129	133.68	0.16
	4	2	114	116.09	0.04
	4	3	116	111.79	0.16
(0,1)	2	1	7	11.30	1.64
	3	1	7	5.94	0.19
	3	2	21	19.42	0.13
	4	1	16	11.99	1.34
	4	2	31	29.58	0.07
	4	3	29	33.88	0.70
(1,0)	2	1	66	69.89	0.22
	3	1	82	80.46	0.03
	3	2	37	35.35	0.08
	4	1	84	79.95	0.21
	4	2	40	38.95	0.03
	4	3	22	27.32	1.04
(1,1)	2	1	159	154.07	0.16
	3	1	159	159.43	0.00
	3	2	204	204.54	0.00
	4	1	150	153.38	0.07
	4	2	194	194.38	0.00
	4	3	212	206.01	0.17

As a rule of thumb, if we consider the residual in each cell as having a χ^2 distribution with one degree of freedom, then a value of the residual greater than 4 is indicative of poor fit at the 5% significance level. To be able to have a better idea of the discrepancies in the margins, given that the value 4 is only indicative, in the examples later in the chapter, we also report residuals greater than 3. A study of the individual margins provides information about where the model does not fit. For the abortion data, all the residuals are very small. On the evidence from the margins, we have no reason to reject the one-factor model. The overall significant result we obtained from the global goodness-of-fit tests cannot, therefore, be attributed to the relationships between the pairs and triplets of items.

Valid test statistics on the lower margins, such as the univariate, bivariate and trivariate are, however, provided in Reiser (1996), Bartholomew and Leung (2002), Maydeu-Olivares and Joe (2005), Reiser (2008) and Cai et al. (2006) and are known as limited information goodness-of-fit test statistics. The M_2 test statistic proposed by Maydeu-Olivares and Joe (2005) is the most widely used and implemented in IRT R packages such as mirt. The M_2 test statistic is a chi-square type statistic calculated by taking the sum of the univariate and bivariate residuals (i.e. the difference between an observed and expected frequency) and standardising them with their corresponding asymptotic covariance matrix. A review of the limited information test statistics can be found in Maydeu-Olivares and Joe (2008). Despite observing no large discrepancies in the bivariate and trivariate margins through the inspection of the chi-squared residuals, the $M_2 = 13.54$ on 2 degrees of freedom and a p-value of 0.0011 reject the one-factor model for the abortion data. That leads to the conclusion that even the limited information goodness-of-fit test statistics should be treated with caution, mainly when sparseness also occurs in the lower-order margins. Inspection of individual cells remains a valuable tool for detecting misfit, particularly in pairs and triplets of items.

iii) *Resampling methods*
When the asymptotic distribution of a test statistic is not known or, in our case, valid, such as the X^2 and G^2 test statistics in the case of sparseness, one can consider obtaining the empirical distribution of the goodness-of-fit test statistic of interest and the corresponding p-value using resampling or bootstrapping methods. Those methods tend to be computationally intensive, but they are available in some of the IRT R Packages such as ltm and could be employed for simpler models.

iv) *Proportion of G^2 explained*
We have remarked at several points in the book that even an incomplete summary of multivariate data can be useful. The same is true of a multivariate model. Even though it may leave something unexplained, it may nevertheless capture some important and interesting features of the data. This is the case with the one-factor model, which is serving as our example in this section.

TABLE 8.5: Chi-squared residuals for the third-order margins for the one-factor model, response (1,1,1) to items (i, j, k), attitude towards abortion

Item i	Item j	Item k	Observed frequency (O)	Expected frequency (E)	$(O - E)^2/E$
1	2	3	153	151.18	0.02
1	2	4	144	145.86	0.02
1	3	4	147	150.15	0.07
2	3	4	185	185.01	0.00

This raises the question of whether we can quantify the degree to which a simple model explains the associations between the binary variables. The same general idea proved useful in PCA and FA, where the proportion of variance explained served a similar purpose. Thus, we observed that the proportion of the total variance accounted for by a set of components might be used as a guide to whether the fs were an adequate summary. The same idea can be used here, but we now talk in terms of the proportion of the likelihood ratio statistic for the independence model, which the model explains with q factors. The independence model would be appropriate if there were no associations between the binary variables $y_1, \ldots, y_p$. The likelihood ratio statistic, G_0^2, for this model can be regarded as a measure of the associations between the ys. The likelihood ratio statistic, G_q^2, for the model with q latent variables is a measure of the residual associations between the ys, which the model has not explained.

The percentage of G^2 explained is given by

$$\%G^2 = \frac{G_0^2 - G_q^2}{G_0^2} \times 100$$

and measures the extent to which the model with q latent variables explains the associations.

For the attitude to abortion data, the percentage of G^2 explained is 98.48%, indicating that the one latent variable model is a much better fit than the independence model or, in other words, there is 98.48% reduction in the likelihood ratio statistic when the one-factor model was fitted. The G^2 test statistic, both for the one-factor model and the independence model, was computed using the observed response patterns only.

The above ways of checking model fit have been discussed in detail in the paper by Bartholomew and Tzamourani (1999).

v) Model selection methods

Another approach, already mentioned in the connection with factor analysis for continuous variables, is based on the use of model selection criteria such as the Akaike information criterion or the Bayesian information criterion (see Sclove 1987).

8.5 Factor scores

Obtaining factor scores for the latent trait model is slightly more complicated than it was for PCA or FA. In PCA, the scores came "ready-made" as linear combinations of the manifest variables. In FA, the position was complicated by the fact that there was no unique value of each f associated with the set of ys. We therefore used a predicted value which turned out to be a linear combination of the ys for which the coefficients were calculated by standard software. Following the same idea for the latent trait model, we would look for a suitable prediction of each f given the ys. Using regression ideas as before, this would suggest using the conditional mean value or conditional expectation:

$$E(f_j \mid y_1, \ldots, y_p) \qquad (j = 1, \ldots, q). \tag{8.7}$$

Unfortunately, these means are not linear combinations of the ys, although they can easily be computed. However, it turns out that (for the logit link function) they are monotonic functions of what we shall call *component scores* which are given by:

$$Y_j = \sum_{i=1}^{p} \alpha_{ij} y_i \qquad (j = 1, \ldots, q). \tag{8.8}$$

In the one-factor case, both the regression function of eq. (8.7) and the components give the same ranking to the individuals in the sample. These components are very simply calculated using the estimated weights obtained from fitting the model. For most practical purposes, it makes no difference whether we use the components or the conditional expectations.

For the logit link function, the component score, Y_j, includes all the information in the data about the latent variables regardless of the assumption made about the distribution of f_j, whereas the posterior mean $E(f_j \mid y_1, \ldots, y_p)$ itself will vary according to whether we assume the distribution of f_j to be normal or some other distribution. This invariance property is a good reason to prefer the component score.

On the other hand, when a distribution is assumed for the fs it is possible to estimate not only the conditional means, $E(f_j \mid y_1, \ldots, y_p)$, but also the conditional standard deviations, $\sigma(f_j \mid y_1, \ldots, y_p)$ for $(j = 1, \ldots, q)$. The estimated standard deviations should be taken into account in judging the ranking of the response patterns when the conditional means are used.

Table 8.6 gives the estimated conditional means and component scores for all the response patterns for the attitude to abortion data. It also gives the expected frequency for each pattern. The sixth column gives the total score of the response pattern. As we can see, the estimated conditional mean, $\hat{E}(f \mid \mathbf{y})$, and the component score gives the same ranking to the individuals. In this particular example, the total score also gives a similar ranking to the individuals, though there are some ties. The reason for this is that all four items have similar discriminating power. There is also a column headed $\hat{\sigma}(f \mid \mathbf{y})$. This is the estimated conditional standard deviation of the latent variable about its conditional mean. This tends to be larger at the extremes but is fairly constant over the middle range. In all cases, it is quite large, indicating that the factor scores are subject to a good deal of uncertainty.

TABLE 8.6: Factor scores listed in increasing order, attitude towards abortion

Observed frequency	Expected frequency	$\hat{E}(f \mid \mathbf{y})$	$\hat{\sigma}(f \mid \mathbf{y})$	Component score (Y_1)	Total score	Response pattern
103	100.0	−1.19	0.55	0.00	0	0000
13	16.6	−0.61	0.32	3.49	1	0001
1	1.7	−0.55	0.30	4.16	1	1000
9	9.1	−0.52	0.29	4.50	1	0100
10	12.3	−0.38	0.26	6.21	1	0010
0	1.3	−0.29	0.24	7.64	2	1001
6	7.4	−0.27	0.24	7.98	2	0101
3	1.0	−0.24	0.24	8.65	2	1100
21	14.8	−0.18	0.24	9.70	2	0011
0	2.0	−0.14	0.25	10.37	2	1010
7	12.3	−0.12	0.26	10.71	2	0110
3	1.9	−0.01	0.28	12.14	3	1101
6	6.2	0.14	0.31	13.85	3	1011
44	41.1	0.17	0.32	14.20	3	0111
12	7.2	0.24	0.34	14.86	3	1110
141	143.9	0.95	0.61	18.35	4	1111

8.6 Rotation

As with the factor analysis model, the solution is not unique when we fit more than one latent variable. An orthogonal rotation of the factors coupled with a corresponding rotation of the estimated loadings $\hat{\alpha}_{ij}$ leaves the likelihood unchanged. We are therefore free to search for a rotation which is more readily interpretable. The cautionary remarks made in Chapter 7 apply with equal force here. In particular, rotation does not produce a new solution so much as express the original solution in a different way. The main use of rotation is to search for "simple structure". In principle, the same kind of rotations could be used for latent trait models as for factor analysis. However, the uncertainties of estimation increase rapidly with the number of factors. It is doubtful whether there is any value in trying to fit more than two factors with the sample sizes that are commonly available. In any case, we have concentrated in this book on solutions which are capable of being represented in up to two dimensions. Our treatment is therefore consistent with this general approach. For practical purposes, rotation can be carried out in two dimensions graphically, as in Chapter 7.

8.7   Underlying variable approach

In this section, we will discuss the alternative approach for constructing and fitting a factor analysis model to binary items. This approach is called the underlying variable (UV) approach. As we explained in Section 8.2, the UV approach is closer in spirit to factor analysis.

In the UV approach, the observed binary variables are assumed to be realisations of fictitious continuous *underlying variables*. Those underlying variables are unobserved, but they should not be confused with the latent variables. They might be better described as *incompletely observed variables*, because all we observe is whether or not they exceed some threshold.

For each binary variable y_i, it is assumed that there is an *incompletely observed* continuous variable y_i^* which is normally distributed with mean μ_i and variance σ_i^2.

The connection between y_i and y_i^* is as follows: when the underlying variable y_i^* takes values below a threshold value τ_i, the binary item y_i takes the value 1, otherwise y_i takes the value 0. The parameters τ_i are called threshold parameters. Since no other information is available about y_i^* ($i = 1, \ldots, p$), its mean and variance are arbitrary and can be set to zero and one, respectively, without loss of information.

The essence of the method is to treat the y_i^*s as if they had been generated by the classical factor analysis model. That is, we suppose that:

$$y_i^* = \alpha_{i1}^* f_1 + \alpha_{i2}^* f_2 + \cdots + \alpha_{iq}^* f_q + e_i \qquad (i = 1, \ldots, p), \qquad (8.9)$$

where the α_{ij}^* are the factor loadings, the f_j are the latent variables, and the e_i are the residuals with zero mean and variance σ_i^2 ($i = 1, \ldots, p; j = 1, \ldots, q$). In factor analysis, the y_i^*s are observable variables, whereas here they are underlying, incompletely observed variables. The parameters of the model are the thresholds τ_i, $i = 1, \cdots, p$ and the factor loadings α_{ij}^*, $i = 1, \cdots, p$, $j = 1, \cdots, q$. As with the logit model, the number of independent parameters is $p + pq - q(q-1)/2$.

All that we need to fit a factor model is the matrix of correlations. The correlation can be estimated from each pair of the binary y_is and hence software such as lavaan, LISREL, and Mplus can be used to fit the factor model. Correlations estimated in this way are called *tetrachoric* correlations and they are explained in Section 5.10.

There are a number of subtle differences between fitting a factor model to tetrachoric correlations and fitting it to product-moment correlations. The thresholds are estimated from the univariate marginal distribution of the underlying variable, y_i^*, and the correlations from the bivariate marginal distributions of the y_i^*s for given thresholds. This amounts to say that the method uses less of the information in the data. The UV approach does assume conditional independence through the independence of the residual terms, e_i, and it also assumes that the univariate and bivariate distributions of the underlying variables are normal.

The results of carrying out a factor analysis on tetrachoric correlations are very similar to those obtained using the logit latent variable model. This is no accident because it can be shown that the two types of models are equivalent for binary data. A mathematical proof of this equivalence can be found in Bartholomew, Knott, and Moustaki (2011), p.87-88. The logit model and the UV probit model give similar results because the normal and logistic distributions are so similar in shape. There is an exact equivalence between the parameter estimates for the probit UV and the probit IRF model given by:

$$\alpha_{i0} = \frac{\tau_i}{\sigma_i}$$

and

$$\alpha_{ij} = -\frac{\alpha_{ij}^*}{\sigma_i}.$$

The same equivalence holds approximately for the probit UV and the logit IRF models. Furthermore, we can standardise the factor loadings α_{ij}s to represent correlations between the latent variables f_js and the binary variables y_is.

The standardised αs are given by:

$$st\alpha_{ij} = \frac{\alpha_{ij}}{\sqrt{\sum_{j=1}^q \alpha_{ij}^2 + 1}} = -\alpha_{ij}^*. \tag{8.10}$$

This is the standardisation we referred to in Section 8.3 and which was given in Table 8.3.

8.8 Example: sexual attitudes

In order to illustrate the full range of analyses, including the fitting of two factors, we shall take an example extracted from the 1990 British Social Attitudes Survey (Brook, Taylor, and Prior 1991). It concerns contemporary sexual attitudes. The questions addressed to 1077 individuals were as follows.

1. Should divorce be easier?
2. Do you support the law against sexual discrimination?

3. View on pre-marital sex: not at all wrong...always wrong.
4. View on extra-marital sex: not at all wrong...always wrong.
5. View on sexual relationship between individuals of the same sex: not at all wrong...always wrong.
6. Should gays teach in school?
7. Should gays teach in higher education?
8. Should gays hold public positions?
9. Should a female homosexual couple be allowed to adopt children?
10. Should a male homosexual couple be allowed to adopt children?

For those items yielding a binary response (1,2,6,7,8,9,10), a positive response was coded as 1 and a negative response as 0. For items 3, 4, and 5 there were five categories: "always wrong", "mostly wrong", "sometimes wrong", "rarely wrong" and "not at all wrong". Responses "sometimes wrong", "rarely wrong", and "not at all wrong" were coded as 1 and responses "always wrong" and "mostly wrong" as 0. With ten variables, there are $2^{10} = 1024$ possible response patterns. Not all of these occur, but with a sample size of 1077 the data matrix takes up a good deal of space. The full dataset is given on the website, but the cases with frequencies greater than ten are given in Table 8.7 in decreasing order of observed frequency as an illustration.

Table 8.8 gives the proportions giving positive and negative responses to each item.

TABLE 8.7: Response frequencies, sexual attitudes data

Response patterns	Frequency	Response patterns	Frequency
0110000000	117	1110000000	17
0110111100	95	0111000000	15
0100000000	93	0111011100	15
0110011100	90	0010011100	14
0110111111	40	0110111110	14
0010000000	35	0110011110	13
0100011100	32	1110011100	13
0000000000	29	1110111111	12
0110000100	27	0110011000	11
0110001100	21	0110100000	11
0111111100	19	0010000100	11
0100000100	18	0000011100	10
0111111111	18	Other patterns	287

Since we come to the data with no preconceived ideas about what the latent variables might be, we begin by fitting a one-factor model to the ten items. The parameter estimates are given in Table 8.9. Items 6, 7, and 8 have large discrimination coefficients, $\hat{\alpha}_{i1}$, indicating that the characteristic curves of those items are very steep. From the $\text{st}\hat{\alpha}_{i1}$ column, we see that item 1 has the weakest relationship with the latent variable, followed by items 2 and 4. The rest of the items show strong relationships with the latent variable f.

TABLE 8.8: Proportions giving positive and negative responses to observed items, sexual attitudes data

Item	Response 1	Response 0
1	0.13	0.87
2	0.82	0.18
3	0.77	0.23
4	0.13	0.87
5	0.29	0.71
6	0.48	0.52
7	0.55	0.45
8	0.59	0.41
9	0.19	0.81
10	0.11	0.89

TABLE 8.9: Estimated difficulty and discrimination parameters with standard errors in brackets and standardised loadings for the one-factor model, sexual attitudes data

Items	$\hat{\alpha}_{i0}$	s.e.	$\hat{\alpha}_{i1}$	s.e.	$\text{st}\hat{\alpha}_{i1}$	$\hat{\pi}_i(0)$
1	−1.94	(0.09)	0.11	(0.10)	0.11	0.13
2	1.65	(0.09)	0.53	(0.10)	0.47	0.84
3	1.46	(0.10)	1.02	(0.11)	0.72	0.81
4	−2.02	(0.10)	0.60	(0.11)	0.52	0.12
5	−1.30	(0.11)	1.79	(0.15)	0.87	0.21
6	−0.11	(0.40)	10.93	(3.12)	1.00	0.47
7	1.94	(0.63)	12.38	(2.69)	1.00	0.87
8	1.00	(0.18)	3.54	(0.31)	0.96	0.73
9	−2.07	(0.14)	1.62	(0.15)	0.85	0.11
10	−3.70	(0.28)	2.40	(0.25)	0.92	0.02

We first investigate the goodness-of-fit of the one-factor model using some of the methods described in Section 8.4. Due to the large number of response patterns with an expected frequency of less than five, we do not use the overall goodness-of-fit to decide on the fit; instead, we check the chi-squared residuals for the two- and three-way margins.

There were large discrepancies between the observed and expected frequencies for many pairs and triplets of items. Table 8.10 gives all the pairs and the (1,1,1) triplets of items where the chi-squared residuals were greater than 3.

TABLE 8.10: Chi-squared residuals greater than 3 for all the second- and (1,1,1) third-order margins for the one-factor model, sexual attitudes data

Response	Items	O	E	$(O-E)^2/E$
(0,0)	5,3	237	210.86	3.21
	9, 10	875	816.22	4.23
(1,0)	9, 10	88	144.34	21.99
	1, 3	19	29.27	3.60
(0,1)	9, 10	2	54.61	50.68
	7, 10	12	2.62	33.66
	6, 10	15	4.38	25.80
	7, 9	36	18.30	17.11
	3, 4	4	22.51	15.22
	8, 10	11	3.89	12.99
	6, 9	46	26.36	14.64
	3, 5	14	36.15	13.57
	2, 5	23	37.64	5.69
	8, 9	29	20.21	3.82
(1,1)	9, 10	112	61.83	40.71
	1, 4	29	18.73	5.63
	6, 9	154	179.81	3.71
	7, 9	164	187.86	3.03
(1,1,1)	3, 9, 10	106	58.85	37.79
	5, 9, 10	89	49.24	32.11
	2, 9, 10	99	57.09	30.77
	8, 9, 10	101	61.27	25.77
	7, 9, 10	100	61.66	23.85
	6, 9, 10	97	61.45	20.57
	1, 9, 10	21	9.09	15.63
	4, 9, 10	33	16.89	15.36
	1, 3, 4	29	15.95	10.68
	1, 5, 10	20	11.96	5.41
	2, 6, 9	137	162.11	3.89
	1, 4, 10	8	4.09	3.74
	6, 7, 9	153	178.58	3.66
	1, 4, 8	22	14.75	3.56
	1, 4, 5	14	8.61	3.38
	2, 3, 4	122	103.60	3.27
	1, 2, 6	50	64.10	3.10

The percentage of G^2 explained is 77.27%, which shows that the model goes a long way in explaining the associations, but taken with the very poor fit indicated by the fit measures, it is desirable to continue by fitting a second latent variable. The two-factor model is a considerable improvement. The percentage of G^2 explained increased from 77.27 to 87.10%.

Comparing the results from the one-factor solution given in Table 8.10, we find that the two-factor solution is a great improvement for predicting the observed two- and three-way margins. The fit was found to be poor (with residuals greater than 3) for the margins given in Table 8.11. Although the fit is

TABLE 8.11: Chi-squared residuals greater than 3 for the first-, second-, and (1,1,1) Third-order margins for the two-factor model, sexual attitudes data

Response	Items	O	E	$(O-E)^2/E$
(1,0)	4, 3	4	18.27	11.15
	5, 2	23	37.53	5.62
	5, 3	14	28.29	7.21
	9, 3	15	9.20	3.66
(1,1)	1, 4	29	20.53	3.50
(1,1,1)	1, 3, 4	29	18.26	6.32
	2, 3, 4	122	103.50	3.31
	1, 4, 8	22	15.00	3.26

still somewhat questionable, the large percentage of G^2 explained encourages us to attempt an interpretation of the two-factor model.

Table 8.12 gives the maximum likelihood estimates together with their asymptotic (estimated using large sample theory) standard errors and the standardised parameters for the factor loadings. The last column shows very striking differences in the response of the median individual to the various questions. The last two items on adoption by homosexual couples show virtually no support for the propositions. There are also small probabilities of responding positively to items 1, 4, and 5. The marginal observed proportions given in Table 8.8 give a similar picture, but they relate to views in the whole sample rather than to the median individual. The unrotated standardised solution presented in Table 8.12 shows that the first factor is a general factor with moderate to high loadings on all items except item 1. The second factor has negative, and of similar magnitude, loadings on items 2, 6, 7, and 8 and positive loadings on items 1, 3, 4, 5, 9, and 10. The interpretation is not entirely clear, but we note that the items in the first group are concerned with views on public matters, whereas items 3 to 5, at least, are concerned with private behaviour. However, the inclusion of items 9 and 10 does not fit with this interpretation. When we rotated the solution using an orthogonal rotation such as Varimax we found that the rotated solution did not produce a simpler pattern than the one revealed from the original factor solution.

The failure to get two clear-cut factors, coupled with the poor fit of the model, suggests that the analysis should be taken further. One issue could be the large estimated factor loadings for some of the items, indicating that the item characteristic curves are very steep for those items. This is a computational issue that needs to be addressed. Another obvious next step would be to try a three-factor model or to re-analyse the data, omitting the last two items, which seem to differ in some fundamental way from the other items. A third possibility, to which we shall return in Chapter 10, is to consider a different kind of model; namely, a latent class model.

TABLE 8.12: Estimated difficulty and discrimination parameters with standard errors in brackets and standardised loadings for the two-factor model, sexual attitudes data

Items	$\hat{\alpha}_{i0}$	s.e.	$\hat{\alpha}_{i1}$	s.e.	$\hat{\alpha}_{i2}$	s.e.	st$\hat{\alpha}_{i1}$	st$\hat{\alpha}_{i2}$	$\hat{\pi}_i(\mathbf{0})$
1	-1.99	(0.10)	0.08	(0.11)	0.37	(0.14)	0.06	0.25	0.12
2	1.65	(0.09)	0.52	(0.10)	-0.23	(0.13)	0.34	-0.16	0.84
3	1.56	(0.12)	1.19	(0.14)	0.54	(0.17)	0.62	0.28	0.83
4	-2.13	(0.12)	0.65	(0.12)	0.47	(0.14)	0.40	0.29	0.11
5	-1.49	(0.13)	2.00	(0.17)	0.34	(0.16)	0.81	0.14	0.18
6	-0.64	(0.47)	11.13	(3.47)	-3.08	(1.15)	0.96	-0.26	0.34
7	1.61	(0.73)	12.44	(4.42)	-3.37	(1.50)	0.96	-0.26	0.83
8	0.87	(0.16)	3.35	(0.29)	-0.72	(0.23)	0.90	-0.19	0.71
9	-6.76	(1.88)	5.19	(1.45)	5.16	(1.64)	0.70	0.69	0.00
10	-29.17	(21.16)	18.09	(13.16)	14.48	(10.41)	0.78	0.62	0.00

8.9   Further examples and suggestions for further work

The Law School Admission Test (LSAT), Section VI

The LSAT example is part of an educational test dataset given in Bock and Lieberman (1970). The LSAT is a classical example in educational testing for measuring ability traits. This is a test that was designed to measure a *single* latent ability scale. The test as given in Bock and Lieberman (1970) consisted of five items taken by 1000 individuals. The main interest is whether the attempt to construct items which are indicators solely of this ability has been successful and, if so, what the parameter estimates tell us about the items. From Table 8.13, you will see that 92% of the students answered item 1 correctly, but only 55% answered item 3 correctly. That makes item 3 the most "difficult" among the five items. The full dataset is given in Table 8.15. To investigate whether the five items form a *unidimensional* scale, you need to test whether the one-factor model is a good fit to the five items. The overall goodness-of-fit measure shows that the one-factor model is a very good fit to the data ($X^2 = 18.13$ on 21 degrees of freedom). In other words, the associations among the five items can be explained by a single latent variable that, in this example, is an ability which the test is designed to measure. Since the overall goodness-of-fit test does not reject the one-factor model, there is no need to check the fit on the two- and three-way margins. X^2 measures how well the model predicts the whole response pattern. The first two columns of

Table 8.15 show small discrepancies between the observed frequencies and the expected frequencies under the one-factor model.

Table 8.14 gives the parameter estimates for the one-factor solution. The last column of the table, $\hat{\pi}_i(0)$, gives the probability that the median individual will respond correctly to any of those five items. The five items have different difficulty levels. However, the median individual has quite a high chance of getting the items correct, indicating that, overall, the items are quite easy.

TABLE 8.13: Proportions of positive and negative responses for observed items, LSAT data

Item	Response 1	Response 0
1	0.92	0.08
2	0.71	0.29
3	0.55	0.45
4	0.76	0.24
5	0.87	0.13

TABLE 8.14: Estimated difficulty and discrimination parameters with standard errors in brackets and standardised loadings for the one-factor model, LSAT data

item	$\hat{\alpha}_{i0}$	s.e.	$\hat{\alpha}_{i1}$	s.e.	st$\hat{\alpha}_{i1}$	$\hat{\pi}_i(0)$
1	2.77	(0.20)	0.83	(0.25)	0.64	0.94
2	0.99	(0.09)	0.72	(0.19)	0.59	0.73
3	0.25	(0.08)	0.89	(0.23)	0.67	0.56
4	1.28	(0.10)	0.69	(0.19)	0.57	0.78
5	2.05	(0.13)	0.66	(0.20)	0.55	0.89

The factor loadings $\hat{\alpha}_{i1}$ are all positive and of similar magnitude with similar standard errors. The same is true for the standardised loadings st$\hat{\alpha}_{i1}$. That implies that all five items have similar discriminating power, and so a similar weight is applied to each response. In that case, the component score that is used to scale individuals on the latent dimension should give close results (similar ranking) to the scores obtained when the total score is used (see columns five and six of Table 8.15). This is an example where the Rasch model might be appropriate, or you might analyse the five items using a latent class model that is discussed in Chapter 10. You could compare the ranking of the individuals obtained from the latent trait model with the allocation of individuals into two distinct classes.

TABLE 8.15: Factor scores in increasing order, LSAT data

Observed frequency	Expected frequency	$\hat{E}(f \mid \mathbf{y})$	$\hat{\sigma}(f \mid \mathbf{y})$	Component score (Y_1)	Total score	Response pattern
3	2.3	−1.90	0.80	0.00	0	00000
6	5.9	−1.47	0.80	0.66	1	00001
2	2.6	−1.45	0.80	0.69	1	00010
1	1.8	−1.43	0.80	0.72	1	01000
10	9.5	−1.37	0.80	0.83	1	10000
1	0.7	−1.32	0.80	0.89	1	00100
11	8.9	−1.03	0.81	1.35	2	00011
8	6.4	−1.01	0.81	1.38	2	01001
29	34.6	−0.94	0.81	1.48	2	10001
14	15.6	−0.92	0.81	1.51	2	10010
1	2.6	−0.90	0.81	1.55	2	00101
16	11.3	−0.90	0.81	1.55	2	11000
3	1.2	−0.88	0.81	1.58	2	00110
3	4.7	−0.79	0.81	1.72	2	10100
16	13.6	−0.55	0.82	2.07	3	01011
81	76.6	−0.48	0.82	2.17	3	10011
56	56.1	−0.46	0.82	2.21	3	11001
4	6.0	−0.44	0.82	2.24	3	00111
21	25.7	−0.44	0.82	2.24	3	11010
3	4.4	−0.42	0.82	2.27	3	01101
2	2.0	−0.40	0.82	2.30	3	01110
28	25.0	−0.35	0.82	2.37	3	10101
15	11.5	−0.33	0.82	2.40	3	10110
11	8.4	−0.30	0.82	2.44	3	11100
173	173.3	0.01	0.83	2.89	4	11011
15	13.9	0.05	0.84	2.96	4	01111
80	83.5	0.13	0.84	3.06	4	10111
61	62.5	0.15	0.84	3.10	4	11101
28	29.1	0.17	0.84	3.13	4	11110
298	296.7	0.65	0.86	3.78	5	11111

Workplace industrial relations data

This example is taken from a section of the 1990 Workplace Industrial Relations Survey (WIRS) dealing with management/worker consultation in firms. A subset of the data is used here that consists of 1005 firms and concerns non-manual workers. The full dataset is given on the website. The questions asked are given below:

Please consider the most recent change involving the introduction of new plant, machinery and equipment. Were discussions or consultations of any of the type on this card held either about the introduction of the change or about the way it was to be implemented?

1. Informal discussion with individual workers.
2. Meetings with groups of workers.
3. Discussions in established joint consultative committee.
4. Discussions in specially constituted committee to consider the change.
5. Discussions with union representatives at the establishment.
6. Discussions with paid union officials from outside.

All six items measure the amount of consultation that occurs within firms at various levels of the firm's structure. Items 1 to 6 cover a range of informal to formal types of consultation. Those firms which place a high value on consultation might be expected to use all or most consultation practices. The six items are analysed here using the latent trait model. We should note that the items discussed here were not initially designed to form a scale, as is the case with the LSAT example and most educational data. Therefore, our analysis is completely exploratory. The proportions giving positive and negative responses to each item are given in Table 8.16. The most common type of consultation among the 1005 firms is the established joint consultative committee.

TABLE 8.16: Proportions giving positive and negative responses to observed items, WIRS data

Item	Response 1	Response 0
1	0.37	0.63
2	0.58	0.42
3	0.28	0.72
4	0.24	0.76
5	0.36	0.64
6	0.15	0.85

The one-factor model gives $M_2 = 163.34$ on 9 degrees of freedom, indicating that the one-factor model is a poor fit to the data. Table 8.17 gives chi-squared residuals greater than 3 for the second- and third-order margins. The largest discrepancies are found between items 1 and 2. As a result, the model fails to explain the associations among the six items, judging by the limited information goodness-of-fit measures, and it also fails to explain the pairwise associations.

You should continue the analysis by fitting one more latent variable that might account for the big discrepancies between the observed and expected frequencies. The percentage of G^2 explained increases from 55.22% for the one-factor model to 78.18% for the two-factor model. The second latent variable contributes substantially to explaining the associations among the six items. The fit of the two-factor model is now very good, judging from the limited information test statistic $M_2 = 2.77$ on 4 degrees of freedom. In addition, the residuals for the two-way margins are all close to zero. The second latent variable accounts for the pairwise associations. The fit is still not satisfactory on the three-way margins. Table 8.18 gives the residuals greater than 3 for the $(1,1,1)$ three-way margins. Item 1 appears in all the triplets that show a bad fit. This is the least formal item, which is also vaguely worded and might be interpreted differently by different respondents.

TABLE 8.17: Chi-squared residuals greater than 3 for the second- and (1,1,1) third-order margins for the one-factor model, WIRS data

Response	Items	O	E	$(O-E)^2/E$
(0,0)	2, 1	186	265.75	23.93
(0,1)	2, 1	233	153.23	41.52
(1,0)	2, 1	444	364.25	17.46
	4, 1	172	145.48	4.84
	4, 2	61	87.00	7.77
(1,1)	2, 1	142	221.77	28.69
	4, 1	69	95.65	7.43
	4, 2	180	154.13	4.34
(1,1,1)	1, 2, 3	37	75.79	19.85
	1 ,2, 4	23	61.75	24.32
	1, 2, 5	53	94.85	18.46
	1, 2, 6	26	40.32	5.08
	1, 3, 4	30	45.69	5.39
	1, 4, 5	35	55.73	7.71
	2, 3, 4	93	75.03	4.31
	2, 4, 5	108	91.39	3.02

TABLE 8.18: Chi-squared residuals greater than 3 for the third-order margins for the two-factor model, response $(1,1,1)$ to items (i,j,k), WIRS data

Item i	Item j	Item k	O	E	$(O-E)^2/E$
1	2	3	37	60.53	9.15
1	2	4	23	40.65	7.66
1	2	5	53	73.99	5.95
1	3	6	31	42.54	3.13
1	5	6	36	49.84	3.84

Although the model is not good at predicting the three-way margins, it does predict well the two-way margins, and therefore it is worth looking at the parameter estimates of the two-factor latent trait model given in Table 8.19. All the loadings $(\hat{\alpha}_{i1})$ of the first factor, except that for item 1 (the least formal item, also not statistically significant), are positive and large, indicating a "general" factor relating to the amount of consultation which takes place. Note that the standard errors for most of the estimated loadings of the second factor are large relatively to the estimated values, resulting in non-statistically significant parameters.

TABLE 8.19: Estimated difficulty and discrimination parameters with standard errors in brackets and standardised loadings for the two-factor model, WIRS data

Items	$\hat{\alpha}_{i0}$	s.e.	$\hat{\alpha}_{i1}$	s.e.	$\hat{\alpha}_{i2}$	s.e.	st$\hat{\alpha}_{i1}$	st$\hat{\alpha}_{i2}$	$\hat{\pi}_i(0)$
1	-0.94	(0.30)	-0.16	(2.07)	-2.38	(1.09)	-0.06	-0.86	0.28
2	0.53	(0.14)	1.07	(1.30)	1.42	(0.99)	0.47	0.62	0.63
3	-1.40	(0.14)	1.62	(0.54)	-0.58	(1.43)	0.73	-0.26	0.20
4	-1.47	(0.11)	1.19	(0.31)	0.31	(1.05)	0.63	0.17	0.19
5	-0.97	(0.14)	1.92	(0.59)	-0.61	(1.69)	0.78	-0.25	0.27
6	-2.40	(0.19)	1.36	(0.58)	-0.62	(1.20)	0.66	-0.30	0.08

The analysis is now repeated with item 1 omitted, so the items used in the analysis are items 2 to 6 (and those names are used here). The one-factor model gives $M_2 = 20.09$ on 5 degrees of freedom. The one-factor model is rejected. The fit of the two-way margins is very good except for two pairs, and there is only one large chi-squared residual in the (1,1,1) three-way margins. These residuals are given in Table 8.20 and all include item 2, which is the second least formal item after item 1 (which is omitted from the current analysis). Further analysis of this dataset can be found in Bartholomew (1998). The same data are analysed with a latent class model in Chapter 10 and a graphical model in Chapter 12.

The parameter estimates of the one-factor model given in Table 8.21 indicate a clear general factor corresponding to the amount of consultation that takes place. Note that item 2 has the smallest factor loading, while items 3 to 6 have similar factor loadings. It is quite apparent that items 3 to 6 can be considered separately to construct a scale measuring the amount of formal consultation which takes place.

TABLE 8.20: Chi-squared residuals greater than 3 for the second- and the (1,1,1) third-order margins for the one-factor model, WIRS data, item 1 omitted

Response	Items	O	E	$(O-E)^2/E$
(1,0)	3, 1	61	84.94	6.75
(1,1)	3, 1	180	156.28	3.60
(1,1,1)	(1, 2, 3)	93	77.09	3.28

TABLE 8.21: Estimated difficulty and discrimination parameters with standard errors in brackets and standardised loadings for the one-factor model with item 1 omitted, WIRS data

item	$\hat{\alpha}_{i0}$	s.e.	$\hat{\alpha}_{i1}$	s.e.	st$\hat{\alpha}_{i1}$	$\hat{\pi}_i(0)$
2	0.35	(0.07)	0.42	(0.10)	0.39	0.59
3	−1.38	(0.14)	1.69	(0.23)	0.86	0.20
4	−1.40	(0.10)	1.05	(0.14)	0.72	0.20
5	−0.95	(0.13)	1.97	(0.31)	0.89	0.28
6	−2.29	(0.16)	1.34	(0.18)	0.80	0.09

Women's mobility

These data are from the Bangladesh Fertility Survey of 1989 (Huq and Cleland 1990). The rural subsample of 8445 women is analysed here. The questionnaire contains a number of items believed to measure different dimensions of women's status. The particular dimension that we shall focus on here is women's mobility or social freedom. Women were asked whether they could engage in the following activities alone (1=yes, 0=no).

1. Go to any part of the village/town/city.
2. Go outside the village/town/city.
3. Talk to a man you do not know.
4. Go to a cinema/cultural show.
5. Go shopping.
6. Go to a cooperative/mothers' club/other club.
7. Attend a political meeting.
8. Go to a health centre/hospital.

First, the one-factor model was fitted to the eight items to investigate whether the variables are all indicators of the same type of women's mobility in society. The one-factor model gives a M_2 equal to 377.92 on 20 degrees of freedom, indicating a bad fit. Table 8.22 shows the chi-squared residuals greater than 3 for the two-way margins and the (1,1,1) three-way margins of the one-factor model.

The two-factor model is still rejected based on a M_2 equal to 117.83 on 13 degrees of freedom. The percentage of G^2 explained increases only slightly from 96.52% to 97.95%. However, although the contribution of the second factor is small, the fit on the two-way margins and the (1,1,1) three-way margins has improved considerably; the margins for which the fit is poor are shown in Table 8.23.

The parameter estimates for the two-factor model are given in Table 8.24. As we can see from the standardised loadings st$\hat{\alpha}_{i1}$ and st$\hat{\alpha}_{i2}$, the eight items

TABLE 8.22: Chi-squared residuals greater than 3 for the second- and the (1,1,1) third-order margins for the one-factor model, women's mobility data

Response	Items	O	E	$(O-E)^2/E$
(0,1)	3, 2	187	229.22	7.78
	6, 2	1986	1900.27	3.87
	7, 6	532	595.83	6.84
	8, 5	194	245.12	10.66
	8, 7	108	134.47	5.21
(1,0)	2, 1	52	117.29	36.35
	5, 1	13	3.02	32.97
	5, 2	98	77.69	5.31
	5, 3	20	12.40	4.67
	5, 4	19	28.74	3.30
	6, 2	274	196.20	30.85
	6, 3	44	32.02	4.48
	7, 1	6	1.13	21.02
	7, 2	62	36.78	17.29
	7, 4	17	8.74	7.80
	7, 6	41	93.66	29.60
	8, 1	28	7.14	60.90
	8, 3	38	22.73	10.25
	8, 4	88	67.80	6.02
	8, 5	340	391.68	6.82
(1,1)	6, 2	665	755.83	10.92
	7, 6	407	356.20	7.24
	8, 5	392	348.06	5.55
(1,1,1)	1, 2, 3	2433	2338.68	3.80
	1, 2, 6	659	750.71	11.20
	1, 5, 8	392	347.22	5.78
	1, 6, 7	403	355.51	6.34
	2, 3, 6	653	736.35	9.44
	2, 4, 6	637	703.82	6.34
	3, 5, 8	389	343.49	6.03
	3, 6, 7	402	352.07	7.08
	4, 5, 8	386	341.51	5.79
	4, 6, 7	396	351.39	5.66
	5, 6, 7	304	271.29	3.94
	5, 6, 8	326	279.56	7.79
	5, 7, 8	276	246.51	3.55
	6, 7, 8	318	266.88	9.79

are positively correlated with the first factor, which can be considered as a general factor measuring women's status. The $\hat{\pi}_i(\mathbf{0})$ values show clearly that a woman who is in the middle of both factors has close to zero chances of responding positively to items 5 to 8.

To facilitate the interpretation, we performed an oblique rotation (oblimin) using the R package mirt (see also Section 7.6). The solution is given in Table 8.25. As we can see from the oblique rotation, items 1 to 3 load heavily on the second factor, while items 5 to 8 load heavily on the first factor. Item 4 contributes similarly to both factors. The oblique rotation gives a substantial correlation of 0.72 between the two factors. The two factors can be interpreted as measuring different dimensions of women's status. Items 5 to 8, and to some extent item 4, indicate a relatively high level of participation in public life; engaging in any of these activities would suggest a high degree of social freedom for a woman in rural Bangladesh. In contrast, items 1 to 3 are less specific, indicating some degree of freedom, but not necessarily in the public life sphere. You should compare the results obtained here with those obtained in Chapter 10 where a latent class model is fitted.

TABLE 8.23: Chi-squared residuals greater than 3 for the second- and the (1,1,1) third-order margins for the two-factor model, women's mobility data

Response	Items	O	E	$(O-E)^2/E$
(0,1)	8, 5	194	237.63	8.01
(1,0)	5, 1	13	6.75	5.78
	5, 4	19	32.86	5.84
	6, 1	15	30.50	7.88
	7, 2	62	78.52	3.48
	7, 6	41	55.57	3.82
	8, 1	28	14.13	13.61
	8, 4	88	71.25	3.94
	8, 5	340	386.41	5.57
(1,1)	8, 5	392	348.38	5.46
(1,1,1)	1, 5, 8	392	346.13	6.08
	2, 5, 8	351	309.13	5.67
	3, 5, 8	389	341.22	6.69
	4, 5, 8	386	340.14	6.18
	5, 6, 8	326	290.26	4.40

TABLE 8.24: Estimated difficulty and discrimination parameters with standard errors in brackets and standardised factor loadings for the two-factor model, women's mobility data

Items	$\hat{\alpha}_{i0}$	s.e.	$\hat{\alpha}_{i1}$	s.e.	$\hat{\alpha}_{i2}$	s.e.	st$\hat{\alpha}_{i1}$	st$\hat{\alpha}_{i2}$	$\hat{\pi}_i(\mathbf{0})$
1	2.64	(0.16)	2.56	(0.19)	0.55	(0.26)	0.86	0.19	0.93
2	−1.59	(0.09)	2.78	(0.18)	0.29	(0.23)	0.89	0.09	0.17
3	1.55	(0.04)	1.51	(0.06)	−0.01	(0.14)	0.73	0.00	0.82
4	−1.19	(0.06)	2.90	(0.13)	−0.76	(0.24)	0.87	−0.23	0.23
5	−6.57	(0.29)	3.64	(0.23)	−1.79	(0.31)	0.85	−0.42	0.00
6	−5.31	(0.30)	3.19	(0.25)	−2.39	(0.33)	0.75	−0.57	0.00
7	−14.83	(3.07)	7.14	(1.53)	−5.37	(1.28)	0.79	−0.59	0.00
8	−4.97	(0.18)	8.23	(0.16)	−1.47	(0.24)	0.81	−0.42	0.01

TABLE 8.25: Rotated unstandardised factor loadings for the two-factor model, oblimin rotation, women's mobility data

Items	rot($\hat{\alpha}_{i1}$)	rot($\hat{\alpha}_{i2}$)
1	−0.05	0.87
2	0.07	0.81
3	0.17	0.53
4	0.52	0.42
5	0.79	0.17
6	0.97	−0.07
7	1.03	−0.07
8	0.78	0.14

8.10 Software

The latent trait model for binary data can be estimated using commercial software such as Stata and R packages, including ltm (Rizopoulos 2006) and mirt (Chalmers 2012). All produce asymptotic standard errors and goodness-of-fit test statistics, but only ltm gives the lower-order margin residuals. The UV approach is implemented in commercial software such as Amos (Arbuckle 2006) built in SPSS, LISREL (Jöreskog and Sörbom 1999), Mplus (Muthén and Muthén 2017), and EQS (Bentler 2008) and the R package lavaan (Rosseel 2012).

8.11 Further reading

Bartholomew, D. J., Knott, M. and Moustaki, I. (2011). *Latent Variable Models and Factor Analysis: A Unified Approach* (3rd ed.). West Sussex, United Kingdom: John Wiley & Sons.

Bock, R. D., and Moustaki. I. (2007). Item response theory. In C. R. Rao and S. Sinharay (Ed.), *Handbook of Statistics* (Vol. 26). Elsevier.

Bollen, K. A. (1989). *Structural Equations with Latent Variables.* New York: Wiley and Sons.

Fischer, G. H. and Molenaar, I. W. (Eds.) (1995). *Rasch Models: Foundations, Recent Developments, and Applications.* New York: Springer-Verlag.

Hambleton, R. K., Swaminathan, H. and Rogers, H. J. (1991). *Fundamentals of Item Response Theory.* Newbury Park, California: Sage Publications.

Heinen, T. (1996). *Latent Class and Discrete Latent Trait Models.* Thousand Oaks: Sage Publications.

Skrondal, A. and Rabe-Hesketh, S.(2004). *Generalized Latent Variable Modelling: Multilevel, Longitudinal, and Structural Equation Models.* Boca Raton, FL: Chapman and Hall/CRC.

van der Linden, W. J. and Hambleton, R. (1997). *Handbook of Modern Item Response Theory.* New York: Springer-Verlag.

9

Factor Analysis for Ordered Categorical Variables

9.1 The practical background

The subject of this chapter lies, in a sense, between that of Chapter 7 on factor analysis and Chapter 8 on factor analysis for binary data. It differs from both of these earlier kinds of factor analysis principally in that the manifest variables are ordered categorical variables; that is, the response will be in one of a number of ordered categories. For example, if we ask someone whether they enjoyed a meal, "very much", "a little", or "not very much", we would be observing an ordered categorical variable. The answer can fall into only one category, and those categories are ordered according to strength of approval. Any ordered categorical variable can be reduced, of course, to a binary variable by amalgamating categories. For example, if we were to amalgamate "a little" and "very much", we would have a binary variable. We did this in the analysis of the sexual attitudes data in Section 8.8, where three of the items had four categories. These were amalgamated into two pairs so that all variables could be treated as binary. In doing this, we are losing information, and that motivates the present chapter. Categorical variables with more than two categories are often referred to as *polytomous*, and where they are ordered they may be called *ordinal*.

In view of these considerations, it might have seemed logical to place this chapter immediately after the factor analysis chapter and then to treat the binary case in Chapter 8 as an important special case. However, work on ordered categorical variables is nearer to the research frontier and is consequently more incomplete and, in some sense, more difficult than the other methods. It is, therefore, better to approach it with the experience gained from the preceding two chapters on latent variable models. For the same reasons, the structure of this chapter will be a little different from its predecessors, though we shall retain the emphasis on practical applications.

Despite all this, we think it is important for social science researchers to be aware of what is available because ordered categorical data are so common in social research. In many social applications, the data collected is coded into a number of ordered categories. Examples of ordered variables are often

DOI: 10.1201/9781003483342-9

attitudinal statements with response alternatives such as "strongly disagree", "disagree", "agree" and "strongly agree" or "very satisfied", "satisfied", "dissatisfied" and "very dissatisfied". These scales are sometimes known as Likert-type scales. In such scales, it is also common to have an alternative which does not fit into the ordering. This is the case with responses such as "neither agree nor disagree", or "don't know". The treatment of scales including such items, is discussed briefly in Section 9.6.

It will be helpful to have an example in mind as we introduce the two main models on which the methods are based.

Example: Attitude to science and technology

The data used in this example come from the Consumer Protection and Perceptions of Science and Technology section of the 1992 Eurobarometer Survey (Karlheinz and Melich 1992) based on a sample from Great Britain. The questions chosen are given below.

1. Science and technology are making our lives healthier, easier and more comfortable [Comfort]

2. Scientific and technological research cannot play an important role in protecting the environment and repairing it [Environment]

3. The application of science and new technology will make work more interesting [Work]

4. Thanks to science and technology, there will be more opportunities for the future generations [Future]

5. New technology does not depend on basic scientific research [Technology]

6. Scientific and technological research do not play an important role in industrial development [Industry]

7. The benefits of science are greater than any harmful effects it may have [Benefit]

All of the above items were measured on a four-point scale, with response categories "strongly disagree", "disagree to some extent", "agree to some extent" and "strongly agree".

Missing values have been excluded from the analysis by listwise deletion, giving a sample of 392 respondents. Listwise deletion implies that response patterns that have missing values for any of the items are omitted from the analysis. Omitting respondents with missing values can bias the results as discussed in Section 7.11.

To start our analysis, we chose the items that were positively worded, namely Comfort, Work, Future, and Benefit. Those four items can be considered as indicators for measuring attitude towards science and technology.

9.2 Two approaches to modelling ordered categorical data

We have remarked that ordered categorical problems can be reduced to factor analysis of binary variables and the analysis of the sexual attitudes in Section 8.8 showed this being done. Dichotomisation of ordinal variables is also applied to the science and technology data in Chapter 10. In the past, however, the approach has often been from the other end. Thus, in applications where the number of ordered categories is large (more than six or seven, say), ordinal categories have often been treated as if they were interval level variables. Having made that assumption, one can go on to compute correlations between these pseudo-continuous variables and carry out a standard factor analysis. Provided that the number of categories is large for all variables, this may not seriously affect the results of the analysis. Even when the number of categories is as low as three or four, it may be acceptable to use this method. We did this for the anxiety data in Chapter 7, though we emphasised that our concern there was to illustrate other aspects of factor analysis. In general, the uncritical factor analysis of categorical data in this way is likely to give biased estimates of the factor loadings and is not recommended. We shall provide empirical evidence of this point in Section 9.5.

One might still wonder whether factor analysis could be used on other types of correlation coefficients specifically designed for ordered categorical data. For example, Kruskal's *gamma*, Somer's *d*, or grouped forms of rank correlation coefficients such as Kendall's *tau* all measure the strength of the relationship between ordered categorical variables. Factor analyses are sometimes carried out on such coefficients and, if they are viewed from a purely descriptive point of view, they may yield useful insights. However, whatever the merits of such *ad hoc* methods, they have been superseded by better, model-based methods.

As already explained in Chapter 8, there are two main approaches for analysing binary data with latent variable models. Each of these can be generalised to the case of variables with more than two ordered categories. They are: the *item response function* (IRF) approach and the *underlying variable* (UV) approach. In this chapter, we will describe these two main methodologies used for multivariate analysis of ordinal items.

The IRF models use a straightforward extension of the logit or probit (normit) models for binary responses discussed in Chapter 8. The logit model for ordinal responses is implemented in commercial software and the R packages ltm and mirt, and it will be used in all the examples in this chapter.

The UV approach is based on the fit of the standard linear factor model using the matrix of polychoric correlation coefficients (which we explain below). The UV and IRF models will be compared through an example. More on the comparison between the two approaches can be found in two research papers by Moustaki (2000) and Jöreskog and Moustaki (2001).

9.3   Item response function approach

In factor analysis for binary items, we were interested in modelling the probability of a randomly selected individual giving a positive response to an item as a function of the latent variables. This was done in terms of a set of probabilities $\{\pi_i(\mathbf{f})\}$. The probabilities of giving the negative response did not appear explicitly because they were simply the complements of the $\pi_i(\mathbf{f})$s. In the ordinal case, where there are more than two categories, we need to specify probabilities for each category. The observed ordinal variables are denoted by $y_1, \ldots, y_p$. Let us suppose that there are m_i categories for variable i labelled $(1, \ldots, m_i)$. For binary items, $m_i = 2$ for each i and the category labels were 0 and 1, but could equally well have been 1 and 2. We now need to define a response probability for each category. Let $\pi_{i(s)}(\mathbf{f})$ be the probability that, given $\mathbf{f}$, a response falls in category s for variable i.

The position with two categories can now be compared with the general case as follows:

Categories	0	1
Response probability	$1 - \pi_i(\mathbf{f})$	$\pi_i(\mathbf{f})$

Categories	1	2	$\ldots$	s	$\ldots$	m_i
Response Probability	$\pi_{i(1)}(\mathbf{f})$	$\pi_{i(2)}(\mathbf{f})$	$\ldots$	$\pi_{i(s)}(\mathbf{f})$	$\ldots$	$\pi_{i(m_i)}(\mathbf{f})$

In both cases, the response probabilities sum to one. In the binary case, we derived the logit model, which expressed the logit of the probability of a response in category one as a linear function of the fs (see eq. 8.3). The question now is how to generalise the argument used there to more than two categories. Suppose we were to divide the categories into two groups with categories $(1, 2, \ldots, s)$ in one group and $(s+1, s+2, \ldots, m_i)$ in the other and were merely to report into which of the two groups the response fell. We would thereby have reduced the polytomous variable to a binary variable. It therefore seems reasonable to require that any model we choose for the polytomous case should be consistent with the one which we have already used for the binary case. We can do this by supposing that *wherever we make the split* the binary logit model will apply. To do this, we need the probabilities of a response falling into the first and second groups, respectively. These may be written:

$$\gamma_{i(s)}(\mathbf{f}) = Pr(y_i \leq s) = \pi_{i(1)}(\mathbf{f}) + \pi_{i(2)}(\mathbf{f}) + \cdots + \pi_{i(s)}(\mathbf{f}),$$

and

$$1 - \gamma_{i(s)}(\mathbf{f}) = Pr(y_i > s) = \pi_{i(s+1)}(\mathbf{f}) + \pi_{i(s+2)}(\mathbf{f}) + \cdots + \pi_{i(m_i)}(\mathbf{f}),$$

where y_i denotes the category into which the ith variable falls. The probabilities $\gamma_{i(s)}(\mathbf{f})$ are referred to as *cumulative response probabilities*. Note that it is only meaningful to describe them in this way because the categories are ordered.

In essence, we now define the model by supposing that the binary logit model holds for all possible divisions of the m_i categories into two groups. We can do this in two equivalent ways according to which group we regard as the "positive" response. That is, we can write the model in terms of $\text{logit}\gamma_{i(s)}(\mathbf{f})$ or of $\text{logit}(1-\gamma_{i(s)}(\mathbf{f}))$. Although it would be natural to take the $(1-\gamma)$ version because it links directly with the model for the binary case, we use here the $\gamma_{i(s)}(\mathbf{f})$ version since that is used more often in the literature. The model is thus written

$$\log\left[\frac{\gamma_{i(s)}(\mathbf{f})}{1 - \gamma_{i(s)}(\mathbf{f})}\right] = \alpha_{i(s)} - \sum_{j=1}^{q} \alpha_{ij} f_j, \tag{9.1}$$

where $s = 1, \ldots, m_i - 1$; $i = 1, \ldots, p$. For a positive factor loading α_{ij} the higher the value of an individual on the latent variable f_j, the higher the probability of that individual responding in the higher categories of item i. Instead of using the logit link function, we could use the probit function. The model that uses the logit as a link is also called the *proportional odds model*. The name proportional odds model comes from the fact that, in the one-factor case, the difference between two cumulative logits, that is, the left side of eq. (9.1), for two persons with factor scores f_1 and f_2 is proportional to $f_1 - f_2$. Note that there is one intercept parameter $\alpha_{i(s)}$ for each category. The ordering of the categories implies that the intercept parameters are also ordered, that is,

$$\alpha_{i(1)} \le \alpha_{i(2)} \le \cdots \le \alpha_{i(m_i)}.$$

However, the factor loadings α_{ij} remain the same across categories of the same variable; in other words, the discriminating power of the item does not depend on where the split into two groups is made. The πs are obtained from the γs by

$$\pi_{i(s)}(\mathbf{f}) = \gamma_{i(s)}(\mathbf{f}) - \gamma_{i,(s-1)}(\mathbf{f}) \qquad (s = 2, \ldots, m_i), \tag{9.2}$$

where $\gamma_{i(1)}(\mathbf{f}) = \pi_{i(1)}(\mathbf{f})$ and $\gamma_{i(m_i)}(\mathbf{f}) = 1$. We refer to $\gamma_{i(s)}(\mathbf{f})$ as the *cumulative response function* and to $\pi_{i(s)}(\mathbf{f})$ as the *category response function*.

We should mention that the regression model used for an observable ordinal response on a set of observable (rather than latent) explanatory variables is known as the *cumulative logit* model for ordinal variables.

Figures 9.1 and 9.2 give the cumulative and category response functions, respectively, for parameter values $\alpha_{i(1)} = -0.5, \alpha_{i(2)} = 0.5, \alpha_{i(3)} = 1.5, \alpha_{i(4)} = 3.5$ and $\alpha_{i1} = 1.0$.

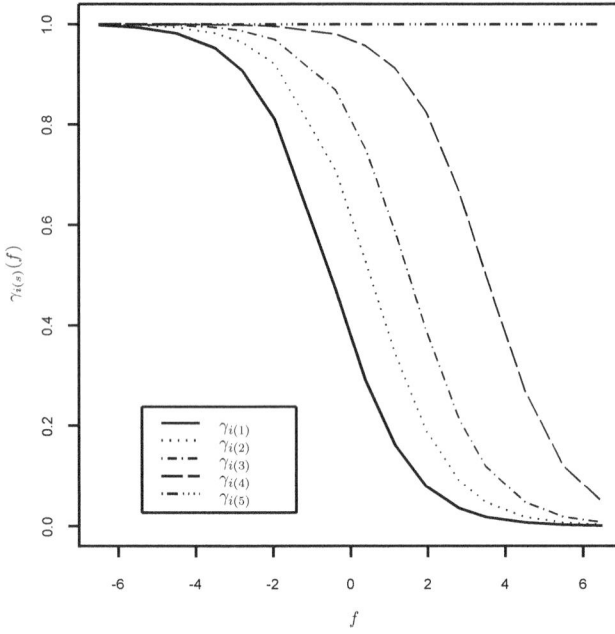

FIGURE 9.1: Cumulative response probabilities

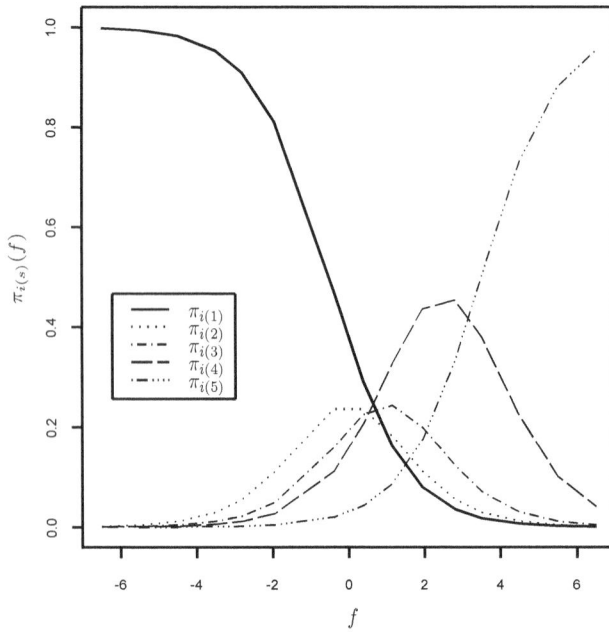

FIGURE 9.2: Category response probabilities

To summarise, the assumptions made under the IRF approach, which are common to other factor models are:

i) the latent variables are independent and normally distributed with mean zero and variance one,

ii) the responses to the ordinal items are independent conditional on the latent variables (conditional independence).

Fitting the model and goodness-of-fit

The model can be fitted in the same manner as the binary latent trait model using the method of maximum likelihood. For example, the R packages ltm and mirt give estimates of the parameters together with their asymptotic standard errors. As in the binary case, the input consists of the list of response patterns.

Goodness-of-fit can, likewise, be judged using the same criteria based on the likelihood ratio and the Pearson chi-squared statistics calculated from the whole response patterns. However, the problems of sparsity are liable to be much more serious in the polytomous case. We can easily see how this comes about by enumerating the number of response patterns. If there are m_i categories for variable i, the total number of response patterns is $(m_1 \times m_2 \times \cdots \times m_p)$. In the case of the science and technology data referred to at the beginning of the chapter, there were four categories for each variable giving $4^4 = 256$ response patterns altogether. This compares with 16 for binary data with four variables. The average expected frequency is therefore likely to be small even when the sample size is large. One way to combat this is to combine response patterns, but there is another alternative which was not available for binary data. This is to amalgamate categories for some of the variables. It often happens that some of the response categories are rarely used, and little is lost if we combine such a category with an adjacent category. This does not destroy the ordering but considerably reduces the number of response categories. We shall do precisely this when we come to the example on attitudes to the environment in the next section.

The number of degrees of freedom for G^2 or X^2 will be equal to the number of response patterns, after any grouping, less the number of independent parameters less one. The number of independent parameters is $(\sum_{i=1}^{p} m_i - p) - pq - q(q-1)/2$. The term $q \times (q-1)/2$ refers to the number of constraints to be imposed on the model parameters (e.g. factor loadings) due to rotational indeterminacy. Thus if there is no grouping, we have:

$$\text{degrees of freedom} = (m_1 \times m_2 \times \cdots \times m_p) - \left(\sum_{i=1}^{p} m_i - p\right) - pq - q(q-1)/2 - 1.$$

If there has been a grouping of response patterns, the relevant m_i will be reduced accordingly. However, grouping is not an optimal solution and grouping itself is arbitrary. The goodness-of-fit can also be assessed by looking at the two-way (or higher) margins. The pairwise distribution of any two variables can be displayed as a two-way contingency table, and chi-squared residuals can be constructed in the usual way by comparing the observed and expected frequencies. We shall illustrate this in the following examples. As noted already in Chapter 8, the residuals computed for each cell are not independent and therefore they cannot be simply summed to give an overall test distributed as chi-squared. Valid tests that combine the individual lower margin residuals are the limited information goodness-of-fit test statistics such as the M_2 discussed in Chapter 8 and can be used also for models with ordinal data (see, for example, Maydeu-Olivares and Joe 2006).

If we fit the one-factor proportional odds model to the science and technology data (the full dataset is given on the website), we obtain the estimates given in Table 9.1. The derivation of the standardised loading, $st\hat{\alpha}_{i1}$ is given in eq. (9.7); it is similar to a correlation coefficient. The high values in Table 9.1 suggest that the single factor model provides a good explanation for all four binary variables, especially for the item relating to the future. Before putting too much weight on this conclusion, we need to look at how well the model fits.

TABLE 9.1: Estimated factor loadings with standard errors in brackets and standardised loadings for the one-factor model for ordinal data, science and technology data

	$\hat{\alpha}_{i1}$	s.e.	$st\hat{\alpha}_{i1}$
Comfort	1.04	(0.19)	0.72
Work	1.23	(0.18)	0.77
Future	2.30	(0.49)	0.92
Benefit	1.09	(0.18)	0.74

Given the sparsity of the data (there is a total frequency of 392 spread over 256 response categories), it is not feasible to carry out global tests. Instead, we look at the fits to the margins. For each pair of items, Table 9.2 gives the sum of the chi-squared residuals over each pair of categories. Sixteen chi-squared residuals are calculated for each pair of items, since each variable has four response categories. Table 9.3 shows how the entry 25.48 of Table 9.2 is computed. The sum of the entries of Table 9.3 is 25.48. Similarly, we compute the sums of chi-squared residuals for the other five two-way tables.

TABLE 9.2: Sums of chi-squared residuals for pairs of items from the two-way margins for the one-factor model for ordinal data, science and technology data

	Work	Future	Benefit
Comfort	25.47	12.01	27.36
Work		9.20	23.21
Future			17.48

TABLE 9.3: Chi-squared residuals for the two-way margins of items Comfort and Work, science and technology data

Categories	1	2	3	4
1	0.87	1.79	0.44	11.45
2	2.42	0.01	0.14	2.08
3	0.02	0.04	0.45	2.40
4	0.65	0.25	1.51	0.95

As discussed in Section 8.4, if the model were correct, the chi-squared residual in a single cell would have approximately a χ^2 distribution with one degree of freedom. A value greater than about 4 would indicate a poor fit. The sum of these residuals over all the cells in a two-way marginal table is analogous to Pearson's chi-squared statistic for goodness-of-fit, but because the model has been fitted to the full multi-way table (rather than to the two-way marginal table), the standard chi-squared test does not apply. We may still use this sum, say S, as a diagnostic guide. A large value of S would suggest that the model does not well explain the associations in that two-way table. But how large is large?

As a rule of thumb, S is too large if it is greater than the upper 1% point of a χ^2 distribution with $[(m_i \times m_j) - 1]$ degrees of freedom. The rationale behind this is that, had the model correctly specified the true cell probabilities for the two-way table, then S would have had this distribution, approximately. Because the cell probabilities are estimated (by fitting a latent variable model to the multi-way table) a value of S larger than this should be even more indicative of a bad fit.

Further work is needed to develop diagnostic procedures for deciding whether or not a given latent variable model well explains associations between pairs of items. However, in practice, a model may still be useful even though it does not fully explain all interrelationships between items.

Returning to the science and technology data, all six entries, values of S, in Table 9.2 are less than 30.58 (the upper 1% point of χ^2 with 15 degrees of freedom). The fit to each two-way marginal table appears satisfactory. However, in Table 9.3 one of the chi-squared residuals is quite large (11.45) even though their sum is only 25.48. But in all six two-way tables, there are only five chi-squared residuals greater than 3. These are listed in Table 9.4. Overall, the one-factor model appears to give an adequate, but not perfect, description of the data.

TABLE 9.4: Chi-squared residuals greater than 3 for a pair of items and categories for the one-factor model, science and technology data

Items	Categories	$(O - E)^2/E$
(1,2)	(1,4)	11.45
(1,3)	(2,4)	6.82
(1,4)	(1,1)	16.53
(2,4)	(1,4)	13.18
(3,4)	(1,2)	4.85

Factor scores

It is possible to calculate factor scores for the proportional odds model, but much of the simplicity of the binary case is lost. In that case, we pointed out that we could either compute the conditional expectations of the latent variable, given the manifest variables, or we could use what we called components, which were linear combinations of the manifest variables formed using the factor loadings as weights. Either method ranked objects in the same order, and in practice, the results from the two methods were very similar. In the general case, this simple correspondence breaks down. There are no components which convey all the information about the latent variables, and we therefore have to rely on the conditional expectations which are more difficult to calculate.

9.4 Examples

Attitudes to the environment

As the first of our main examples, we take another dataset on the environment, this time extracted from the Environment section of the 1990 British Social Attitudes Survey (Brook, Taylor, and Prior 1991). A sample of 291 individuals were asked whether they were "very concerned", "slightly concerned", "not very concerned", or "not at all concerned" with the following environmental issues.

1. Lead from petrol [LeadPetrol]
2. River and sea pollution [RiverSea]
3. Transport and storage of radioactive waste [RadioWaste]
4. Air pollution [AirPollution]
5. Transport and disposal of poisonous chemicals [Chemicals]
6. Risks from nuclear power station [Nuclear]

Since the proportion of individuals falling into the "not very concerned" category was less than 10%, categories "not very concerned" and "not at all concerned" were amalgamated. We treat the above six items as indicators of individuals' attitudes towards environmental issues. The one-factor model with the logit link function was fitted to the six items each with three ordered response categories. The parameters of interest are the factor loadings α_{i1} and their standardised form stα_{i1} and estimates of these are given in Table 9.5.

TABLE 9.5: Estimated factor loadings with standard errors in brackets and standardised loadings for the one-factor model, environment data

Items	$\hat{\alpha}_{i1}$	s.e.	st$\hat{\alpha}_{i1}$
LeadPetrol	1.38	(0.21)	0.81
RiverSea	2.34	(0.39)	0.92
RadioWaste	3.12	(0.52)	0.95
AirPollution	3.28	(0.58)	0.96
Chemicals	2.95	(0.48)	0.95
Nuclear	1.76	(0.25)	0.87

The standardised parameters are all positive and close to one, indicating that the six items are all strong indicators of attitude towards the environment.

The goodness-of-fit of the model is again investigated through the chi-squared residuals computed for the two-way margins. The chi-squared residuals for the two-way margins are smaller than 3 for most of the cells. The sums of these residuals for the 15 two-way marginal tables are given in Table 9.6. These are smaller than 20.09 (the 1% point of χ_8^2). The limited information test statistic, $M_2 = 12.84$, on 3 degrees of freedom, rejects the one-factor model. However, since the bivariate residuals are all small, we can use this factor with some confidence as a summary measure of attitude towards environmental issues.

TABLE 9.6: Sums of chi-squared residuals for pairs of items from the two-way margins for the one-factor model, environment data

Items	2	3	4	5	6
1	13.15	3.89	6.66	2.92	2.31
2		8.97	9.77	4.92	5.00
3			3.81	9.32	11.04
4				4.65	4.83
5					3.58

Science and technology data

This is the same set of data that we used to introduce the chapter, but now we take all seven items from the Science and Technology section of the 1992 Eurobarometer Survey. This dataset will also be analysed in Chapter 10 in binary form using a three latent class model.

The first step is to fit a one-factor model to the seven items. The chi-squared residuals for the two-way margins are smaller than 3 for most of the cells, but there are a few cells where the model shows a very bad fit. We report in Table 9.7 the sums of these chi-squared residuals for all pairs of items.

TABLE 9.7: Sums of chi-squared residuals for pairs of items from the two-way margins for the one-factor model, science and technology data (seven items)

Items	Items					
	2	3	4	5	6	7
1	13.36	26.11	12.45	18.96	9.27	24.75
2		27.48	24.05	98.39	90.86	23.49
3			10.16	21.02	35.73	23.05
4				23.20	26.94	17.07
5					103.33	17.07
6						20.28

Values as big as 103.33, 98.39, and 90.86 indicate a very poor fit indeed. It is clear that we need to look at the two-factor model without inspecting the parameter estimates for the one-factor model. Adding a second factor improves the fit of the two-way margins considerably. As we can see from Table 9.8, the sums of the chi-squared residuals for the two-way marginal tables are all smaller than 30.58 (the 1% point of χ^2_{15}), apart from those involving item 6 (Industry). Thus, the two-factor model appears to be adequate at least for the other five items.

TABLE 9.8: Sums of chi-squared residuals for pairs of items from the two-way margins for the two-factor model, science and technology data

Items	Items					
	2	3	4	5	6	7
1	12.22	24.93	11.98	18.52	7.23	24.78
2		21.35	23.04	20.73	31.71	22.62
3			9.32	15.11	30.41	23.75
4				21.65	26.51	17.33
5					34.57	16.33
6						20.66

The parameter estimates with the standard errors in brackets for the two-factor model are given in Table 9.9. Note that parameter α_{72} is set to zero by the R package mirt for identification purposes.

The two-factor solution obtained is not unique since orthogonal transformation of the factors leaves the value of the likelihood unchanged. We performed both orthogonal (varimax) and oblique (oblimin) rotation to the two-factor solution but both rotations produced the same solution due to the factor correlation being close to zero. The varimax solution is given in Table 9.10

The factor loadings of items Comfort, Work, Future and Benefit load heavily on the first factor and items Environment, Technology, and Industry on the second factor. It looks as if, on the one hand, people vary in the extent to which they believe that technology can give answers to society's problems, and on

TABLE 9.9: Estimated intercepts and factor loadings with standard errors in brackets for the two-factor model, science and technology data

Items	Cat	$\hat{\alpha}_{i(s)}$	s.e.	$\hat{\alpha}_{i1}$	s.e.	$\hat{\alpha}_{i2}$	s.e.
1. Comfort	1	−1.53	(0.17)	−1.12	(0.20)	−3.60	(0.19)
	2	2.73	(0.24)				
	3	4.50	(0.50)				
2. Environment	1	−1.01	(0.17)	0.04	(0.24)	−1.64	(0.25)
	2	1.25	(0.18)				
	3	3.45	(0.32)				
3. Work	1	−2.29	(0.21)	−1.23	(0.19)	0.30	(0.20)
	2	0.91	(0.15)				
	3	2.95	(0.24)				
4. Future	1	−1.90	(0.29)	−2.18	(0.42)	0.13	(0.29)
	2	2.16	(0.32)				
	3	5.08	(0.64)				
5. Technology	1	−1.09	(0.18)	0.05	(0.25)	−1.73	(0.26)
	2	1.48	(0.20)				
	3	4.17	(0.39)				
6. Industry	1	−0.46	(0.15)	−0.43	(0.24)	−1.58	(0.25)
	2	2.51	(0.25)				
	3	4.69	(0.44)				
7. Benefit	1	-1.70	(0.17)	-1.13	(0.18)	0.00	(—)
	2	1.00	(0.14)				
	3	3.38	(0.28)				

TABLE 9.10: Rotated unstandardised factor loadings for the two-factor model, varimax rotation, science and technology data

Items	rot$(\hat{\alpha}_{i1})$	rot$(\hat{\alpha}_{i2})$
1. Comfort	0.53	0.19
2. Environment	-0.06	0.69
3. Work	0.59	-0.12
4. Future	0.79	-0.02
5. Technology	-0.07	0.71
6. Industry	0.14	0.68
7. Benefit	0.55	0.02

the other hand they vary according to how important they judge science and technology to be. An alternative explanation of why the items Environment, Technology, and Industry form a scale by themselves is that they are the ones that have been expressed negatively. This second explanation suggests that the result could be an artefact of the questionnaire rather than of people's attitudes. Possible problems with question wording can often be discovered in the course of a latent variable analysis. The analysis carried out here agrees with the results found later when the latent class model with three classes is fitted to a dichotomised version of the seven ordinal items. We shall return to the interpretation of this set of data when we meet it again in Chapter 10.

9.5   The underlying variable approach

The underlying variable (UV) approach follows the same lines as in the case of binary data. The essential assumption we made there was that we had a factor analysis problem in which we merely observed whether or not each variable exceeded a threshold. The problem then was how to carry out a linear factor analysis using the binary data.

When responses fall into m_i categories rather than two, we have rather more information about the incompletely observed variables. We now assume that the category in which a response occurs is determined by where the manifest variable falls in relation to a series of thresholds. Let us denote, as before, the observed ordinal variables by $y_1, y_2, \ldots, y_p$ and the incompletely observed variables or underlying variables by $y_1^*, y_2^*, \ldots, y_p^*$. For variable y_i with m_i categories, there are $m_i - 1$ threshold parameters. Let the thresholds for variable i be denoted by,

$$\tau_{i(1)}, \tau_{i(2)}, \ldots, \tau_{i(m_i-1)}.$$

This divides the scale of the underlying variable into m_i segments. The connection between the ordinal variable y_i and the underlying variable y_i^* is then that

$$y_i = s \quad \text{if} \quad \tau_{i(s-1)} < y_i^* \leq \tau_{i(s)} \qquad (s = 1, 2, \ldots, m_i).$$

The extreme lower and upper threshold parameters, $\tau_{i(0)}$ and $\tau_{i(m_i)}$, are $-\infty$ and $+\infty$, respectively. Since ordinal information only is available about y_i^*, the mean and variance of y_i^* are not identified and are therefore set to zero and one, respectively.

The model to be fitted is the classical linear factor analysis model

$$y_i^* = \alpha_{i1}^* f_1 + \alpha_{i2}^* f_2 + \cdots + \alpha_{iq}^* f_q + e_i \qquad (i = 1, 2, \ldots, p), \qquad (9.3)$$

where e_i is a residual term and y_i^* is an unobserved continuous variable *underlying* the ordinal variable y_i. In classical factor analysis, y_i^* is directly observed but here it is only partially observed through y_i. The number of independent parameters is the same as the proportional odds model in Section 9.3.

The reader who finds that the mathematical content is already becoming rather heavy needs only to note at this point that if we could estimate the correlations between the underlying variables, y_i^*, we would have all that we needed to carry out a standard factor analysis. It is possible to estimate these correlations from the ordinal variables, and the resulting estimates are called *polychoric* correlations. In the special case when there are only two categories, the corresponding coefficients are known, as noted in Chapters 6 and 8, as *tetrachoric* correlations. The "tetra" part refers to the four categories formed by cross-classifying two binary variables; the "poly" covers all cases where there are more than four such categories. For those who would like to delve a little deeper into the method, we provide an outline in the following three paragraphs; however, nothing essential will be missed if these are passed over.

To fit the model, it is clear that there are three different sets of parameters to be estimated, namely the thresholds, the polychoric correlations between the underlying variables, and, finally, the factor loadings of eq. (9.3). Those three sets of parameters can, in theory, be estimated in one step. Under the assumptions of the linear factor model that the latent variables are independent and normally distributed with mean 0 and variance 1 and that the residual terms are independent and normally distributed with mean 0 and variance σ_i^2, it follows that the underlying variables $y_1^*, \ldots, y_p^*$ have a multivariate normal distribution with zero means, unit variances, and some correlation matrix $\mathbf{P} = \{\rho_{ij}\}$, where $\rho_{ij} = \sum_{l=1}^q \alpha_{il}^* \alpha_{jl}^*$. Although in principle we could estimate all the parameters simultaneously, the assumption of multivariate normality requires the evaluation of multiple integrals that make the estimation computationally unfeasible when the number of ordinal variables is greater than about five.

One solution to that computational problem is to estimate the three sets of parameters in two or three stages. Software such as lavaan, LISREL, Mplus, and EQS use a three-step procedure assuming underlying normally distributed variables y_i^*. In the first step, the thresholds are estimated from the univariate margins of the observed variables. In the second step, the polychoric correlations are estimated from the bivariate margins of the observed variables for given thresholds. In the third step, the factor model is estimated from the polychoric correlations by weighted least squares (WLS) using a weight matrix. The different weightings take into account the differing precisions of the correlation estimates. If one is only interested in obtaining consistent parameter estimates, then any positive definite weight matrix can be used. However, suppose one is interested in obtaining asymptotically correct chi-squared measures of goodness-of-fit and standard errors for the parameter estimates. In that case, an estimate of the inverse of the asymptotic covariance matrix of the polychoric correlations needs to be used as the weight matrix which is often unstable in small samples, and so it can only be used for large samples. When the sample size is small "maximum likelihood estimation" is used instead. Note that ML estimation treats the matrix of polychoric correlations as if it were the covariance matrix of a sample from a multivariate normal distribution. The assumptions made under the three-stage estimation procedure are:

i) the latent variables are independent and normally distributed with mean zero and variance one,

ii) the residuals $(e_1, \ldots, e_p)$ are independent and normally distributed with mean zero and variance σ_i^2,

iii) univariate and bivariate normality of the underlying variables y_i^*. The bivariate normality assumption can be tested using a large sample likelihood ratio or goodness-of-fit chi-squared test in each marginal two-way table. The software available provides those tests.

From the description of the three-stage estimation procedure, we can see that the model is not fitted to the whole response pattern (as is the case in the one-stage formulation) but rather to the univariate and bivariate distributions. This approach is called a limited information method because it does not use all the information in the data. The advantage of the three-stage method is that it is implemented in widely available software such as lavaan, LISREL, Mplus, and EQS, and it allows researchers to fit a model with a large number of ordinal items and a large number of latent variables. Pairwise maximum likelihood estimation (PML) for binary and ordinal variables (Katsikatsou et al. 2012) is another limited information method that, contrary to WLS, estimates all the parameters (thresholds and factor loadings) in one step. It is available in the R Package lavaan and readers are encouraged to try both WLS and PML when fitting factor models to categorical data.

To illustrate the results of calculating polychoric correlations and also to demonstrate that it is not sufficient to use product moment correlations calculated from the ordinal variables, Table 9.11 gives the Pearson correlation matrix and the polychoric correlations for four science and technology variables. The Pearson correlations are smaller than the polychoric correlations for all pairs of items. The polychoric correlations are more accurate when the underlying variable model holds and are generally larger than the Pearson product-moment correlations.

TABLE 9.11: Correlation matrices for science and technology data

	Pearson correlations				Polychoric correlations			
	Comfort	Work	Future	Benefit	Comfort	Work	Future	Benefit
Comfort	1.00				1.00			
Work	0.15	1.00			0.20	1.00		
Future	0.28	0.40	1.00		0.35	0.48	1.00	
Benefit	0.33	0.17	0.31	1.00	0.41	0.21	0.38	1.00

Science and technology data

The one-factor model (eq. (9.3) for $q = 1$) has been fitted to the polychoric correlation matrix of the four items from the science and technology dataset using lavaan. The parameters of the model are estimated using two estimation methods, namely, maximum likelihood (ML) and weighted least squares (WLS) (where the weight matrix is the inverse of the asymptotic covariance matrix of the polychoric correlations). Table 9.12 gives the estimated factor loadings $\hat{\alpha}_{i1}^{*}$ with standard errors in brackets and the chi-squared statistics under the two estimation methods. The factor loadings for the two methods differ only slightly. The factor loadings are all positive, indicating a strong correlation between the items and the latent variable. The chi-squared statistic is large under both ML and WLS. As we have already argued, the chi-squared statistic should not be the only criterion for judging the goodness-of-fit of the model when, as in this case, the data are sparse. It would therefore be desirable to apply some of the other methods based on chi-squared residuals.

TABLE 9.12: UV approach, estimated factor loadings with asymptotic standard errors in brackets and chi-squared measures of fit, science and technology data (four items)

	ML		WLS	
	$\hat{\alpha}_{i1}^{*}$	s.e.	$\hat{\alpha}_{i1}^{*}$	s.e.
Comfort	0.48	(0.10)	0.51	(0.08)
Work	0.55	(0.07)	0.52	(0.07)
Future	0.79	(0.09)	0.76	(0.07)
Benefit	0.51	(0.09)	0.54	(0.07)
chi-squared	28.56		29.51	
p-value	0.00		0.00	

If we accept the fit as a first approximation to the truth, the conclusion to be drawn from fitting the UV model is essentially the same as that for the fit of the IRF model reported in Table 9.1. It supports the view that there is a single common factor underlying the data, of which the "Future" variable is the most important indicator. The factor loadings for the UV model are correlations between a normal latent variable and the normal underlying variables, whereas for the IRF logit model the standardised loadings are correlations between the normal latent variable and underlying variables that are not normally distributed. Had we used the probit IRF model instead of the logit, they would have been even closer. The important point is that the *relative* values are much the same, and that is what matters for interpretation. More information about the relationship between the numerical values of the loadings for the logit and probit models can be found in Bartholomew et al. (2011).

Relationship between the UV and IRF approaches

Although the UV and the IRF models look quite distinct in the sense that the model fitting procedures and some of the model assumptions are different, an equivalence has been noticed between the two approaches by Bartholomew et al. (2011). We made the same point in Chapter 8 when comparing the two approaches for binary data. The equivalence in the general case implies the following relationships between the parameters of the probit IRF and the probit UV models:

$$\alpha_{ij} = \frac{\alpha_{ij}^*}{\sigma_i} \, , \tag{9.4}$$

$$\alpha_{i(s)} = \frac{\tau_{i(s)}}{\sigma_i} \, , \tag{9.5}$$

where $\tau_{i(s)}$, α_{ij}^* and σ_i^2 are the thresholds, the factor loading of the jth latent variable and the variance of the error term in the linear factor model for the ith ordinal variable. The equivalence holds approximately for the logit IRF and the probit UV since the logit is a good approximation to the probit.

For the factor analysis model of eq. (9.3), the correlation between a latent variable f_j and an underlying variable y_i^* is

$$\text{Corr}(y_i^*, f_j) = \frac{\alpha_{ij}^*}{\sqrt{\sum_{j=1}^{q} \alpha_{ij}^{*2} + \sigma_i^2}} \, . \tag{9.6}$$

Substituting eq. (9.4) into eq. (9.6), we get the same correlation in terms of the IRF parameters α_{ij}. Those will be called standardised parameters and they will be denoted, as before, by $\text{st}\alpha_{ij}$:

$$\text{st}\alpha_{ij} = \frac{\alpha_{ij}}{\sqrt{1 + \sum_{j=1}^{q} \alpha_{ij}^2}} \, . \tag{9.7}$$

When results are presented from both the UV and the IRF approaches, it is best to produce the standardised version of the model parameters to allow for comparisons.

Given the equivalence between the two types of model, it is pertinent to ask why we have presented both and which is to be preferred in practice. We prefer the IRF version, and that is why we have used it for the analysis of the examples. It is a "full information" method in that the estimation method makes use of the full distribution over all score patterns. The UV method is a "partial information" method, using information only from the pairwise distributions of the ordinal variables. Our preference is therefore based partly on the grounds of efficiency. It is also based on the close link with the logit model for binary data, which we have already given powerful reasons for adopting. There is also a link with the unordered model, which we shall look at briefly in Section 9.6. There is a comprehensive and appealing simplicity about including so much under a single umbrella.

Nevertheless, at the present stage, there are two arguments in favour of including the UV approach. It gives an alternative way of looking at the model, which places it firmly in the factor analysis tradition and allows the standardised alphas to be interpreted as correlations. From this perspective, the method is merely a way of extending the scope of the factor model. Secondly, the software necessary to fit the model using polychoric correlations is widely available and familiar to social scientists.

9.6 Unordered and partially ordered observed variables

We have already remarked that categories such as "don't know" do not fit into a sequence of ordered categories obtained from attitude questions. Some method is therefore needed to deal with the case of partial ordering of categories. There are other cases where we suspect an ordering but would prefer the method to be flexible enough to allow us to check whether or not this is so. There are also many situations where some or all of the categorical variables are nominal. All of these cases can be handled by an alternative generalisation of the binary latent trait model of Chapter 8. A full exposition of this model requires technical resources beyond the limits we have set ourselves for this book, but a brief introductory account will help to complete the story of the factor analysis of categorical data.

The IRF approach, which we used for binary data, can easily be extended to cover the polytomous case. To do this, we treat each category of a nominal item as a binary item. An individual either belongs to that category or does not. A binary response function model is then specified for each category of the nominal item, having its own difficulty (intercept) and discrimination (slope) parameter.

We have already shown in Section 4.8 how to represent a polytomous variable by a vector of binary elements. In the present notation, the variable y_i is replaced by a vector-valued indicator function with its sth element defined as:

$$y_{i(s)} = \begin{cases} 1, & \text{if the response falls in category } s, \text{ for } (s = 1, \ldots, m_i), \\ 0, & \text{otherwise,} \end{cases}$$

where m_i denotes the number of categories of variable i and $\sum_{s=1}^{m_i} y_{i(s)} = 1$.

We now introduce a response function for each category of each variable, exactly as in the ordinal case, and using the same notation. Thus, the single response function of the binary case is now replaced by a set of response functions, one for each category, $\pi_{i(s)}(\mathbf{f})$ $(s = 1, \ldots, m_i)$ where $\sum_{s=1}^{m_i} \pi_{i(s)}(\mathbf{f}) = 1$.

Constructing a model now requires us to express these response probabilities as functions of the latent variables. In Chapter 8, we explained why we could not choose a linear function and that led to the search for some function of the response probability which could be expressed as a linear function of the fs. Of the several possibilities, we favoured the logit function. The logit is the logarithm of the ratio of the response probability for one category to the corresponding probability for the other. In the polytomous case, it is not immediately clear how to extend this idea to the case of several categories. Whatever choice we make, the probabilities must add up to one across all categories. It turns out that this can be achieved if one category is designated as what we shall call the *reference* category and if the response probabilities for all other categories are expressed in terms of that probability. Therefore, we define $(m_i - 1)$ logits for pairs of categories where the reference category can be selected to be any of the m_i categories. If the first category of the nominal variable is selected to be the reference category, the generalised logit model for variable i may be written as:

$$\log \left[\frac{\pi_{i(s)}(\mathbf{f})}{\pi_{i(1)}(\mathbf{f})} \right] = \alpha_{i0(s)} + \alpha_{i1(s)} f_1 + \cdots + \alpha_{iq(s)} f_q , \qquad (9.8)$$

where $s = 2, 3, \ldots, m_i$. Eq. (9.8) is the one used in multinomial logistic regression in the case where the variables on the right-hand side of the equation are known. If we compare this with the binary model, the only difference on the right-hand side is the subscript s appearing on all the parameters. The intercepts $\alpha_{i0(s)}$ are difficulty parameters not for the whole item i but for a specific category s of item i. The factor loadings or discrimination parameters, $\alpha_{ij(s)}$, measure the effect of the latent variable f_j on the log of the odds of being in category s rather than the reference category. We refer to this as the *nominal* model. We reanalyse the environment data already given in Section 9.4. The results are given in Table 9.13.

TABLE 9.13: Estimated difficulty and discrimination parameters with standard errors in brackets for the one-factor model for nominal data, environment data

Items	Category	$\hat{\alpha}_{i0(s)}$	s.e.	$\hat{\alpha}_{i1(s)}$	s.e.
LeadPetrol	2	−0.75	(0.17)	1.33	(0.40)
	3	−3.05	(0.45)	2.06	(0.22)
RiverSea	2	−2.32	(0.34)	2.11	(0.39)
	3	−8.83	(2.44)	5.04	(1.81)
RadioWaste	2	−2.57	(0.50)	3.30	(0.69)
	3	−5.98	(1.18)	6.50	(1.40)
AirPollution	2	−1.38	(0.34)	3.10	(0.56)
	3	−8.49	(2.10)	6.50	(2.29)
Chemicals	2	−2.81	(0.56)	3.59	(0.80)
	3	−5.40	(0.99)	4.73	(1.21)
Nuclear	2	−0.33	(0.18)	1.54	(0.38)
	3	−1.95	(0.37)	2.81	(0.33)

Each variable has three categories and, since category 1 was chosen as the reference category, parameter estimates are given only for categories 2 and 3. The difficulty parameters $\alpha_{i0(s)}$ are all negative and relatively large, which means that the median individual is most likely to be in category 1 [very concerned], less likely to be in category 2 [concerned] and least likely to be in category 3 [not very or not at all concerned]. The factor loadings given in the last column are the most interesting for the comparison with the proportional odds model. Although they have relatively large standard errors, it appears that in every case, the loading contrasting the first and the third category is greater than that contrasting the first and the second. This indicates that the items are measured on an ordinal scale. Fitting the nominal model has confirmed the ordering of the categories. In this particular example, this is little more than proving the obvious, but in less clear-cut examples, such confirmation would be useful.

The goodness-of-fit of the nominal model can also be investigated by computing the chi-squared residuals for the two-way margins. Table 9.14 gives the sums of the chi-squared residuals for the two-way marginal tables. Again, the values are all smaller than 20.09 (the 1% point of χ_8^2). The nominal model with one factor adequately explains the pairwise associations between the six items. Comparing the fit of the nominal model with the proportional odds model (Table 9.6), we see that for some pairs of items one model fits slightly better than the other. However, both models give a very good fit.

TABLE 9.14: Sums of chi-squared residuals for pairs of items from the two-way margins for the one-factor model for nominal data, environment data

	Items				
Items	2	3	4	5	6
1	13.48	4.91	7.93	2.48	1.68
2		4.98	8.42	3.67	4.03
3			4.43	8.22	12.98
4				1.99	6.00
5					6.08

In Chapter 8, we noted that for the binary response model of eq. (8.8), the linear combinations of the manifest variables, which we called components, were "sufficient" in that they contained all the information in the data about the latent variable. This property was lost when we moved to the ordinal model, but it continues to hold for the nominal model. These "sufficient statistics" are weighted sums of the responses with weights equal to the discrimination coefficients: thus the score for the hth individual on the jth factor is

$$Y_{hj} = \sum_{i=1}^{p} \sum_{s=2}^{m_i} \alpha_{ij(s)} y_{ih(s)},$$

where $j = 1, \ldots, q$ and $y_{ih(s)}$ takes the value 1 when individual h responds to the sth category and 0 otherwise. These sufficient statistics can be used as factor scores to scale individuals in the latent space.

The nominal model is the "natural" generalisation of the binary model in the sense that most of the latter's attractive properties are preserved. It is broader than the ordinal (proportional odds) model in that it covers ordinal and nominal categories as well as mixtures of the two. It can also be used when missing values are present since "missing" can be included as a separate response category. The model will then predict the place of the missing value among the response categories. The model, therefore, lets the data "speak for itself" in a way that the ordinal model fails to do.

On the other hand, there are two strong arguments for using the ordinal model whenever its assumptions are met. It makes use of relevant prior information about the ordering of the categories, and when this is incorporated, we might expect to get more precise estimates and more powerful tests of fit. More importantly, however, the flexibility of the nominal model is bought at a considerable price. The number of parameters to be estimated in the nominal model for variable i is:

$$(m_i - 1) + q \times (m_i - 1),$$

where $(m_i - 1)$ denotes the number of difficulty parameters and the number of loadings estimated under the nominal model, which compares with

$$(m_i - 1) + q,$$

for the ordinal model. Both models involve a large number of parameters, which can lead to numerical problems with the fitting, but this is much worse for the nominal model. When, for example, $p = 7$, $q = 2$ and $m_i = 4$ for all i, there will be 63 parameters to estimate in the nominal model but only 35 in the ordinal model. Often it will be found that the standard errors become very large and there are liable to be problems of identification. It is technical difficulties of this kind that make this more general approach hazardous.

9.7   Additional examples and further work

Government data

The data analysed here, using the ordinal IRF model, relate to 786 respondents from the 1996 British Social Attitudes Survey (Jowell et al. 1996). The data are given on the website. The items selected for the analysis are:

On the whole, do you think it should or should not be the government's responsibility to

1. provide a job for everyone who wants one [JobEvery]
2. keep prices under control [PriceCont]
3. provide a decent standard of living for the unemployed [Living]
4. reduce income differences between the rich and the poor [Income]
5. provide decent housing for those who can't afford it [Housing]

The four response alternatives are: "definitely should be", "probably should be", "probably should not be", and "definitely should not be".

You might expect that there would be a range of opinions on the extent to which the government should provide for the basic needs of its citizens through social security so it is reasonable to try a model with one latent variable. Table 9.15 gives the sums of chi-squared residuals for the two-way margins and using our rule of thumb these should be compared with 30.58 (the 1% point of χ^2_{15}). There is only one pair of items (item 1 and item 2) that is found not to be adequately fitted (chi-squared residual 40.31). The estimated factor loadings for the one-factor model with their standard errors are given in Table 9.16. The standardised loadings are all close to one, indicating a strong association between the latent variables and the five items. The one-factor model for ordinal responses provides a good, but not perfect, fit to the five items.

You could also repeat the analysis by fitting a latent class model as discussed in Chapter 10.

TABLE 9.15: Sums of the chi-squared residuals for pairs of items from the two-way margins for the one-factor model for ordinal responses, government data

	Item 2	Item 3	Item 4	Item 5
Item 1	40.31	22.92	7.91	14.42
Item 2		26.66	16.67	24.41
Item 3			9.58	17.90
Item 4				15.68

TABLE 9.16: Estimated factor loadings with standard errors in brackets and standardised loadings for the one-factor model, government data

Items	$\hat{\alpha}_{i1}$	s.e.	st$\hat{\alpha}_{i1}$
JobEvery	1.91	(0.15)	0.89
PriceCont	1.30	(0.11)	0.79
Living	2.65	(0.22)	0.94
Income	2.31	(0.18)	0.92
Housing	2.57	(0.22)	0.93

Voter study in Flanders, Belgium, 1991-1992

The set of questions analysed here, again using the IRF approach, is from an opinion poll survey on political attitudes after general legislative elections

(Carton et al. 1993). The survey was carried out between December 1991 and April 1992 in Flanders. The items chosen for analysis are expected to form a scale for measuring "Ethnocentrism". The following questions concern immigrants, by which we understand primarily Turks and Moroccans.

Please tell me whether or not you agree with the following statements. Response alternatives: "completely agree", "agree", "neither agree nor disagree", "disagree", "completely disagree" and "no opinion".

1. Belgium shouldn't have brought in guest workers
2. Generally speaking, immigrants can't be trusted
3. Guest workers are a threat to the employment of Belgians
4. Guest workers come here to exploit our Social Security
5. In some neighbourhoods, government is doing more for immigrants than for the Belgians who live there

The sample size was 2691. "No opinion" is treated here as missing. The percentage of missing values for items 1, 2, 3, 4, and 5 is 2.1, 3.4, 2.0, 1.7, and 13.9, respectively. After missing values are eliminated using listwise deletion, we are left with 2227 respondents. A more correct analysis, including all respondents and using a model that takes account of missing values was carried out by O'Muircheartaigh and Moustaki (1999). The results showed that missing values were not related to the "ethnocentrism" attitude, and so we believe that their elimination does not lead to substantial bias. Table 9.17 gives the frequency distribution of the five items.

TABLE 9.17: Frequency distribution of observed items, Flanders data

			Items		
Categories	1	2	3	4	5
1	0.15	0.08	0.10	0.20	0.16
2	0.24	0.18	0.28	0.37	0.32
3	0.22	0.29	0.25	0.21	0.24
4	0.34	0.37	0.33	0.18	0.25
5	0.05	0.07	0.05	0.04	0.04

The response alternatives include a middle category "neither agree nor disagree", which might disrupt the ordering of the response categories. However, you could still start with five categories for each item by fitting the one-factor model for ordinal responses. The sums of chi-squared residuals are given in Table 9.18 and are all greater than 42.98 (the 1% point of χ^2_{24}) indicating that the one-factor model is a poor fit. The fit is not substantially improved when the two-factor model is fitted. Most of the large chi-squared residuals are found between pairs of items for categories (1,2) and (4,5).

There are a number of possible reasons why the two-factor model for ordinal responses is not a good fit, and these are given below.

TABLE 9.18: Sum of the chi-squared residuals for pair of items from the two-way margins for the one-factor model for ordinal responses, Flanders data

	Item 2	Item 3	Item 4	Item 5
Item 1	171.52	169.01	161.53	149.20
Item 2		229.87	163.87	169.47
Item 3			143.08	194.28
Item 4				128.70

i) The assumption of conditional independence is not satisfied, and so more than two factors are needed to explain the associations among the five items.

ii) The logit model is not appropriate; a different link function is needed.

iii) The categories are not ordered. In that case, the model for partially ordered responses or nominal variables is likely to be more appropriate.

As far as point i) is concerned, a model with more than two factors cannot be fitted when there are only five items.

Instead of the two-factor logit model, the two-factor probit model could have been used. Still, since the response functions are so similar in shape, no significant differences are expected. You could try fitting one- and two-factor models using the UV approach for ordinal response categories for comparison.

The point that remains to be investigated is whether the five response categories can be treated as ordered for each of the five attitude items. Close examination of the chi-squared residuals for each pair of categories (before they are added to produce the sum of residuals) suggests that problems occur for other categories as well as for "neither agree nor disagree". These chi-squared residuals are not given here, but you can calculate them yourself when you repeat the analysis.

The analysis is continued by fitting the model for nominal responses given in eq. (9.8). The one-factor model was a bad fit, judging by the large chi-squared residuals observed for the two-way margins. When the one-factor model for nominal items is fitted, we would expect that, if the response categories are ordered, the factor loadings $\alpha_{i1(s)}$ should also be ordered across categories. This is because the factor loadings indicate an increase in the odds of falling into a category as the individual's position on the latent variable increases. Remember that the first category of each item is treated as a reference category. The factor loadings are given in Table 9.19, and you can see that they are ordered across categories for all items, indicating that the middle category is in the middle of the scale for each item. However, when the five items were analysed using the proportional odds model for ordinal responses, neither the one-factor nor the two-factor model was a good fit. You can also see that the factor loadings for the last two categories for each item are big, indicating that the response function fitted to those categories is very steep.

TABLE 9.19: Estimated factor loadings with standard errors in brackets for the one-factor model for nominal responses, Flanders data

Item	Category	$\hat{\alpha}_{i1(s)}$	s.e.
1	2	1.53	(0.88)
1	3	2.28	(0.07)
1	4	2.98	(0.12)
1	5	4.37	(0.13)
2	2	2.64	(0.89)
2	3	3.73	(0.16)
2	4	5.11	(0.19)
2	5	6.77	(0.18)
3	2	4.35	(1.19)
3	3	5.63	(0.33)
3	4	6.82	(0.29)
3	5	8.65	(0.25)
4	2	3.38	(0.97)
4	3	5.70	(0.49)
4	4	7.85	(0.29)
4	5	10.79	(0.36)
5	2	1.98	(0.85)
5	3	2.44	(0.12)
5	4	3.65	(0.14)
5	5	5.67	(0.16)

From the analysis so far, we have concluded that the five items do not seem to be indicators of a single latent dimension called "ethnocentrism". On the other hand, when the two-factor model for ordinal responses was fitted, the fit was not improved. Also, not much information will be lost by grouping the first two and last two categories, and there is no evidence that the middle alternative is not in the middle of the response categories.

Suppose you continue the analysis by fitting a one-factor model for ordinal responses to the five items with three categories each. In that case, you will obtain the sums of the chi-squared residuals given in Table 9.20. Six out of these ten sums are greater than 20.09 (the 1% point of χ_8^2). Investigation of the chi-squared residuals across categories for pairs of items shows that the poor fit occurs in the middle category. Again, when the two-factor model is fitted, the fit is not improved considerably, indicating that the poor fit is not due to the existence of a second factor. There is no reason to suppose that a second factor is needed to explain the associations among the five items.

Table 9.21 gives the parameter estimates for the one-factor model for three response categories. The loadings are close to one, indicating a strong association between the single factor and each item. This example illustrates the sad truth that statistical analyses do not always produce clear-cut results.

TABLE 9.20: Sums of the chi-squared residuals for pairs of items from the two-way margins for the one-factor model for ordinal responses (three categories), Flanders data

	Item 2	Item 3	Item 4	Item 5
Item 1	24.44	27.97	17.18	21.31
Item 2		20.95	17.82	36.14
Item 3			24.54	13.86
Item 4				10.46

TABLE 9.21: Estimated factor loadings with standard errors in brackets and standardised loadings for the one-factor model for ordinal responses (three categories), Flanders data

Items	$\hat{\alpha}_{i1}$	s.e.	st$\hat{\alpha}_{i1}$
1	1.53	(0.08)	0.84
2	2.17	(0.11)	0.91
3	2.09	(0.11)	0.90
4	3.08	(0.20)	0.95
5	1.49	(0.08)	0.83

9.8 Software

The UV models can be fitted using commercial software such as Amos (Arbuckle 2006), EQS (Bentler 2008), LISREL (Jöreskog and Sörbom 1999) and MPlus (Muthén and Muthén 2017) and the R package lavaan. The IRF logit model can be fitted with commercial software such as Stata and R packages, including ltm (Rizopoulos 2006) and mirt (Chalmers 2012).

9.9 Further reading

Bartholomew, D. J., Knott, M. and Moustaki, I.(2011). *Latent Variable Models and Factor Analysis: A Unified Approach* (3rd ed.). West Sussex, United Kingdom: John Wiley & Sons.

Bock, R. D., and Moustaki. I. (2007). Item response theory. In C. R. Rao and S. Sinharay (Ed.), *Handbook of Statistics* (Vol. 26). Elsevier.

Krzanowski, W. J. and Marriott, F. H. C. (1995b). *Multivariate Analysis, Part 2: Classification, Covariance Structures, and Repeated Measurements*. London: Arnold.

Skrondal, A. and Rabe-Hesketh, S. (2004). *Generalized Latent Variable Modelling: Multilevel, Longitudinal, and Structural Equation Models*. Boca Raton, FL: Chapman and Hall/CRC.

10

Models with Categorical Latent Variables

10.1 Introduction

In this chapter, we focus on the two cells at the bottom of Table 7.1 and discuss models with categorical latent variables. These models are known as latent profile models or latent class models, depending on whether the manifest variables are continuous or categorical. The situations in which we might wish to apply these models are very similar to those mentioned in connection with cluster analysis. Here, however, we shall usually have a clearer idea of how many clusters there might be and what they might represent. In addition, cluster analysis methods are typically descriptive, while the methods introduced in this chapter are model-based. The following examples illustrate some possibilities.

i) *Educational assessment.* There has been a good deal of research on how children acquire new concepts. One hypothesis is that a child has either mastered the concept or has not. This may be judged by asking the child to perform tasks which depend for their successful completion on having grasped the concept. This is known as *criterion referenced* testing because performance is judged by reference to an external criterion. If the hypothesis is correct we might expect the "masters" to get all the items correct and the "non-masters" to get them all wrong. In practice, non-masters will sometimes get items correct by chance and masters will sometimes get items wrong by making silly mistakes. As the observed data are categorical (correct or incorrect answers), a latent class model may help us to decide whether or not the hypothesis is supported by the data allowing for any errors which have been made. In this case, we would fit a model with two classes - representing the "masters" and the "non-masters", respectively. An example of this type is discussed in Section 10.8 based on a dataset analysed in Macready and Dayton (1977).

DOI: 10.1201/9781003483342-10

ii) *Medical diagnosis.* Many medical conditions cannot be diagnosed directly, or without invasive surgery, because the root of the condition is deep-seated. However, many symptoms can be easily observed, some of which may point towards one cause and some to another. It would be useful if we could use observations of an individual's symptoms to estimate the probability that the patient has any of the possible conditions. As the observed data are categorical (having symptoms or not), a latent class model may help us to do this.

iii) *Selection methods.* Aptitude for performing a complex task, like flying an aircraft, can only be inferred in advance by testing the candidate's performance on a variety of tests designed to give an indication of the required skills. One might anticipate that there existed three groups of candidates: those who were ready for immediate acceptance, those who should be rejected, and those who, in time and with additional preparation, might make the grade. With continuous test scores, a latent profile analysis would enable us to investigate this hypothesis and provide a rule for assigning individuals to the groups.

Models with categorical latent variables are commonly used in real-world applications. Their principal objectives are:

i) To reduce the complexity of a dataset by explaining the associations between the observed variables in terms of membership of a small number of unobservable groups (latent profiles/classes), and hence to gain understanding of the interrelationships between the observed variables.

ii) To be able to allocate an object to one of these groups — or sometimes just to estimate the probabilities of belonging to each group — on the basis of the values of the observed variables for that object.

In this chapter, we shall focus on two popular models – a latent profile model for continuous data (referred to as a Gaussian mixture model) and a latent class model for categorical data. The main difference between these models and the factor analysis models discussed in the previous three chapters is that FA assumes that the latent variable(s) are continuous, and possibly normally distributed, whereas the models in this chapter assume a *single* latent variable that is categorical, representing the unobserved groups. For simplicity, we will refer to the categories of the latent variable as the latent classes in the rest of this chapter, while noting that they are often referred to as the latent profiles in latent profile models. In a model with J latent classes, the latent variable, c, can be defined to take the value 1 for an object in class 1, 2 for an object in class 2, ..., and J for an object in class J. (When $J = 2$, the classes might be labelled 0 and 1 rather than 1 and 2. In any case, the precise labelling is irrelevant.) We also consider models with latent classes for the analysis of longitudinal data in Chapter 14.

10.2   The Gaussian mixture model for continuous data

Suppose we have an n by p data matrix, where each row corresponds to an object or individual, and each column represents a continuous variable, denoted by $y_1, \ldots, y_p$. A Gaussian mixture model makes the following assumptions:

    i) The n individuals are a random sample from some population, and every individual in that population belongs to just one of the J latent classes.

    ii) Given an individual's latent class membership, the distribution of the manifest variables $(y_1, \ldots, y_p)$ is a multivariate normal distribution. This distribution is the same for all the individuals in the same class.

Notation

The Gaussian mixture model assumes that the conditional distribution of $(y_1, \ldots, y_p)$ given $c = j$ is multivariate normal, for each $j = 1, \ldots, J$. We denote this conditional distribution given $c = j$ as $N(\boldsymbol{\mu}_j, \Sigma_j)$, where $\boldsymbol{\mu}_j = (\mu_{j1}, \ldots, \mu_{jp})$ is the mean vector and Σ_j is a $p \times p$ covariance matrix. Let η_j be the proportion of the population in latent class j or equivalently the probability that a randomly selected individual from the population belongs to latent class j, for $j = 1, \ldots, J$. Sometimes η_j is referred to as the *prior* probability of belonging to class j.

    Note that in FA models, we conditioned on the values of q factors or latent traits, $f_1, \ldots, f_q$, and modelled the conditional distribution of $(y_1, \ldots, y_p)$ given $(f_1, \ldots, f_q)$. Here, we have a *single, categorical* latent variable c taking values $1, \ldots, J$.

Fitting the model and possible constraints

The Gaussian mixture model involves unknown parameters η_j, $\boldsymbol{\mu}_j$, and Σ_j, $j = 1, 2, \ldots, J$. These parameters are most commonly estimated by the maximum likelihood estimator, which is obtained using an iterative optimisation algorithm. One should be cautious here that the likelihood function of a Gaussian mixture model could have several local maxima, and the iterative algorithm may result in a local maximum rather than a global one. Running the optimisation algorithm from a good starting point makes it more likely to find the global optimum, where a good starting point may be obtained using a clustering algorithm, such as the K-means method. It is also recommended to run the algorithm multiple times with different starting points, hoping that if they all give the same solution, then it is likely the global maximum. If not, then we should choose the solution that gives the largest likelihood function value among all the solutions. Previous research suggests that local maxima are unlikely to be a serious problem when only two or three classes are fitted but that the risk is greater for larger numbers of classes.

We also note that each Σ_j contains $p(p+1)/2$ parameters. When p is large, there may be too many parameters in these covariance matrices. To keep the model parsimonious, it is common to impose constraints on Σ_js to reduce the number of model parameters. The following constraints are commonly used:

i) $\Sigma_1 = \Sigma_2 = \cdots = \Sigma_J = \Sigma$, i.e. the covariance matrices are assumed to be the same across classes.

ii) Σ_js are class-specific diagonal matrices, for all $j = 1, \ldots, J$. That is, given the latent class to which an object or individual belongs, the p variables are uncorrelated or, equivalently, independent due to the normality assumption.

iii) $\Sigma_1 = \Sigma_2 = \cdots = \Sigma_J = \Sigma$, and Σ is a diagonal matrix. This combines the previous two sets of constraints.

Interpretation of the latent classes

In some cases, we may have an idea about what classes we expect to find. We shall then wish to ask whether those uncovered by the analysis correspond to what we expected. In other cases, our approach is exploratory, and then we do not know in advance how many classes are needed or what they might represent. In the latter case, we usually start by fitting a model with two classes and then proceed to add further classes as necessary. In both approaches, we need to be able to use estimates of the class-specific parameters, $\hat{\mu}_j$ and $\hat{\Sigma}_j$, to infer something about the nature of the classes to which they relate.

For illustration, we consider a dataset consisting of cost-of-living information for 888 cities worldwide. We focus on two variables: the logarithm of the average price per square meter to buy an apartment in the city centre (in US dollars) and the logarithm of the average monthly net salary (after tax; in US dollars). Figure 10.1 gives a scatter plot of these two variables. It does not seem to be unimodal, and thus a bivariate normal distribution may not fit the data well.

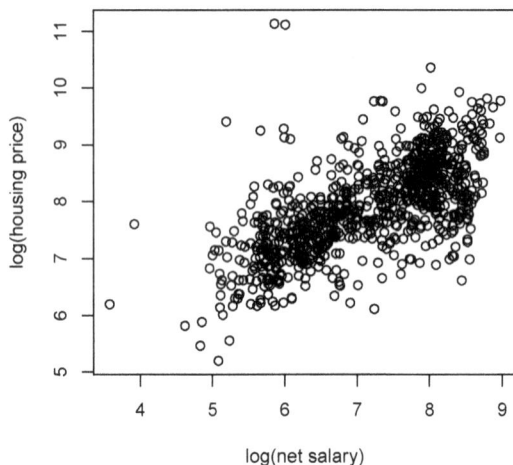

FIGURE 10.1: Scatterplot of the logarithm of the average monthly net salary versus the logarithm of the average price per square meter to buy an apartment in the city centre; cost-of-living data

Panel (a) of Figure 10.2 gives the result of estimating a Gaussian mixture model with two latent classes, for which no constraint is imposed on the class-specific parameters. In the plot, the two points marked by an asterisk symbol represent the mean vectors of the two classes, and the contour of the estimated density function is also given. Table 10.1 gives the values of the estimated parameters, including $\hat{\eta}_1$, $\hat{\eta}_2$, $\hat{\boldsymbol{\mu}}_1$, $\hat{\boldsymbol{\mu}}_2$, $\hat{\Sigma}_1$ and $\hat{\Sigma}_2$. From these results, we note that each class contains about half of the cities, with $\hat{\eta}_1 = 0.53$ and $\hat{\eta}_2 = 0.47$. Cities in the second class tend to have both higher income and higher housing prices than those in the first. The estimated mean value of the average price per square metre to buy an apartment in the city centre is about $\exp(7.51) = \$1826$ and $\exp(8.41) = \$4492$ for the two classes, respectively. The mean average monthly net salary is about $\exp(6.39) = \$596$ and $\exp(8.05) = \$3134$ for the two classes, respectively. In addition, the logarithm of housing price and that of net salary tend to be less correlated for the second class, with the estimated correlation being $0.25/\sqrt{0.48 \times 0.53} = 0.50$ for the first class and $0.05/\sqrt{0.38 \times 0.13} = 0.22$ for the second class.

For illustration purposes, we also fit a two-class model with a common co-variance matrix for the two classes. The resulting contour plot is given in Panel (b) of Figure 10.2, and the parameter estimates are given in Table 10.2. We notice that the estimated centroids of the two classes do not change substantially, but the covariance matrices for the two classes are quite different from the results in Table 10.1. Consequently, the two contour plots look different in Figure 10.2.

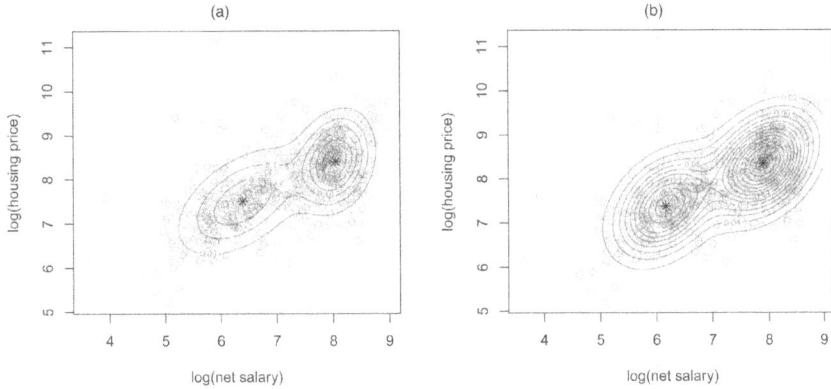

FIGURE 10.2: Contour plots for estimated Gaussian mixture models. Panel (a): model with unconstrained class-specific covariance matrices. Panel (b): model with a common covariance matrix; cost-of-living data

TABLE 10.1: Estimated parameters of a Gaussian mixture model with unconstrained class-specific covariance matrices; cost-of-living data

Variable	$\hat{\boldsymbol{\mu}}_1$	$\hat{\Sigma}_1$		$\hat{\boldsymbol{\mu}}_2$	$\hat{\Sigma}_2$	
log(net salary)	6.39	0.48	0.25	8.05	0.13	0.05
log(housing price)	7.51	0.25	0.53	8.41	0.05	0.38
$\hat{\eta}_j$		0.53			0.47	

TABLE 10.2: Estimated parameters of a Gaussian mixture model with a common class-specific covariance matrix; cost-of-living data

Variable	$\hat{\boldsymbol{\mu}}_1$	$\hat{\boldsymbol{\mu}}_2$	$\hat{\Sigma}$	
log(net salary)	6.17	7.90	0.27	0.12
log(housing price)	7.38	8.34	0.12	0.44
$\hat{\eta}_j$	0.53	0.47		

Model selection

For a given dataset, we can apply multiple Gaussian mixture models – models with different numbers of latent classes and different constraints. Models are most commonly compared using the Bayesian information criterion (BIC), as it is theoretically guaranteed to detect the correct number of latent classes with high probability when the sample size is large. Given several estimated models, the one with the smallest BIC is preferred.

For the cost-of-living dataset, we compare eight models with two possible choices of latent classes, $J = 2, 3$, and four choices of covariance matrices. The BIC values of these models are given in Table 10.3, where the models are ordered according to their BIC values, from the smallest to the largest. We see that the two-class model with class-specific unconstrained covariance matrices fits the best among these models, followed by the three-class model with class-specific unconstrained covariance matrices. The rest of the models fit substantially worse than these two models.

TABLE 10.3: BIC values for eight Gaussian mixture models with different numbers of latent classes and covariance matrix assumptions, cost-of-living data

Model	J	Covariance	BIC
1	2	Class-specific, unconstrained	4030
2	3	Class-specific, unconstrained	4036
3	2	Common, unconstrained	4059
4	3	Class-specific, diagonal	4079
5	3	Common, unconstrained	4080
6	2	Class-specific, diagonal	4120
7	2	Common, diagonal	4133
8	3	Common, diagonal	4142

Allocation to classes

Having selected a latent class model as a useful simplification of our data, we may want to allocate the individuals to the identified classes using their observed data $y_1, \ldots, y_p$. This can be seen as model-based clustering. It is also closely related to the "factor scores" problem in factor analysis in Section 7.7 of Chapter 7 and latent trait analysis in Section 8.5 of Chapter 8. We solve the problem by estimating the conditional probability that an individual falls into a particular class, given their observed data. This conditional probability, sometimes called the *posterior* probability, is:

$$\Pr(\text{individual is in class } j \mid y_1, \ldots, y_p), \qquad j = 1, \ldots, J,$$

which can be easily computed under the estimated model.

Table 10.4 gives the estimated allocation probabilities for 15 randomly selected cities from the 888 cities in the data. Most of the probabilities in Table 10.4 are close to either zero or one, indicating that there is little uncertainty with their allocations. However, there are also cases whose allocations have higher uncertainty; for example, the allocation probabilities of the two classes are 0.58 versus 0.42 for Bari of Italy and 0.21 versus 0.79 for Girona of Spain. The last column gives the allocation of the cities to the two classes based on the mode of the estimated allocation probabilities; a city is allocated to the first class if $\hat{\Pr}(\text{Class 1} \mid y_1, y_2) > 0.5$ and to the second class otherwise. We also note that most of the cities allocated to the second class are in developed countries, while those allocated to the first class are in less developed countries. This is consistent with the interpretation of the estimated model that cities in the second class tend to have both higher income and higher housing prices than those in the first.

TABLE 10.4: Estimated allocation probabilities for 15 randomly selected cities; cost-of-living data

City	Country	$\hat{\Pr}(\text{Class 1} \mid y_1, y_2)$	$\hat{\Pr}(\text{Class 2} \mid y_1, y_2)$	Class allocation
Kranj	Slovenia	0.94	0.06	1
Ramat Gan	Israel	0.05	0.95	2
Accra	Ghana	1.00	0.00	1
Bari	Italy	0.58	0.42	1
Arlington	United States	0.01	0.99	2
Gaborone	Botswana	1.00	0.00	1
Pretoria	South Africa	0.92	0.08	1
Tabriz	Iran	0.99	0.01	1
Luxembourg	Luxembourg	0.02	0.98	2
Ashdod	Israel	0.13	0.87	2
Quebec City	Canada	0.03	0.97	2
Hua Hin	Thailand	1.00	0.00	1
Nice	France	0.07	0.93	2
Fort Worth	United States	0.01	0.99	2
Girona	Spain	0.21	0.79	2

Gaussian mixture models for longitudinal data

Variants of Gaussian mixture models are widely used in the analysis of longitudinal data where there are repeated measurements of a continuous variable

y over time. In Chapter 14, we consider two generalisations of the Gaussian mixture model, referred to as latent class growth analysis (or group-based trajectory models) and growth mixture models. Both approaches assume that individual response patterns in y (or "trajectories") can be grouped into latent classes such that individuals in the same class follow the same average trajectory.

10.3 The Gaussian mixture model versus the linear factor model

Both the Gaussian mixture model and the linear factor model introduced in Chapter 7 apply to continuous data. How do we decide which model to use? The main consideration should be the purpose of the analysis.

Suppose the goal is to gain an understanding of the observed variables through one or multiple latent variables. In that case, we should consider whether the phenomenon is best interpreted using continuous latent traits or discrete latent classes (i.e. unobserved groups). The relevant substantive theory and models used in relevant studies may help us make this decision. If it remains unclear which model to use, we can fit both models and then decide based on their interpretation and goodness-of-fit results.

If the goal is to learn the joint distribution of the observed variables for making predictions, then we should choose the model depending on its prediction performance, for example, based on some cross-validation scores. We note that the Gaussian mixture model is a more flexible family of distributions than the linear factor model with the assumption of normality. That is, a normal linear factor model is a submodel of an unconstrained multivariate normal model for the observed variables, and the latter can further be seen as a special case of a Gaussian mixture model with only $J = 1$ mixing component. When the sample size is sufficiently large, we may afford to estimate a more flexible model, and thus, a Gaussian mixture model may be preferred. In contrast, when the sample size is small, a model with fewer parameters tends to perform better, so we may prefer the linear factor model or a Gaussian mixture model with constraints (e.g. constraining Σ_js to be diagonal).

10.4 The latent class model for binary data

We now consider latent class analysis for binary data. Suppose that, as before, we have an n by p data matrix of values of p binary variables, $y_1, \ldots, y_p$ taking the values 0 or 1 for n objects or individuals. For example, consider the attitude to abortion data analysed using cluster analysis in Chapter 2 and again using factor analysis for binary data in Chapter 8. There are four binary variables or items, and 379 individuals who have responded to the items. A latent class model might have better explained these data. The frequency distribution for the observed response patterns is given in Table 8.1. Response patterns (1111) and (0000) have the largest frequencies with 141 and 103 respectively, which suggests they might be used as the nuclei of two classes, one consisting of those tending to favour abortion and the other of those tending to oppose abortion. Part of the interest in trying to fit a latent class model is to see whether the remaining response patterns could be allocated to one or other of these classes. In that connection, it is interesting that 44 respondents had the pattern (0111), indicating that they agreed to all the items except the first one. The first item was the one found to have the largest "difficulty" coefficient in the latent trait analysis described in Chapter 8.

The latent class model for binary variables with J latent classes makes the following assumptions, of which the first two assumptions are also made in the Gaussian mixture model.

i) The n individuals are a random sample from some population, and every individual in that population belongs to just one of the J latent classes.

ii) The probability of giving a positive response to a particular item is the same for all individuals in the same class but may be different for individuals in different classes.

iii) Given the latent class to which an individual belongs, its responses to different items are conditionally independent.

The third assumption is typically known as the conditional or local independence assumption. It is not assumed in a Gaussian mixture model unless the covariance matrices Σ_j are all constrained to be diagonal matrices. A similar assumption is also made in the factor models in Chapters 7–9.

Notation

We again let η_j be the proportion of the population in latent class j, as in the Gaussian mixture model. In addition, let $\pi_{ij} = \Pr(y_i = 1 \mid c = j)$ be the probability that a randomly selected individual from class j will answer positively to item i, for $(i = 1, \ldots, p; \ j = 1, \ldots, J)$. Thus, π_{ij} is the conditional probability of a positive response to item i, given (or conditional on) membership of class j.

Fitting the model

The model is fitted iteratively to obtain maximum likelihood estimates, similar to the fitting of a Gaussian mixture model. The algorithm for the latent class model may also result in a local rather than a global maximum. Our comments about the computation for the Gaussian mixture model still apply here.

There is another problem that can arise in fitting latent class models known as *under-identification*. Roughly speaking, this means that there are too many parameters to be estimated from the data available. An important difference between LCA and FA models for binary variable models is that the latter impose explicit functional relationships (such as the logit-linear relationship) between the probability of a correct/positive response and the latent variable or factor. The latent class models do not impose any such restriction on the form of the probabilities. This means that there are many more parameters to be estimated in a typical latent class model, and this has practical implications for the complexity of the models that can be fitted.

Interpretation of the latent classes

The process of interpreting the latent classes from a latent class model is similar to that of a Gaussian mixture model. We will use the estimated class-specific response probabilities $\hat{\pi}_{ij}$s to infer something about the nature of the classes to which they relate. This process will be illustrated using the attitude to abortion data and other examples.

The estimates of the model parameters with their standard errors in brackets for the attitude to abortion data are given in Table 10.5. Individuals in Class 2 have much higher estimated probabilities of agreeing with all the propositions about abortion than those in Class 1. The estimated standard errors are sufficiently small for the probabilities to be taken at face value. Item 1, which says that the law should allow abortion when the woman decides on her own, has rather less support among members of Class 2 than other items ($\hat{\pi}_{12} = 0.71$).

Class 1 comprises individuals who have a close to zero estimated probability of agreeing with item 1 and rather larger, though still small, estimated probabilities of agreeing with items 2, 3 and 4 ($\hat{\pi}_{21} = 0.09$, $\hat{\pi}_{31} = 0.12$, $\hat{\pi}_{41} = 0.15$). The more practical matters of marriage and affordability seem to be less critical in defining the attitudes of this class than the first two items. The last row of the table shows that about 39% of the individuals in the sample belong to Class 1 and 61% to Class 2.

TABLE 10.5: Estimated conditional probabilities, $\hat{\pi}_{ij}$, and prior probabilities, $\hat{\eta}_j$, with standard errors in brackets for the two-class model, attitude to abortion data

Item (i)	$\hat{\pi}_{i1} = \hat{\text{Pr}}(y_i = 1 \mid 1)$	$\hat{\pi}_{i2} = \hat{\text{Pr}}(y_i = 1 \mid 2)$
WomanDecide	0.01 (0.01)	0.71 (0.03)
CoupleDecide	0.09 (0.03)	0.91 (0.02)
NotMarried	0.12 (0.04)	0.96 (0.02)
CannotAfford	0.15 (0.04)	0.91 (0.02)
$\hat{\eta}_j$	0.39 (0.03)	0.61 (0.03)

Goodness-of-fit

The methods of assessing goodness-of-fit described in Chapter 8 on FA for binary data can also be applied to latent class models. The only new feature concerns the degrees of freedom for the two goodness-of-fit statistics, G^2 and X^2. The degrees of freedom equal the number of different response patterns (2^p) minus the number of independent parameters $(J - 1 + Jp)$ minus 1 which is equal to $(2^p - J(p + 1))$. For a test to be possible, the degrees of freedom must be greater than zero. Problems will arise if p is small and J is large. For example, for the attitude to abortion data $p = 4$, so when $J = 2$ there are six degrees of freedom, but this will be reduced if response patterns have to be grouped to ensure that the expected frequencies are not less than 5. For the attitude to abortion data, the likelihood-ratio statistic is $G^2 = 37.02$ and the chi-squared test statistic is $X^2 = 44.81$ on six degrees of freedom, indicating that the two-class model is not a good fit to the data. The values of G^2 and X^2, after grouping the response patterns with expected frequency less than 5, become 34.84 and 31.09, respectively, on one degree of freedom, which also indicates a poor fit.

The chi-squared residuals for the attitude to abortion data given in Table 10.6 show that the two-class model predicts the two-way margins very well for most pairs of items. There is only one pair with a large residual for response $(0,1)$ to items 2 and 1. Judging the model by how well it predicts the two- and three-way margins, we have no reason to reject the two-class model. The chi-squared residuals for the $(1,1,1)$ three-way margins are all close to zero.

The percentage of G^2 explained is 94.2% indicating that the two-class model is a much better fit than the independence model. If we decide that the fit is not good enough, we could go on to fit a three-class model, but it is here that we run into trouble with identifiability. When $J = 3$, the number of degrees of freedom becomes one, and these may be lost if grouping takes place. With more than three classes, there will certainly not be enough degrees of freedom to carry out a test. We therefore have to do the best we can with a less well-fitting model. In the case of the attitude to abortion data, it is possible to make a sensible interpretation of the data using the two-class model (see Table 10.7), even though the requirement of conditional independence (assumption iii) may not be fully satisfied.

TABLE 10.6: Chi-squared residuals for the second-order margins for a two-class model, attitude to abortion data

	Items	O	E	$O - E$	$(O - E)^2/E$
Response (0,0)	2, 1	147	137.79	9.21	0.62
	3, 1	131	130.16	0.84	0.05
	3, 2	117	117.58	−0.58	0.00
	4, 1	129	129.12	−0.12	0.00
	4, 2	114	114.61	−0.61	0.00
	4, 3	116	109.97	6.03	0.33
Response (0,1)	1, 2	66	75.21	−9.21	1.13
	1, 3	82	82.84	−0.84	0.01
	1, 4	84	83.88	0.12	0.00
	2, 1	7	16.21	−9.21	5.24
	2, 3	37	36.42	0.58	0.01
	2, 4	40	39.39	0.61	0.01
	3, 1	7	7.84	−0.84	0.09
	3, 2	21	20.42	0.58	0.02
	3, 4	22	28.03	−6.03	1.30
	4, 1	16	15.88	0.12	0.00
	4, 2	31	30.39	0.61	0.01
	4, 3	29	35.03	−6.03	1.04
Response (1,1)	2, 1	159	149.79	9.21	0.57
	3, 1	159	158.16	0.84	0.01
	3, 2	204	204.58	−0.58	0.00
	4, 1	150	150.12	−0.12	0.00
	4, 2	194	194.61	−0.61	0.00
	4, 3	212	205.97	6.03	0.18

Allocation to classes

Similar to the Gaussian mixture model, we can allocate the individuals to the identified classes based on their responses. Again, we estimate the conditional probability of an individual falling into a particular class given their responses, also known as the *posterior* probability:

$$\Pr(\text{individual is in class } j \mid y_1, \ldots, y_p) \qquad j = 1, \ldots, J.$$

Table 10.7 gives the estimated allocation probabilities for the attitude to abortion data. Most of the probabilities are close to either zero or one, indicating that there is little doubt as to the class to which each individual should be allocated. In a few cases, the position is more ambiguous, especially for response pattern 0101, where the probability is 0.55. However, there are only six respondents in this category. Most individuals can be allocated to one or other latent classes with little uncertainty. The last column allocates response patterns to clusters 1 and 2 based on a cluster analysis performed in Chapter 2. The allocation based on LCA and CLA do not match. Which do you think is best?

TABLE 10.7: Estimated posterior probabilities of class membership for the attitude to abortion data

Response pattern	$\hat{\text{Pr}}(\text{Class } 1 \mid y_1, \ldots, y_4)$	$\hat{\text{Pr}}(\text{Class } 2 \mid y_1, \ldots, y_4)$	Class allocation	CLA cluster allocation
0000	1.00	0.00	1	1
0001	0.99	0.01	1	1
0010	0.96	0.04	1	2
0100	0.98	0.02	1	1
1000	0.95	0.05	1	1
0011	0.31	0.69	2	2
0101	0.45	0.55	2	1
0110	0.21	0.79	2	2
1100	0.16	0.84	2	1
0111	0.00	1.00	2	2
1011	0.00	1.00	2	2
1101	0.00	1.00	2	1
1110	0.00	1.00	2	2
1111	0.00	1.00	2	2

Example: attitude to science and technology data

The seven variables used in this example have already been given in Chapter 9 in their original form. Originally, all the items were measured on a four-point scale with response categories "strongly disagree", "disagree to some extent", "agree to some extent", and "strongly agree". For the present analysis, response categories were dichotomised. As already noted in Section 9.4, items 1 [Comfort], 3 [Work], 4 [Future], and 7 [Benefit] express positive attitude to science and technology whereas items 2 [Environment], 5 [Technology], and 6 [Industry] express negative attitudes. Therefore, for items 1, 3, 4, and 7, we have used 1 for "agree to some extent" and "strongly agree" and 0 for "strongly disagree" and "disagree to some extent" and we have used the reverse coding for items 2, 5 and 6. Thus, the higher code, 1, corresponds to a more positive attitude to science and technology.

The number of respondents who answered all seven items is 392. There are $2^7 = 128$ possible response patterns, so each pattern occurs, on average, about three times, but there is a great deal of variation, and many patterns do not occur at all. The full dataset is given on the website.

First, we fit the two-class model. Table 10.8 gives the estimated probabilities and their asymptotic standard errors for the two-class model. Class 2, containing an estimated 79% of the population, consists of those individuals who are likely to have a positive attitude to all seven items. This may be described as a pro-science class. The way to describe Class 1 is less clear. On some items, Environment, Technology, and Industry, the probabilities are very similar to those for Class 2, so one cannot describe Class 1 as anti-science. They are, however, markedly less likely to respond positively about the prospects for Work, the Future, and about the benefits outweighing the harmful effects of science. Thus, whereas all tend to agree about the "technical" advances of science, Class 1 members are more sceptical about the social benefits. Although this seems a reasonable way of interpreting the analysis, the fit of the two-class model is, in fact, not very good, at least as judged by the global tests. The X^2 and G^2 statistics are 235.28 and 156.01, respectively, on 112 degrees of freedom. Due to the excessive sparseness, neither statistic can be trusted. The percentage of G^2 explained by the two-class model is 27.7%. Table 10.9 gives the chi-squared residuals for pairs of items where the residual was greater than 3. There was also a large residual, 5.19, for the positive responses to items 2, 5, and 6 among the three-way margins. As already mentioned in Chapter 8, we use as a rule of thumb a value of the residuals greater than 4 as an indication of poor fit. However, since that value is only indicative in the examples, we also report residuals greater than 3.

TABLE 10.8: Estimated conditional probabilities, $\hat{\pi}_{ij}$, and prior probabilities, $\hat{\eta}_j$, with standard errors in brackets for the two-class model, science and technology data

Item (i)	$\hat{\pi}_{i1} = \hat{\Pr}(y_i = 1 \mid 1)$	$\hat{\pi}_{i2} = \hat{\Pr}(y_i = 1 \mid 2)$
Comfort	0.75 (0.06)	0.95 (0.02)
Environment	0.76 (0.06)	0.68 (0.03)
Work	0.36 (0.09)	0.75 (0.04)
Future	0.20 (0.18)	0.94 (0.04)
Technology	0.74 (0.06)	0.72 (0.03)
Industry	0.84 (0.05)	0.86 (0.02)
Benefit	0.41 (0.09)	0.77 (0.03)
$\hat{\eta}_j$	0.21 (0.07)	0.79 (0.07)

The fit revealed by the margins suggests that the model captures an important part of the data structure. It also suggests that something more might be learnt by looking at a three-class model.

TABLE 10.9: Chi-squared residuals greater than 3 for the second-order margins for the two-class model, science and technology data

	Items	O	E	$O - E$	$(O - E)^2/E$
Response (0,0)	5, 1	17	10.05	6.95	4.81
	5, 2	52	33.18	18.82	10.67
	6, 1	10	5.56	4.44	3.54
	6, 2	31	17.23	13.77	11.00
	6, 5	26	15.83	10.17	6.53
Response (0,1)	2, 5	67	85.82	−18.82	4.13
	5, 2	57	75.82	−18.82	4.67
	6, 2	26	39.77	−13.77	4.77

The parameter estimates for the three-class model are given in Table 10.10. Note that the asymptotic standard errors for one of the estimated conditional probabilities is at the boundary and it is not computed.

TABLE 10.10: Estimated conditional probabilities, $\hat{\pi}_{ij}$, and prior probabilities, $\hat{\eta}_j$, with standard errors in brackets for the three-class model, science and technology data

Items	$\hat{\pi}_{i1}$	$\hat{\pi}_{i2}$	$\hat{\pi}_{i3}$
Comfort	0.80 (0.05)	1.00 (—)	0.85 (0.05)
Environment	0.72 (0.05)	0.80 (0.04)	0.39 (0.09)
Work	0.43 (0.07)	0.74 (0.04)	0.90 (0.06)
Future	0.38 (0.09)	0.98 (0.05)	0.98 (0.06)
Technology	0.74 (0.06)	0.94 (0.06)	0.14 (0.13)
Industry	0.84 (0.04)	0.95 (0.03)	0.63 (0.08)
Benefit	0.47 (0.07)	0.82 (0.04)	0.75 (0.06)
$\hat{\eta}_j$	0.33 (0.07)	0.48 (0.07)	0.18 (0.06)

Individuals in Class 2 tend to have strong positive attitudes towards science and technology, and we may continue to refer to them as the pro-science group. What appears to have happened in the three-class solution is that the original Class 2 is now split into two separate classes. The new Class 2 is much the same as Class 2 in the two-class solution (tend to have strong positive attitudes towards science and technology, and we may continue to refer to them as the pro-science group), except that it is now much smaller (48%). The new Class 3 is distinguished by a relatively smaller probability of agreeing to Environment and Technology items. The new Class 1 remains the same as Class 1 in the two-class solution. Similar patterns were found when items were analysed as four-point scales with ordinal items as we saw in Chapter 9.

The overall X^2 and G^2 statistics are 110.82 and 93.29, respectively, on 104 degrees of freedom but they are not useful due to the sparseness issue. The fit of the model judged by the chi-squared residuals on the two-way margins and the (1,1,1) three way margins is greatly improved. There are no discrepancies greater than 1 for any pair or triple of items. The amount of G^2

explained increases from 27.7% to 56.8% as we go from two to three classes. The AIC and BIC also selected the three-class model over the two-class model (two-class: AIC=2881.47, BIC=2941.03, and three-class: AIC=2831.75 and BIC=2923.01).

It therefore appears that the two-class solution did convey the essence of the situation, but that the three-class solution has enabled us to describe the minority group in more detail.

The sexual attitudes data revisited

The questions on sexual attitudes from the 1990 British Social Attitudes Survey were analysed in Chapter 8 using a two-factor latent trait model. We found that the two-factor solution was unsatisfactory both in terms of fit and by failing to produce two clear-cut factors. Here we reanalyse the data using a latent class model. The same set of questions has been thoroughly analysed in de Menezes and Bartholomew (1996).

Latent class models with two, three, and four classes were fitted to the ten items. The AIC and BIC values are given in Table 10.11. The four-class model is selected over the two- and three-class model.

TABLE 10.11: AIC and BIC, sexual attitudes data

	Two-class	Three-class	Four-class
AIC	9328.43	8947.53	8849.90
BIC	9433.05	9106.95	9064.12

The parameter estimates for the two, three, and four-class solution are given in Table 10.12. The two-class solution divides the population into two groups of similar size (49 and 51%). Individuals in Class 1 have low probabilities of agreeing with any of the ten items except items 2, 3 and 7. Members in Class 2 have higher probabilities of agreeing with all the items except item 1. The two-class solution provides a permissive/non-permissive dichotomisation of the population. However, the two-class solution was not a good fit. We proceed to the three-class solution, which has the first class to be the same in terms of size and response probabilities as Class 1 of the two-class solution. Class 2 remains, to some extent, as the permissive one, but with low probabilities on the adoption items 9 and 10. Class 3 comprises 11% of the population, and members in this class have the highest probability of agreeing with each item. In the four-class solution, the first two classes remain the same as the first two classes in the three-class solution. The 11% in the third class is now split into two classes. One is identical to the third class of the three-class solution, and the other one consists of only 1% of the population. That class reveals a completely different group of individuals showing a positive attitude towards the adoption items but opposing gay people teaching in schools and universities, and holding public positions.

TABLE 10.12: Estimated conditional probabilities, $\hat{\pi}_{ij}$, and prior probabilities, $\hat{\eta}_j$, sexual attitudes data

Item	Two-class		Three-class			Four-class			
	$\hat{\pi}_{i1}$	$\hat{\pi}_{i2}$	$\hat{\pi}_{i1}$	$\hat{\pi}_{i2}$	$\hat{\pi}_{i3}$	$\hat{\pi}_{i1}$	$\hat{\pi}_{i2}$	$\hat{\pi}_{i3}$	$\hat{\pi}_{i4}$
1	0.13	0.12	0.13	0.09	0.21	0.14	0.09	0.21	0.07
2	0.76	0.88	0.76	0.87	0.92	0.77	0.87	0.92	0.60
3	0.64	0.88	0.64	0.86	0.96	0.63	0.86	0.96	0.87
4	0.09	0.17	0.09	0.13	0.30	0.08	0.13	0.31	0.27
5	0.08	0.49	0.08	0.38	0.82	0.07	0.38	0.83	0.60
6	0.01	0.92	0.01	0.87	0.97	0.01	0.87	1.00	0.00
7	0.08	0.99	0.06	0.98	1.00	0.06	0.98	1.00	0.20
8	0.23	0.94	0.21	0.91	0.99	0.21	0.91	1.00	0.27
9	0.07	0.30	0.07	0.10	0.98	0.05	0.11	0.98	1.00
10	0.02	0.19	0.02	0.00	0.84	0.00	0.00	0.85	1.00
$\hat{\eta}_j$	0.49	0.51	0.47	0.41	0.11	0.46	0.42	0.11	0.01

In the sexual attitudes example, the latent class analysis is more revealing than the factor analysis.

10.5 The latent class model for ordinal and nominal data

As noted earlier, ordinal (ordered) and nominal (unordered) manifest variables are commonly encountered in social science research. For example, a situational judgement test is a type of psychological test widely used in professional recruitment processes to assess how candidates might behave in a work-related situation. An item in such a test may present a work-like situation and a number of possible actions they could take, where the possible actions do not have a natural ordering. In this case, candidates' item responses are unordered categorical manifest variables, and a latent class analysis could reveal different behavioural patterns in work-related situations. For another example, multiple-choice items are common in educational testing. When the goal is to measure students' performance, their item responses are binary-scored as correct and incorrect based on the answer key, which can then be analysed using a latent trait model or a latent class model for binary data. However, a latent class analysis of the unscored raw responses may also be of interest. By identifying the specific choices the students are likely to make, this analysis may provide insights into the subgroups of students and the misconceptions that each subgroup has. An example of multiple-choice data is given in Section 10.8.

To deal with ordinal and nominal manifest variables, we describe an extension of the latent class model in Section 10.4. This model can be applied to both ordered and unordered manifest variables. However, when applied to ordered variables, the model ignores the ordering information and treats them as unordered variables.

Similar to the setting in Chapter 9, suppose we have an $n \times p$ data matrix, where each column i records an ordinal/nominal variable y_i with possible categories $1, \ldots, m_i$, where m_i is the total number of categories for variable i. These categories do not need to have an order. The key assumptions of a latent class model for ordinal/nominal variables with J classes are the same as those of a model for binary variables. They are:

i) The n individuals are a random sample from some population and every individual in that population belongs to just one of the J latent classes.

ii) The probability of giving a positive response to a particular item is the same for all individuals in the same class but may be different for individuals in different classes.

iii) Given the latent class to which an individual belongs, its responses to different items are conditionally independent.

The difference between these models lies in the conditional distribution of each y_i given an individual's latent class membership. In the model for binary variables, this conditional distribution follows a Bernoulli distribution, while in the model for ordinal/nominal variables, it becomes a categorical distribution. As a result, more parameters are needed. More specifically, we denote $\pi_{ijk} = \Pr(y_i = k \mid c = j)$, $k = 1, \ldots, m_i$, $i = 1, \ldots, p$, and $j = 1, \ldots, J$. These parameters satisfy the constraints

$$\sum_{k=1}^{m_i} \pi_{ijk} = 1$$

for each $i = 1, \ldots, p$ and $j = 1, \ldots, J$, as one has to choose one from the m_i categories. Therefore, the current model has $m_i - 1$ free parameters for each pair of item i and latent class j, while the model for binary data only has one free parameter. We again let η_j be the proportion of the population in latent class j.

The model fitting, interpretation, assessment of goodness-of-fit, and allocation to classes of this model follow the same principles as the latent class model for binary data. We, therefore, do not repeat all the details here. However, we note that the interpretation of the latent classes is now based on the estimated π_{ijk} parameters. While the orders of the categories are not accounted for in the latent class model and its fitting when the manifest variables are ordered, this information could be useful when interpreting the latent classes. We illustrate this point using the attitude to science and technology data that we have analysed earlier in this chapter in the binary form and in Chapter 9 with a latent trait model. In this example, $m_i = 4$ for all the items. Similar to the binary data example, we have used the reverse coding for the negatively worded items, 2, 5, and 6. Thus, the code of 1 corresponds to the most positive attitude towards science and technology, and the code of 4 corresponds to the most negative attitude. Table 10.13 gives the estimated parameters for a two-class model. The two classes are estimated to have about equal sizes. Overall, individuals in the second class have a more positive attitude towards science and technology than those in the first class. This can be seen from the fact that this class is more likely to give a response in a higher category for all the items. While the interpretation of these two classes is similar to that of the two-class model for the binary version of the same data, there are some notable differences. First, the estimated population proportions are quite different. The two classes are of almost equal sizes in the current model, while one class is much bigger than the other in the model for binary data. Second, the model-implied conditional probabilities are quite different. For example, according to the result in Table 10.13, the conditional probabilities of answering "agree to some extent" or "strongly agree" are 0.90 and 0.91 for the two latent classes, while these conditional probabilities are 0.75 and 0.95 under the estimated model for binary data. These differences are likely due to the information loss when converting the ordinal/nominal variables into binary ones.

TABLE 10.13: Estimated conditional probabilities, $\hat{\pi}_{ijk}$, and prior probabilities, $\hat{\eta}_j$, for the two-class model with categorical manifest variables, science and technology data

Items	$\hat{\pi}_{i11}$	$\hat{\pi}_{i12}$	$\hat{\pi}_{i13}$	$\hat{\pi}_{i14}$	$\hat{\pi}_{i21}$	$\hat{\pi}_{i22}$	$\hat{\pi}_{i23}$	$\hat{\pi}_{i24}$
Comfort	0.01	0.09	0.79	0.11	0.02	0.07	0.57	0.34
Environment	0.04	0.33	0.51	0.11	0.11	0.12	0.23	0.54
Work	0.04	0.34	0.58	0.04	0.12	0.16	0.47	0.24
Future	0.02	0.22	0.66	0.10	0.06	0.15	0.41	0.39
Technology	0.02	0.32	0.55	0.11	0.07	0.14	0.25	0.54
Industry	0.01	0.19	0.73	0.07	0.04	0.05	0.15	0.76
Benefit	0.02	0.33	0.58	0.07	0.09	0.18	0.40	0.33
$\hat{\eta}_j$			0.50				0.50	

10.6 How can we distinguish the latent class model from the latent trait model?

One reason for using the attitude to abortion data to illustrate both the latent trait model and the latent class model was to show, incidentally, that the two models are often equally successful in fitting a set of data. In this particular case, neither model provided a particularly good fit, but both were good enough to provide reasonable interpretations. The results for the attitude to abortion data are given in Table 10.14.

TABLE 10.14: Observed and expected frequencies under the latent trait and the latent class models, attitude to abortion data

Observed	Expected frequency for latent trait	Expected frequency for latent class	Response pattern
103	100.0	98.0	0000
13	16.6	17.8	0001
1	1.7	1.1	1000
9	9.1	10.3	0100
10	12.3	14.2	0010
6	7.4	4.1	0101
3	1.0	0.6	1100
21	14.8	7.8	0011
7	12.3	6.7	0110
3	1.9	5.4	1101
6	6.2	13.0	1011
44	41.1	54.2	0111
12	7.2	12.8	1110
141	143.9	130.9	1111

In other examples where the fit is much better, one often finds that the expected frequencies for the two models are virtually identical. At first sight, it seems surprising that models which make such different assumptions about the distribution of the latent variable should give such similar fits. In the case of the latent trait model, the latent variable is assumed to have a standard normal distribution. We have already remarked that the choice of the form of this distribution is not critical. It now appears that even if we replace the continuous prior by a distribution of the latent variable which is concentrated on a small number of discrete points (which is what the latent class model amounts to), the fit is not much affected. Another way of putting this is to say that the distribution of the latent variable is poorly determined by the data. What, then, are we to do about interpretation when our various methods for judging fit have little chance of distinguishing between the models?

Two things may be said. First, we should be wary of pressing our conclusions to the point where the distinction between the models has practical implications. For example, our analysis of the attitude to abortion data makes it very clear that people differ in their attitude to abortion. It is less clear whether that variation is best described by a distribution of attitudes along a continuum or by a polarisation into two groups. There would therefore be little empirical justification for claiming that about 61% of the population were pro-abortion, because that depends on the selection of the latent class model as the more appropriate. It might, however, be the appropriate model to use if one wanted to predict the outcome of a vote on the question since that would force people to adopt one or other stance.

Secondly, background knowledge, which cannot easily be quantified, may favour one model over the other. The sexual attitudes data illustrates this point. Having fitted a model with two continuous factors, we still did not have a very satisfactory fit, although it was clear that the latent trait model did capture some important aspects of the variation between individuals. The latent class model seemed to be more successful in providing an interpretable solution by identifying minority groups whose existence seemed credible on general grounds.

The relationship between the latent trait and latent class models can be explored theoretically, and the interested reader may refer to Bartholomew, Knott, and Moustaki (2011), (p.159-161). There is one aspect of this comparison which we mention here because it links back to the "threshold" effect which we met in Chapter 8. We noted there that estimated values of the discrimination parameter α_{i1} were sometimes very large. This implied that the item response function was almost vertical at its central point, which means that the probability of a positive response switches from a low value to a high value at that point. This is why it is called a threshold effect. But this is exactly what happens with the two-class latent class model — as we move from one class to the other, the probability of a positive response also changes abruptly. The appearance of large discrimination parameters in a latent trait model is therefore an indication that a latent class model might be more appropriate.

10.7 Latent profile analysis, latent class analysis and cluster analysis

Both latent profile analysis and latent class analysis involve a class allocation step, which can be viewed as a form of cluster analysis (CLA). They differ from the methods given in Chapter 2 principally in that they start from a statistical model. It is worth pointing out some of the similarities and differences. CLA typically begins by constructing a similarity (or distance) matrix and then goes on to look for clusters of objects or individuals which are close together. Latent profile/class analysis begins with specifying a categorical latent variable, which assumes that there are J clusters in the population, and then allocates objects to clusters based on the estimated model. Unlike CLA, this allocation is probabilistic; instead of knowing which cluster an object belongs to, we merely have probabilities of belonging to the various clusters. The rationale behind CLA is rooted in the similarities between rows of the data matrix. On the other hand, latent profile/class analysis is based on the distribution of the elements in the rows.

10.8   Additional examples and further work

Five more examples will be considered in this section, of which one applies the Gaussian mixture model, three apply the latent class model for binary responses, and one applies the latent class model for nominal responses. Among the examples of latent class analysis for binary data, two datasets have been analysed in Chapter 8 using the latent trait model for binary responses.

Big-five personality data

We consider an example of a big-five personality test based on a dataset studied in Johnson (2014). It contains the scores of 16489 UK participants on five personality scales, where a participant's continuous score for each scale was calculated based on their responses to 24 items. These five scales measure the big-five personality factors of neuroticism (N; sensitive/nervous vs. resilient/confident), extraversion (E; outgoing/energetic vs. solitary/reserved), openness to experience (O; inventive/curious vs. consistent/cautious), agreeableness (A; friendly/compassionate vs. critical/judgmental), and conscientiousness (C; efficient/organised vs. extravagant/careless). The data are pre-standardardised so that each variable has a mean of zero and a variance of one.

We apply Gaussian mixture models to the data, with the number of classes J varying from 2 to 10. Models with different constraints on the covariance matrices Σ_j are considered, including those without constraints. The top five models with the smallest BIC values all have unconstrained covariance matrices. The BIC values of these models are given in Table 10.15. Below, we interpret the model with the smallest BIC value, i.e. the model with $J = 5$ latent classes. However, we should note that the BIC values of the top models are close to each other, meaning that the other models in this table are all competitive. In real-life scenarios, we may want to examine these models to gain further insights.

TABLE 10.15: The number of latent classes (J) and BIC values for the five Gaussian mixture models with the smallest BIC values, big-five personality data

J	5	4	6	3	7
BIC	221975	221993	222016	222040	222108

Table 10.16 gives the estimated mean vectors, i.e. the centres, of the five latent classes. From this table, we see that the classes are of comparable sizes, with the estimated population proportions taking values between 0.1 and 0.3. The classes may be interpreted based on the pattern of the estimated mean vectors. For example, Class 1 has the most negative mean score for the N factor, the most positive mean scores for the A and C factors, and moderate mean scores for the other two factors. Its most extreme mean scores suggest that individuals in this class tend to be the most resilient, confident, friendly, compassionate, efficient, and organised among all the classes. The remaining classes can be interpreted similarly.

TABLE 10.16: Estimated mean vectors and population proportions of the five-class Gaussian mixture model with unconstrained covariance matrices, big-five personality data

Variable	$\hat{\mu}_1$	$\hat{\mu}_2$	$\hat{\mu}_3$	$\hat{\mu}_4$	$\hat{\mu}_5$
N	-0.93	0.26	0.08	0.33	-0.20
E	0.49	0.09	-0.09	-0.77	0.89
O	0.04	-0.52	0.71	-0.33	-0.12
A	0.43	0.25	0.30	-0.70	-0.35
C	0.77	0.01	-0.10	-0.22	-0.27
$\hat{\eta}_j$	0.14	0.23	0.28	0.23	0.12

Macready and Dayton data

This dataset arises from educational testing where one wishes to study the learning process in children. Macready and Dayton (1977) used a two-class model on a set of four tests to identify two distinct classes; that of "masters" and "non-masters". The frequency distribution of the response patterns together with the expected frequencies for each response pattern under the two-class model and the probability of belonging to the "master" class are given in Table 10.17. The estimated posterior probabilities show that most of the individuals can be allocated to the "master" class or the "non-master" class with high confidence. For example, individuals who have responded to at least two items correctly have probabilities greater than 0.9 of being allocated into the "master" class. On the other hand, individuals who got all items wrong (0000) are allocated to the "non-master" class with probability 0.98. There are, however, three response patterns (0100, 0001, 0010) that have probabilities close to a half. These include only 11 individuals out of 142. Therefore, the latent class model has been able to classify with high confidence most of the response patterns in the two classes.

TABLE 10.17: Observed and predicted frequencies and estimated class probabilities for the two-class model, Macready and Dayton data

Observed frequency	Expected frequency	$\hat{Pr}$(master | y)	Class	Response pattern
15	14.96	1.00	2	1111
23	19.72	1.00	2	1101
7	6.19	1.00	2	1110
4	4.90	1.00	2	0111
1	4.22	1.00	2	1011
7	8.92	0.91	2	1100
6	6.13	0.90	2	1001
5	6.61	0.98	2	0101
3	1.93	0.90	2	1010
2	2.08	0.97	2	0110
4	1.42	0.97	2	0011
13	12.91	0.18	1	1000
6	5.62	0.47	1	0100
4	4.04	0.45	1	0001
1	1.31	0.44	1	0010
41	41.04	0.02	1	0000

The $X^2 = 9.5$ and the $G^2 = 9.0$ on six degrees of freedom indicate a near perfect fit to the data. The percentage of G^2 explained is 91%. The parameter estimates and standard errors of the model are given in Table 10.18. Members of the first class have small estimated probabilities of answering items correctly. This class is clearly the "non-master" one. Members in the second class have for all items much higher probabilities of answering correctly. This class is the "master" class.

TABLE 10.18: Estimated conditional probabilities, $\hat{\pi}_{ij}$, and prior probabilities, $\hat{\eta}_j$, with standard errors in brackets for the two-class model, Macready and Dayton data

Item (i)	$\hat{\pi}_{i1}$	$\hat{\pi}_{i2}$
1	0.21 (0.06)	0.75 (0.06)
2	0.07 (0.06)	0.78 (0.06)
3	0.02 (0.03)	0.43 (0.06)
4	0.05 (0.05)	0.71 (0.06)
$\hat{\eta}_j$	0.41 (0.06)	0.59 (0.06)

You could reanalyse the data using a latent trait model and compare the expected frequencies obtained under the latent class model and the latent trait model.

Women's mobility

The eight items on women's mobility analysed in Chapter 8 can be analysed using latent class analysis. The items indicate whether a woman living in rural Bangladesh could engage in various activities alone.

The aim is to identify groups of women with similar patterns of mobility. You should be able to show that the two- and three-class models give poor fits to the second and third order margins. For the two-class model, chi-squared residuals greater than 100 occurred for many pairs and triplets of items. The three-class model improves the fit considerably, but there are still pairs of items with residuals greater than 20. The four-class model improves the fit even further. The percentage of G^2 explained increases from 72 to 93.7% as we go from the two to three classes, and to 96.7% as we increase the number of classes to four. Table 10.19 shows the two-way margins and the (1,1,1) three-way margins for which the chi-squared residuals were greater than 3 under this model.

TABLE 10.19: Chi-squared residuals greater than 3 for the second and the (1,1,1) third-order margins for the four-class model, women's mobility data

	Items	$O - E$	$(O - E)^2/E$
(0,1)	2, 6	40.45	7.00
	2, 8	−25.37	3.80
	5, 8	−42.82	4.79
	6, 7	−14.69	3.88
	8, 5	−42.82	7.74
(1,1)	8, 5	42.82	5.25
(1,1,1)	1, 5, 8	45.77	6.05
	2, 5, 8	45.86	6.89
	3, 5, 8	44.70	5.80
	4, 5, 8	44.81	5.88

Table 10.20 shows the estimated conditional probabilities of responding positively to each item given class membership, and the prior probabilities of belonging to each of the four classes. The classes have a clear interpretation and appear in increasing order of social freedom. Women in the first class have the lowest degree of mobility; the only activities in which a reasonable proportion of women in this class can engage are moving within their locality and talking to a man they do not know (items 1 and 3). Women in the second class have a greater level of social freedom than those in the first class since most of them responded positively to items 1 and 3, and, in addition, more than a quarter reported that they could go outside their village or town, or visit the cinema. It is estimated that the majority (73%) of women fall into one of the first two classes. Women in latent class 3 have still more freedom; most women in this class can engage in the first four activities. Finally, the fourth class contains a small group of women with a very high level of mobility.

TABLE 10.20: Estimated conditional probabilities, $\hat{\pi}_{ij}$, and prior probabilities, $\hat{\eta}_j$ for the four-class model, women's mobility data

Items	$\hat{\pi}_{i1}$	$\hat{\pi}_{i2}$	$\hat{\pi}_{i3}$	$\hat{\pi}_{i4}$
1	0.44	0.98	0.98	0.99
2	0.01	0.27	0.72	0.89
3	0.46	0.86	0.96	0.99
4	0.04	0.26	0.92	0.98
5	0.00	0.00	0.11	0.78
6	0.00	0.03	0.20	0.96
7	0.00	0.00	0.03	0.80
8	0.01	0.00	0.15	0.82
$\hat{\eta}_j$	0.34	0.39	0.21	0.06

Workplace Industrial Relations Survey, WIRS

The WIRS data were analysed in Chapter 8 with a factor analysis model for binary variables. The same set of six items can be analysed using latent class analysis.

The six items, given in Chapter 8, measure the amount of consultation that takes place in firms at different levels of the firm structure. Items 1 to 6 cover a range of informal to formal types of consultation. The first two items are less formal practices, and items 3 to 6 are more formal. The factor analysis for binary items model revealed that the two-factor model was a good fit to the data.

The latent class analysis aims to group firms according to the patterns of consultation they are adopting.

The two-class model fitted to the six items is rejected not only by the overall goodness-of-fit measures ($X^2 = 369.70, G^2 = 306.45$ on 50 degrees of freedom) but also by the large chi-squared residuals for some of the two- and three-way margins. All the chi-squared residuals with values greater than 3 include item 1. The three-class model is still rejected ($X^2 = 76.71, G^2 = 76.15$ on 43 degrees of freedom). However, you should be able to show that the fit to the two- and three-way margins is very good. The estimated probabilities for this model are given in Table 10.21. Class 1 represents those firms that mainly use informal policies (items 1 and 2). Class 3 includes those firms that use all the methods, except the first informal one. Lastly, firms in Class 2 use all methods, including those under item 1 (with lower probabilities than in Class 3 for items 2 to 6). The last row of the table estimates that the majority of the firms (55%) are in Class 1.

Both the latent class analysis and the factor analysis for binary variables had problems fitting item 1 (the most informal one). In Chapter 8, the analysis was repeated without item 1. We suggest that you repeat the latent class analysis with item 1 omitted.

TABLE 10.21: Estimated conditional probabilities, $\hat{\pi}_{ij}$, and prior probabilities, $\hat{\eta}_j$, for the three-class model, WIRS data

Items	$\hat{\pi}_{i1}$	$\hat{\pi}_{i2}$	$\hat{\pi}_{i3}$
1	0.21	0.95	0.06
2	0.59	0.27	1.00
3	0.08	0.43	0.68
4	0.14	0.19	0.62
5	0.11	0.53	0.85
6	0.02	0.25	0.37
$\hat{\eta}_j$	0.55	0.26	0.19

Multiple-choice item response data

We consider a dataset containing 472 students' raw responses to eight multiple-choice test items. Each item has four candidate choices, only one of which is correct. A student can only choose one of the four candidate choices. Through latent class analysis, we aim to identify subgroups of students and the common mistakes within each subgroup, from which the teacher can identify the misconceptions each subgroup has and develop tailored instructions accordingly. The raw data of multiple choices provides more information than the binary-scored data, allowing us to identify the specific distractors that each subgroup is likely to choose.

In what follows, we present the result of a three-class model for the nominal response data. You may try models with different numbers of latent classes. Table 10.22 gives the estimated conditional probabilities $\hat{\pi}_{ijk}$ and population proportions $\hat{\eta}_j$. The correct choices are indicated by the boldface $\hat{\pi}_{ijk}$s. For example, the correct choices for the first two items are 2 and 3, respectively. We see that the first class is the largest class among the three, with an estimated population proportion of 0.62. This is also the most capable subgroup of students. Students in this group have a relatively high chance of answering each question correctly.

Students in both the second and third classes may have some misconceptions about the knowledge tested by these questions. Specifically, the second class has an estimated population proportion of 0.34, which covers about one-third of the population. Based on the result in Table 10.22, students in this class are likely to choose one or two distractors for each item. For example, besides the correct choices, they are likely to choose the first candidate choice for the first item and choose the first and fourth candidate choices for the second item. For items 2 and 6, the probabilities of selecting the correct answer are even smaller than choosing some specific distractors. This information, when combined with substantive knowledge about the items and their distractors, may provide insight into the misconceptions these students hold.

TABLE 10.22: Estimated conditional probabilities, $\hat{\pi}_{ijk}$, and prior probabilities, $\hat{\eta}_j$, for the three-class model, multiple-choice data

Items	$\hat{\pi}_{i11}$	$\hat{\pi}_{i12}$	$\hat{\pi}_{i13}$	$\hat{\pi}_{i14}$	$\hat{\pi}_{i21}$	$\hat{\pi}_{i22}$	$\hat{\pi}_{i23}$	$\hat{\pi}_{i24}$
1	0.13	**0.87**	0.01	0.00	0.31	**0.64**	0.02	0.02
2	0.10	0.01	**0.83**	0.06	0.31	0.10	**0.21**	0.38
3	**0.89**	0.03	0.08	0.00	**0.43**	0.36	0.14	0.06
4	**0.85**	0.00	0.06	0.09	**0.61**	0.01	0.30	0.09
5	0.04	0.02	0.00	**0.94**	0.21	0.12	0.02	**0.64**
6	**0.79**	0.05	0.05	0.12	**0.29**	0.37	0.23	0.10
7	0.03	**0.74**	0.12	0.11	0.13	**0.51**	0.25	0.10
8	**0.89**	0.10	0.00	0.01	**0.41**	0.36	0.19	0.05
$\hat{\eta}_j$		0.61				0.35		

Items	$\hat{\pi}_{i31}$	$\hat{\pi}_{i32}$	$\hat{\pi}_{i33}$	$\hat{\pi}_{i34}$
1	0.47	**0.48**	0.05	0.00
2	0.17	0.48	**0.19**	0.16
3	**0.16**	0.48	0.29	0.07
4	**0.00**	0.21	0.41	0.38
5	0.81	0.14	0.05	**0.00**
6	**0.00**	0.72	0.06	0.22
7	0.10	**0.18**	0.36	0.36
8	**0.00**	0.74	0.00	0.26
$\hat{\eta}_j$		0.04		

While the third latent class is small (with an estimated population proportion of 0.04), students in this subgroup likely have the most misconceptions and thus need the teacher's attention. They have higher probabilities of choosing the distractors for all the items than the second latent class, and their probabilities of choosing the correct answers for items 4, 5, 6, and 8 are almost zero.

10.9 Software

A number of software packages such as Latent Gold (Vermunt and Magisdon 2005), Mplus (Muthén and Muthén 2017), and WinLTA (Collins et al. 1999) are available for estimating latent class models. There is also the R Package PoLCA (Linzer and Lewis 2011) for fitting latent class models to categorical data and the R package Mclust (Scrucca et al. 2023) for Gaussian mixture models. For a review of R packages for Gaussian mixture models in R see Chassagnol et al. (2023).

10.10 Further reading

Bartholomew, D. J., Knott, M. and Moustaki, I. (2011). *Latent Variable Models and Factor Analysis: A Unified Approach* (3rd ed.). West Sussex, United Kingdom: John Wiley & Sons.

Hagenaars, J. A. and McCutcheon, A. L. (eds.) (2002). *Applied Latent Class Analysis*. Cambridge: Cambridge University Press.

Heinen, T. (1996). *Latent Class and Discrete Latent Trait Models*. Thousand Oaks: Sage Publications.

Krzanowski, W. J. and Marriott, F. H. C. (1995b). *Multivariate Analysis, Part 2: Classification, Covariance Structures, and Repeated Measurements*. London: Arnold.

11

Confirmatory Factor Analysis and Structural Equation Models

11.1 Introduction

This chapter will introduce the main ideas of confirmatory factor analysis (CFA) and structural equation modelling (SEM) with latent variables. As will become clear from the discussion later, CFA is mainly concerned with testing hypotheses about the values of factor loadings (usually, that some are zero), SEM is concerned with estimating (linear and non-linear) relationships between factors (latent variables).

The reader should not be deterred by noticing that many of the models discussed in this chapter are more complicated than those that have gone before. An appropriate software package will take care of all the technicalities. SEM software packages even allow one to specify the model graphically rather than in terms of equations. As always, our emphasis here will be on the formulation of the model and the interpretation of the results – not on the mathematical analysis which lies between the two.

Chapters 7–9 presented factor analysis for continuous and categorical observed responses. The number of factors was unknown and was estimated using the available goodness-of-fit statistics and measures of fit. The choice of the number of factors may also be guided by PCA. Latent variables have been taken to be independent except where an oblique rotation was applied to the factor solution to simplify the interpretation. That type of analysis is known as exploratory factor analysis (EFA).

It is important at this stage to make a clear distinction between exploratory and confirmatory factor analysis. EFA analyses a set of correlated observed variables without knowing in advance either the number of factors that are required to explain their interrelationships or their meaning or labelling. Depending on the q-factor model finally chosen (based on goodness-of-fit criteria and fit measures) and the rotation applied (orthogonal or oblique), one names the factors according to the indicators to which each factor is related. In addition, residuals are not allowed to be correlated with each other. Finally, note that the choice of the number of factors depends on statistical criteria or rules that might not be clearly defined or valid.

DOI: 10.1201/9781003483342-11

CFA (Jöreskog 1969) postulates certain relationships among the observed and the latent variables assuming a pre-specified pattern for the model parameters (factor loadings, structural parameters, residual variances). CFA is mainly used for testing a hypothesis arising from theory. Therefore, the number of latent variables and the indicators that will be used to measure each latent variable are known in advance. This implies that the expert has sufficient knowledge to define the model hypothesis, that is, the relationships between latent variables or *constructs* and the observed variables or *indicators* that they explain. To run a CFA, one needs to impose constraints on the parameters in such a way that not all items are linked to all latent variables. Setting a factor loading equal to zero, such as $\alpha_{ij} = 0$, implies that factor j does not have a direct influence on item i. In CFA, the structure must be determined before the data are examined. For example, an educationalist might believe that success in humanities depends on one factor and in mathematics, on another. She might therefore specify a model for the data giving the marks of the subjects (Table 11.1) in which the loadings of factor 1 on Arithmetic, Algebra, and Geometry are zero and, similarly, the loadings of factor 2 on Gaelic, History and English are zero. The analysis would then estimate the remaining factor loadings and other parameters of the model. The goodness-of-fit of this model could be evaluated. Therefore, the focus is on testing an existing theory. The factors may be correlated and, if so, estimating their correlation is part of the analysis.

TABLE 11.1: Factor loadings with standard errors, estimated variances of the error terms, and coefficients of reliability /communalities for the two-factor confirmatory model, subject marks data (parameters in blank cells are constrained to equal zero)

Item	$\hat{\alpha}_{i1}$	s.e.	$\hat{\alpha}_{i2}$	s.e.	$\widehat{Var}(\delta_i)$	R^2
Gaelic	0.69	0.08			0.52	0.48
English	0.67	0.08			0.55	0.45
History	0.53	0.08			0.72	0.28
Arithmetic			0.76	0.07	0.42	0.58
Algebra			0.76	0.07	0.42	0.58
Geometry			0.62	0.07	0.62	0.38

We should note here that there is not always a clear distinction between an EFA and a CFA. Researchers improve the fit of a confirmatory factor analysis model by reducing the number of constraints and by allowing for more parameters to be estimated. Also, results from an EFA might lead a researcher to exclude certain items from the analysis. Both of those strategies will result in mixing elements of EFA and CFA. SEM develops from CFA by studying the relationships between the latent variables.

In Chapter 7, the linear factor model with two factors (EFA) was fitted to the subject marks data described in Section 5.5. The first factor was found to measure overall ability in the six subjects, while the second contrasted humanities and mathematics subjects. An orthogonal rotated factor solution revealed one factor, which we identified with mathematics (M) and a second factor for humanities (H). An oblique rotation brought this out more clearly, showing a strong positive correlation between the two factors. An educationalist would have probably started the analysis by fitting a CFA model allowing for the subjects Gaelic, History and English to be indicators of ability in humanities, and the subjects Arithmetic, Algebra and Geometry to be indicators of mathematical ability. In the CFA, the two latent variables would be treated as correlated.

Examples of problems involving confirmatory factor analysis

i) Psychologists are interested in estimating how the verbal ability of children is related to visual perception and writing ability. A set of observed indicators is required for measuring each one of those three constructs or latent variables (e.g. tests of writing, recognising and distinguishing among different shapes, and reading tests). The relationships between the indicators and the constructs are expressed through a measurement model. There is also interest in estimating the correlations among those three constructs.

ii) Measuring changes in political and social attitudes across time. For example, a set of indicators is used to measure "political efficacy" at different points in time (an example of measurement model). Since latent variables measure the same construct or constructs across time, we could estimate the correlation between them. Furthermore, one is also interested in studying changes in the means and variances of the latent variables across time.

11.2   Path diagram

A path diagram is a pictorial representation of how the different kinds of variables entering into the model are related. We have already met a simple example of a path diagram in our brief introduction to Path Analysis in Section 6.15. Here we describe a path diagram in a sufficiently general way to cover both CFA and SEM, to which we come later in the chapter. The path diagram shows the relationships between the observed and the latent variables (measurement model), among the latent variables (structural model) and between covariates and the latent variables (structural model). It also includes residual terms also known variously in the literature as *measurement errors*, *disturbances*, *specific* or *unique* factors. In the path diagram, observed variables appear in boxes, latent variables appear in circles, and disturbance or error terms appear unenclosed. Covariances or correlations between latent variables and between disturbances are represented by curved lines with arrowheads at both ends. Relationships between observed variables and latent variables, as indicated in the measurement part of the model, and between latent variables, as indicated by the structural part of the model, are represented by straight lines with an arrowhead pointing towards the dependent variable. Two straight lines with headed arrows connecting two variables denote a reciprocal relationship (see Figure 11.5 and accompanying text). The models discussed in this chapter will be represented both by path diagrams and mathematical equations.

Figures 11.1 and 11.2 give the conceptualised path diagrams for the exploratory factor analysis with uncorrelated factors and a confirmatory factor model with correlated factors respectively for the subject marks data of Chapter 7. As we see from both figures, the indicator variables are included in boxes, and the latent variables appear inside circles. The terms δ_i $(i = 1, \ldots, p)$ show that each indicator is measured with error. This is usually referred to as an item-specific error or measurement error. It is clear that there is no need to use a path diagram to represent an EFA model because EFA does not impose any constraints on the factor loadings.

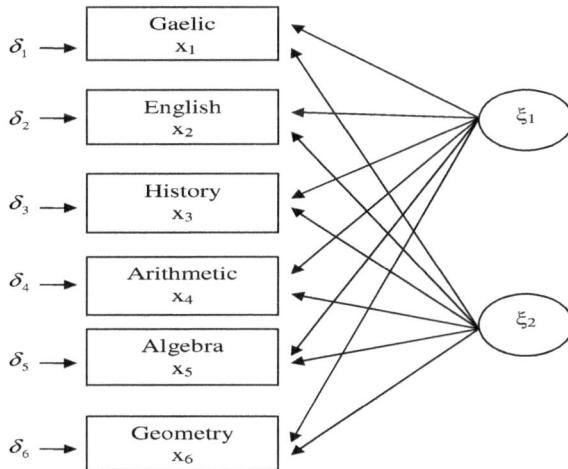

FIGURE 11.1: Path diagram, exploratory factor analysis, subject marks data

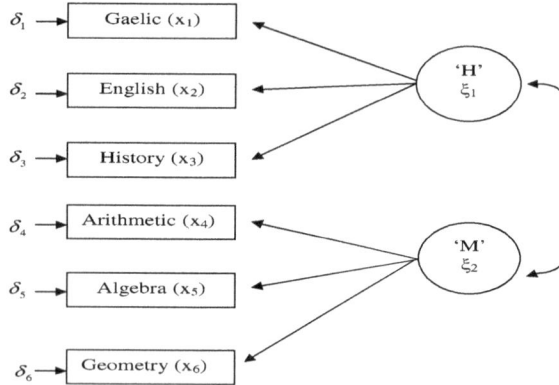

FIGURE 11.2: Path diagram, confirmatory factor analysis, subject marks data

11.3 Measurement models

Measurement equations are already known to us from Chapters 7–9. Throughout this chapter, we adopt the standard notation used in SEM. The notation used in the measurement equations given in eqs. (7.3), (8.3) and (9.1) will now be changed. Latent variables will now be denoted by the Greek letter ξ_j instead of the letter $f_j (j = 1, \ldots, r)$ and the residual term will be denoted by the Greek letter $\delta_i (i = 1, \ldots, p)$ instead of $e_i (i = 1, \ldots, p)$. We retain the same notation as in Chapters 7–9 for the factor loadings $(\alpha_{ij}, i = 1, \ldots, p; j = 1, \ldots, r)$ and intercept parameters $(\alpha_{i0}, i = 1, \ldots, p)$. In SEM notation, Greek letters are used to denote latent variables and residual terms, and Latin letters are used for the observed variables. In Chapters 7–9, we used y for the observed variables, but in this chapter, we will need two different notations for the observed variables. Therefore, we first introduce the observed variables denoted by x $(x_i, i = 1, \ldots, p)$ as the indicators for the latent variables ξ. More about notation will be given later in the chapter.

In Figure 11.1, all items are taken to be indicators of the two latent variables and no correlation is allowed between the latent variables. In the notation of the present chapter, the measurement equations may be written as:

$$
\begin{aligned}
x_1 &= \alpha_{10} + \alpha_{11}\xi_1 + \alpha_{12}\xi_2 + \delta_1 \\
x_2 &= \alpha_{20} + \alpha_{21}\xi_1 + \alpha_{22}\xi_2 + \delta_2 \\
x_3 &= \alpha_{30} + \alpha_{31}\xi_1 + \alpha_{32}\xi_2 + \delta_3 \\
x_4 &= \alpha_{40} + \alpha_{41}\xi_1 + \alpha_{42}\xi_2 + \delta_4 \\
x_5 &= \alpha_{50} + \alpha_{51}\xi_1 + \alpha_{52}\xi_2 + \delta_5 \\
x_6 &= \alpha_{60} + \alpha_{61}\xi_1 + \alpha_{62}\xi_2 + \delta_6,
\end{aligned}
$$

where x_i denotes an observed variable, ξ_1 and ξ_2 are the overall ability factor and the factor contrasting humanities and mathematics subjects respectively and δ_i is the item specific residual. The parameters α_{i0} are needed when the observed variables are not centred. The parameters α_{i1} and α_{i2} are the factor loadings.

In Figure 11.2, the factor loadings of the first latent variable, "Humanities", on the mathematical subjects are set to zero and corresponding values are assigned for the factor loadings of the second latent variable "Mathematics" on the humanities subjects. The correlation between the two latent variables is indicated by the curved line. The straight arrow-headed lines from the x variables to the ξ variables represent one-way effects of the latent variables ξ on the observed variables x. As has been emphasised for all the factor models in the book, the existence of the latent variables ξ_1 and ξ_2 is taken to be responsible for the association between the six observed variables. When those two correlated latent variables are held fixed, the answers to the six items are independent (known as *conditional independence*). This assumption implies that the two latent variables account for the interrelationships among the observed variables. The measurement equations for the confirmatory factor analysis model of Figure 11.2 are:

$$
\begin{aligned}
x_1 &= \alpha_{10} + \alpha_{11}\xi_1 + & \delta_1 \\
x_2 &= \alpha_{20} + \alpha_{21}\xi_1 + & \delta_2 \\
x_3 &= \alpha_{30} + \alpha_{31}\xi_1 + & \delta_3 \\
x_4 &= \alpha_{40} + & \alpha_{42}\xi_2 + \delta_4 \\
x_5 &= \alpha_{50} + & \alpha_{52}\xi_2 + \delta_5 \\
x_6 &= \alpha_{60} + & \alpha_{62}\xi_2 + \delta_6.
\end{aligned}
$$

Observed variables are usually centred, that is measured as deviations from the mean, implying that $E(x_i) = 0$ for all i. If that is the case, the intercept parameters α_{i0} are omitted from eq. (11.1). In addition, $E(\delta_i) = 0$ for all i. It is also assumed that the error terms δ are uncorrelated with the latent variables ξ_1 and ξ_2. Finally, the origin of the latent variables ξ_1 and ξ_2 is set to zero $(E(\xi_1) = E(\xi_2) = 0)$.

Figure 11.3 gives the standardised factor solution for the confirmatory factor model with two correlated factors fitted to the subject marks data. The standardised solution is obtained when both the observed and the latent variables have been standardised to have zero means and unit variances. Although all variables have been standardised, the factor loadings are not correlation coefficients between the observed and the latent variables because the latent variables are correlated. In CFA, factors are taken to be correlated, and the standardised factor loadings should be considered as standardised regression coefficients.

The parameter estimates, together with their standard errors, are given in Table 11.1. Later in the chapter, we shall introduce methods for fitting the models. The estimated factor loadings are all positive and have small standard

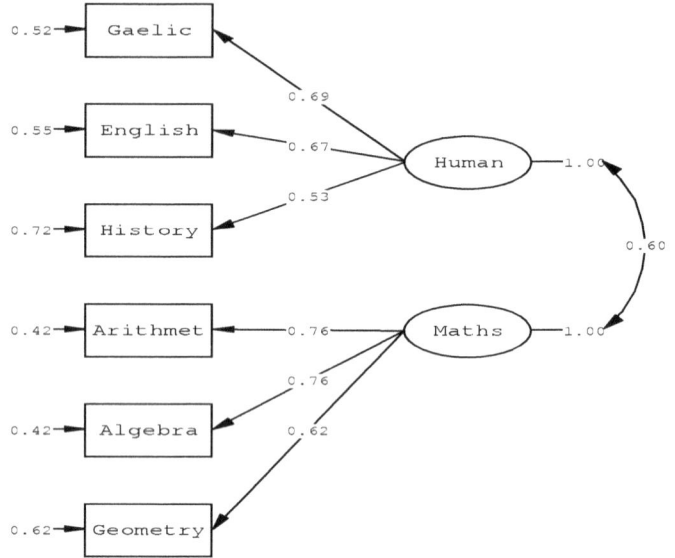

FIGURE 11.3: Path diagram for confirmatory factor analysis, standardised solution, subject marks data

errors. Since the correlation matrix of the six observed variables has been analysed (standardised observed variables), the estimated factor loadings obtained are standardised factor loadings. In the standardised solution, the latent variables, ξ, have unit variance. The R^2 values given in the last column of the table are the squared multiple correlations for each measurement equation, also known as communalities (see also Chapter 7). The latter can be considered as measures of item reliability as well as measures of the strength of the linear relationship imposed by the measurement equations. In our example, items Arithmetic and Algebra are found to have the largest reliability (being the more reliable indicators for the latent variable "Mathematics") and the item Gaelic is the most reliable indicator for the latent variable "Humanities".

Scaling the latent variables

The latent variables are unobserved constructs and therefore there is no natural scaling for them. For example, variables such as aspiration, political attitudes, conservatism, stress, discrimination, performance and motivation do not have universally accepted scales of measurement like weight or height. Both the origin and the unit of measurement are arbitrary, and therefore, they must be chosen in advance. The origin of the latent variables is usually set to zero.

There are two alternative ways of setting the scale of a latent variable, and they lead to equivalent solutions.

1. *Standardised latent variables*

   One approach, already used in Chapters 7–9, is to standardise the latent variables, assuming that they have zero means and unit variances in the population. When the latent variables are standardised and the correlation matrix of the observed variables is analysed, the confirmatory factor analysis model estimates the correlations between the latent variables. The path diagram for the subject marks in Figure 11.3 provides the solution for the case where the latent variables have been standardised. The estimated value of 0.60 that appears on the curved line that links the two latent variables is the estimated correlation between the two latent variables. This was found to be statistically significant (i.e. we reject the null hypothesis that the true population correlation is zero) with a standard error of 0.07 and a t-value of 8.32.

2. *Unstandardised latent variables using a reference variable*

   An alternative way to set the scale of a latent variable is to force its scale to be the same as one of the observed variables. Usually, we choose the variable that best represents the latent variable and set its factor loading equal to one. The selected variable is then known as a *reference variable*. When a good reference variable is available, it is better to use it for assigning the scale of the latent variable since it provides a more interpretable solution. Figure 11.4 gives the solution for the two-factor confirmatory model when the scale of the first latent variable, ξ_1, is set to be the same as the metric of item Gaelic (the item with the highest loading) and the scale of the second latent variable, ξ_2, is set to be the same as the metric of item Arithmetic (the item with one of the highest loadings). Setting the scale of a latent variable in this way implies that, on average, a unit change in ξ_1 and in ξ_2 will change x_1 (Gaelic) and x_4 (Arithmetic) respectively by one unit too. Goodness-of-fit statistics and fit indices remain the same as for the solution obtained from the standardised model. The only difference in the solutions given in Figure 11.3 and Figure 11.4 is that the units of measurement of the latent variables ξ_1 and ξ_2 are different. The estimated covariance between the two latent variables is 0.32, and the estimated correlation is obtained by dividing the estimated covariance by the product of the estimated standard deviations giving $\frac{0.32}{\sqrt{0.48}\sqrt{0.58}} = 0.60$, as in the standardised solution.

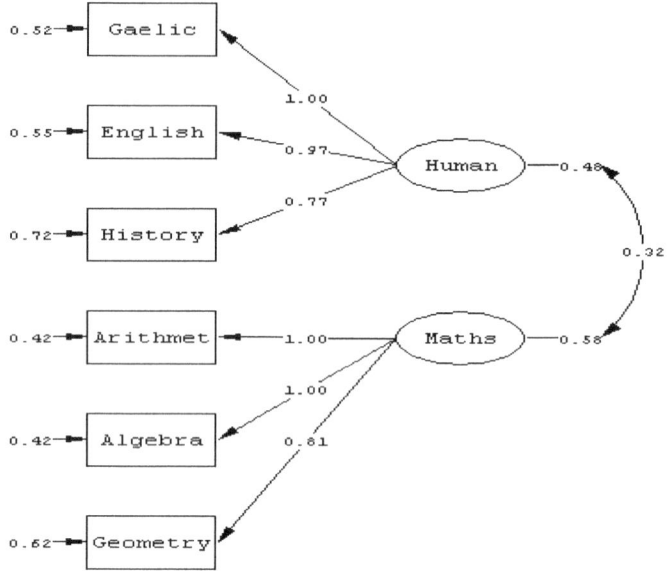

FIGURE 11.4: Path diagram, confirmatory factor analysis, unstandardised solution (using Gaelic and Arithmetic as reference variables), subject marks data

Identification conditions for CFA

Before proceeding with fitting the model, we need to check whether the hypothesised model is identified by the data. A model is said to be *identified* when a unique solution exists for all the model parameters. When different solutions for the model parameters produce the same covariance matrix, and those solutions are not identical, the model is not identified.

If no restrictions are imposed on the model parameters (factor loadings, regression coefficients, structural parameters, or error variances) the model is *under-identified*. Sometimes, under-identified models can be made identifiable by the addition of extra constraints. We have already seen that CFA postulates constraints on some of the factor loadings, allowing only certain items to be indicators of a latent variable (e.g. set some of the α_{ij} parameters equal to zero). Constraints are also imposed to define the scale of the latent variable (i.e. for each factor j, one of the α_{ij} parameters is set equal to one). Also the error terms $\delta_i (i = 1, \ldots, p)$ are usually assumed to be uncorrelated with each other (that will set all the off-diagonal elements of the covariance matrix of δs equal to zero).

Finally, when a model parameter can be obtained from different elements of the covariance matrix then those elements should be the same under a correctly specified model and the parameter is over-identified. Detailed discussion of the rules required for identifiability can be found in Bollen (1989). Note also that rules of identifiability depend on the model and therefore no general rule can be established. Here, we will provide some general guidelines for checking model identifiability and discuss those rules briefly. The standard commercial software will give a warning or will fail to provide a solution if the proposed model is not identifiable.

A necessary, but not sufficient, condition which applies to any type of CFA model is the t-rule (Bollen 1989). This indicates that for a model to be identified

$$t \leq \frac{1}{2}p(p+1),$$

where p is the number of observed variables and t is the total number of unknown and unconstrained parameters. For the subject marks data, the number of observed items is 6 so that the total number of distinct elements in the sample covariance matrix of the six items $(x_1, \ldots, x_6)$ is $\frac{1}{2}(6)(6+1) = 21$ and the total number of free parameters to be estimated is $t = 13$ (six regression coefficients, six error variances and one correlation between the two latent variables). Since t is smaller than 21 the necessary condition for identifiability is satisfied.

A second necessary condition is that the scale of the latent variable must be set, for example, by standardising to unit variance or by scaling to a reference variable. Either method reduces the number of free parameters to be estimated by one for each factor.

In addition, sufficient conditions have been introduced for the case where each item loads only on one factor, or, in other words, where an item cannot be an indicator for more than one factor, and the measurement errors or residual terms are not correlated. These rules are known as the *two-* and *three*-indicator rules and that are sufficient, but not necessary, conditions for identifiability.

1. The two-indicator rule is for the case where there are two or more factors fitted, and it requires that each latent variable has at least two indicators, and that the covariance matrix for the latent variables does not have zero elements (variances and covariances of the latent variables are freely estimated).

2. The three-indicator rule is for the case of a model with a single factor, and it requires that there are at least three indicators for the single factor.

More complicated rules can be established for the case where the same item is an indicator of more than one factor.

11.4   Adequacy of the model

As has already been stated in this chapter, CFA is concerned with testing a theory or a hypothesis by fitting a specified model. There are several ways of judging the statistical adequacy of the model.

Goodness-of-fit test
The Pearson chi-squared test statistic can be obtained with a variety of estimation methods provided by the standard packages. These are usually known by their initials – ML for maximum likelihood, GLS for generalised least squares and WLS for weighted least squares. The test statistic is defined as $(n-1)$ times the value of the minimised fit function used for estimating the model parameters. All estimation methods provide parameter estimates that minimise the distance, in some sense, between the sample and the model covariance matrix. The WLS method requires the asymptotic covariance matrix of the variances and covariances of the observed variables. As already mentioned, the fit function is a scaled discrepancy measure between the observed sample covariance matrix and that estimated under the model. Under the assumption that the model is correct (i.e. holds exactly in the population), and for large sample sizes, the test statistic follows a χ^2 distribution with $p(p+1)/2 - t$ degrees of freedom, where p is the total number of observed variables and t is the total number of unknown and unconstrained model parameters. For a specified level of significance, one can test whether there is evidence against the null hypothesis that the specified model is correct. More specifically, the null hypothesis (H_0) states that the observed covariance matrix is generated by the hypothesised model whereas the alternative hypothesis (H_1) allows the covariance matrix to be unrestricted. In practice, researchers are more interested in whether the model holds approximately than in testing whether it holds exactly. For a model that holds approximately, one can compare the chi-squared statistic obtained with an appropriate non-central χ^2 distribution. Our experience with fitting these models and the published literature recommends that one needs at least 100 observations per parameter to be estimated. Violation of the normality assumption for the observed variables results in invalid standard errors and goodness-of-fit tests. Alternative robust procedures of estimation that provide robust goodness-of-fit tests have been proposed in the literature, and these include scaled chi-squared test statistics due to Satorra and Bentler which are available in standard software. Browne (1984) has also suggested

alternative parameter estimators and test statistics that account for non-normality in the data.

The Pearson chi-squared test statistic for the subject marks data is 8.05 on 8 degrees of freedom ($6 \times (6 + 1)/2 - 13 = 8$, parameters to be estimated are the six factor loadings, the six error variances and the correlation between the two factors). The p-value of 0.43 suggests that the model is a very good fit. It should be noted that failing to reject a model does not mean that there are no alternative models which would also provide an adequate fit to the data. This is an important point, often overlooked in practice. The difference between two chi-squared statistics can be used for selecting among two nested models.

The Pearson chi-squared test statistic is known to be sensitive to sample size. Large samples will tend to produce large chi-squared values that will lead to the rejection of a model, even if the small differences which occur between the observed and reproduced co-variance matrices are not practically important. On the other hand, small sample sizes will tend to produce small chi-squared values that will lead to failure to reject purely because of lack of evidence.

Standard errors and t-tests

Available software for CFA provides standard errors for all the parameters being estimated. t-values obtained by dividing the estimated value by its standard error (see also Section 6.4) are used for judging the statistical significance of individual model parameters. Table 11.1 gives the standard errors for the estimated factor loadings for the subject marks data. The t-values are all greater than 2, indicating that there is evidence that the population coefficients are different from zero at the 5% significance level. The t-test statistic requires the assumption of normality for the data and therefore should be used with some caution.

Reproduced correlation matrix

As suggested in Chapter 7, a good way of assessing the fit of any model is to compare the fitted (reproduced) correlation matrix of the xs with the correlation matrix computed from the sample data. The sample correlation matrix of the subject marks data is given in Table 5.2. Table 11.2 shows the reproduced correlation matrix obtained from fitting the confirmatory factor model of Figure 11.2 to the subject marks data. The lower section of the table shows the differences (discrepancies or residuals) between the reproduced correlations and the sample correlation matrix given in Table 5.2. All differences are very small, suggesting a very good fit. The reproduced correlation matrix for the EFA model is given in Table 7.5.

The fit of the CFA model is almost as good as the fit of the EFA model despite having five fewer parameters.

Standardised Residuals (discrepancies)

The adequacy of the model or its badness of fit can be examined by using the standardised residuals. Table 11.3 gives the standardised residuals for the confirmatory model fitted to the subject marks data. Standardised residuals are the residuals divided by their standard errors. Standardised residuals smaller than -2 or greater than 2 in magnitude indicate a bad fit. In our example, all standardised residuals are small, indicating a very good fit. A large residual for two variables suggests that the model has not adequately explained the correlation between those two variables.

TABLE 11.2: Reproduced correlations for a confirmatory two-factor model fitted to the subject marks data, and discrepancies between observed and reproduced correlations (bottom section), subject marks data

Correlation	Gaelic	English	History	Arithmetic	Algebra	Geometry
Gaelic	1.00					
English	0.46	1.00				
History	0.37	0.36	1.00			
Arithmetic	0.32	0.31	0.24	1.00		
Algebra	0.32	0.31	0.24	0.58	1.00	
Geometry	0.25	0.25	0.20	0.48	0.47	1.00
Residual						
Gaelic	0.00					
English	-0.02	0.00				
History	0.04	-0.01	0.00			
Arithmetic	-0.03	0.04	-0.08	0.00		
Algebra	0.01	0.01	-0.05	0.01	0.00	
Geometry	-0.00	0.08	-0.02	-0.00	-0.01	0.00

TABLE 11.3: Standardised residuals or discrepancies for a confirmatory two-factor model fitted to the subject marks data

Correlation	Gaelic	English	History	Arithmetic	Algebra	Geometry
Gaelic						
English	-1.90					
History	1.78	-0.26				
Arithmetic	-0.75	1.21	-1.99			
Algebra	0.43	0.38	-1.21	0.82		
Geometry	-0.09	1.81	-0.31	-0.05	-0.70	

11.5 Introduction to structural equation models with latent variables

Structural equation models (SEM) provide a general framework that brings together simultaneous equation models developed in econometrics, factor analysis theory developed mainly in psychometrics and path analysis developed mainly in sociology. What is the difference between structural equation modelling and factor analysis? Factor analysis concentrates on the relationship between observed variables and latent variables. That relationship is expressed through a measurement model such as those discussed in Chapters 7–9 and in Section 11.3. Measurement models can be defined in either an exploratory or a confirmatory model setting.

SEM allows the simultaneous estimation of a measurement model and a structural model. The methodology allows us to test complex hypotheses on the relationships between observed variables and latent variables via the measurement model (measurement part) as well as relationships among latent variables via the structural model (structural part). The structural model can estimate the effects of latent variables and observed explanatory variables, such as demographic and socio-economic variables, on other latent variables. Structural equation modelling is a tool for translating an assumed causal theory into a graphical model with directed paths and estimating the implied relationships using a statistical model to assess the theory with empirical data. For example, when academic performance of students is studied at two or more different time points, one is interested in estimating whether ability changes over time but also the differences in ability between girls and boys, as well as differences defined by variables related to the educational level of the parents, family income and so on.

The new thing here is that we express the covariances of the observed variables in terms of parameters defined in both the measurement part of the model and the structural part of the model. Interest is more on the structural part of the model ("causal" relationships between constructs) but the structural part cannot stand without the measurement model that defines the constructs through the observed indicators.

11.6 The linear structural equation model

Unlike the models presented in Chapters 7–9, SEMs are confirmatory factor analysis models. SEM originated in the work of Jöreskog (1970) who described a structural equation model as having:

i) a measurement part, linking the observed variables to the latent variables, usually in a confirmatory way and

ii) a structural part, linking the latent variables to each other.

A structural equation model can be expressed by a set of algebraic equations and assumptions, or equivalently as a path diagram. Path diagrams are easier to use for modelling the complex structures encountered in real applications. Path analysis for observed variables is introduced in Section 6.15.

The confirmatory nature of SEM requires very good knowledge of the area under study. Constructs need to be clearly defined in the measurement part of the model, both in terms of dimensions and indicators to be used for their measurement. A distinction needs to be made between two sorts of latent variables: *exogenous*, *independent* or explanatory latent variables and *endogenous* or *dependent* latent variables. The terms exogenous and endogenous variables come from economics where they are widely used in regression and time series modelling.

The majority of structural equation models are based on some theory of the system under study. For this reason they can become very complex and it is sensible to break the testing into two parts. The measurement model is tested first and, when an adequate or satisfactory fit is obtained, the structural part is added in and interpreted.

Figure 11.5 gives the path diagram of a hypothetical structural equation model.

Because SEM involves specifying, simultaneously, the measurement and structural equations, many parameters are required and therefore some new notation needs to be introduced and explained. The notation used in this chapter is in accordance with the standard notation used in SEM. The path diagram in Figure 11.5 contains two different types of latent variable inside circles namely the η and the ξ variables. The η variables are dependent or endogenous and the ξ are independent or exogenous latent variables. The exogenous latent variables (ξ_1 and ξ_2) are explanatory variables not "explained" by other variables in the system. As can been seen from the diagram their correlation, ϕ_{21}, is a free parameter to be estimated. In all cases, the first subscript of a parameter indicates where the arrow points in the path diagram and the second subscript where the arrow comes from. The latent variables denoted by η are named dependent or endogenous, indicating that they are response variables partially explained or "caused" by other latent variables in the system. They can also affect other endogenous variables in the model. More generally, latent variables are classified into one of the following two categories:

1. r ξ-type independent or exogenous latent variables denoted by the vector $\boldsymbol{\xi} = (\xi_1, \ldots, \xi_r)'$ and

2. s η-type dependent or endogenous latent variables denoted by the vector $\boldsymbol{\eta} = (\eta_1, \ldots, \eta_s)'$.

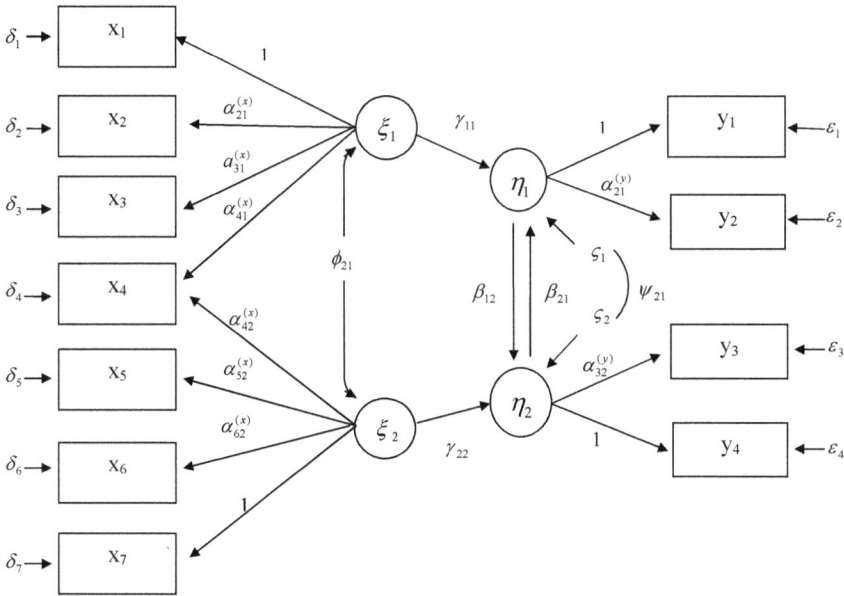

FIGURE 11.5: Path diagram, structural equation model

The observed variables are classified into two categories as follows:

1. p x-type observed variables denoted by the vector $\mathbf{x} = (x_1, \ldots, x_p)'$ and

2. q y-type observed variables denoted by the vector $\mathbf{y} = (y_1, \ldots, y_q)'$.

The x variables measure or serve as indicators for the independent or exogenous latent variables and the y variables measure the dependent or endogenous latent variables.

The disturbances, specific errors or residuals are also classified into two categories as follows:

1. the δ-type denoted by the vector $\boldsymbol{\delta} = (\delta_1, \ldots, \delta_p)'$ being the disturbances for $x_1, \ldots, x_p$ and

2. the ϵ-type denoted by the vector $\boldsymbol{\epsilon} = (\epsilon_1, \ldots, \epsilon_q)'$ being the disturbances for $y_1, \ldots, y_q$.

The reader will also have noticed that in Figure 11.5 the factor loadings, α, have superscripts indicating whether they relate to xs or to ys.

Examples of problems involving structural equation models

i) Duncan, Haller, and Portes (1971) fitted a structural equation model to investigate peer influence on educational and occupational aspiration. They assumed that a respondent's educational and occupational aspiration (dependent latent variables: η_1 and η_2) were affected by independent latent variables such as respondent's parental aspirations (ξ_1), respondent's intelligence (ξ_2), and the respondent's and their friend's socioeconomic status (ξ_3 and ξ_4 respectively). In addition, for each respondent, a friend's educational and occupational aspiration (η_3 and η_4) were assumed to be affected by the independent latent variables: friend's parental aspirations (ξ_5), friend's intelligence (ξ_6), respondent's and friend's socioeconomic status (ξ_3 and ξ_4 respectively). Finally, the respondent's and friend's aspirations were assumed to affect each other. The above description refers only to the structural part of the model. The measurement part of the model defines the relationships between all the constructs (dependent: $\eta_j, j = 1 \ldots, 4$ and the independent: $\xi_i, i = 1, \ldots, 6$ latent variables) and their corresponding indicators.

ii) Another example is that of Wheaton et al. (1977). They studied the stability of attitudes such as alienation over the years 1967 and 1971 and their relation to explanatory variables such as education and socioeconomic status. The dependent or endogenous latent variable "Alienation" in years 1967 and 1971 was measured through the Powerlessness and the Anomia scale. The independent or exogenous latent variable, socioeconomic status, is measured using as indicators education and a socioeconomic index. In addition, "Alienation" measured in year 1971 is affected by "Alienation" in year 1967.

iii) A lot of current research focuses on defining the profile of the green consumer. The research is trying to establish the causes of a more environmentally concerned behaviour. Theory suggests that environmental attitudes and environmental values (independent latent variables) affect ecological behaviour intention (a dependent latent variable). Moreover, additional independent observed variables such as income, education, and social status can be used as explanatory variables for ecological behaviour intention (a dependent latent variable). Such a model is known as the MIMIC (Multiple Indicator Multiple Cause) structural equation model where explanatory variables that affect the dependent or endogenous latent variables are also included in the structural part of the model.

Algebraic representation of a structural equation model

The model given in Figure 11.5 can also be represented by mathematical equations. The dependent and independent latent variables are measured indirectly through a measurement model, as described in Section 11.3. From the path diagram, it is clear that there are two measurement models. One measurement model defines the relationships between the observed variables x and the exogenous latent variables ξ. If the x and the y variables are centred around their mean, the first measurement model becomes:

$$
\begin{aligned}
x_1 &= \xi_1 + \delta_1 \\
x_2 &= \alpha_{21}^{(x)}\xi_1 + \delta_2 \\
x_3 &= \alpha_{31}^{(x)}\xi_1 + \delta_3 \\
x_4 &= \alpha_{41}^{(x)}\xi_1 + \alpha_{42}^{(x)}\xi_2 + \delta_4 \\
x_5 &= \alpha_{52}^{(x)}\xi_2 + \delta_5 \\
x_6 &= \alpha_{62}^{(x)}\xi_2 + \delta_6 \\
x_7 &= \xi_2 + \delta_7.
\end{aligned}
$$

The second measurement model, that defines the relationship between the y observed variables and the endogenous latent variables, is given by:

$$
\begin{aligned}
y_1 &= \eta_1 + \epsilon_1 \\
y_2 &= \alpha_{21}^{(y)}\eta_1 + \epsilon_2 \\
y_3 &= \alpha_{32}^{(y)}\eta_2 + \epsilon_3 \\
y_4 &= \eta_2 + \epsilon_4.
\end{aligned}
$$

All of the equations correspond to relationships shown in the path diagram given in Figure 11.5.

From Figure 11.5 we see that the values of the factor loadings $\alpha_{11}^{(x)}$ and $\alpha_{72}^{(x)}$ are set equal to one. Those constraints are used for fixing the scale of the latent variables. More specifically, they assign to the latent variables ξ_1 and ξ_2 the same scale of measurement as the observed variables x_1 and x_7 respectively. The same is being done when defining the scale of the endogenous latent variables η_1 and η_2. The factor loadings $\alpha_{11}^{(y)}$ and $\alpha_{42}^{(y)}$ have been set equal to one so that latent variable η_1 has the same scale as the observed variable y_1 and η_2 the same scale as the observed variable y_4. The random variables δ and ϵ are the measurement residuals, or specific errors.

The structural model combines the dependent with the independent latent variables:

$$
\begin{aligned}
\eta_1 &= \beta_{12}\eta_2 + \gamma_{11}\xi_1 + \zeta_1 \\
\eta_2 &= \beta_{21}\eta_1 + \gamma_{22}\xi_2 + \zeta_2.
\end{aligned}
$$

The above two equations model the reciprocal relationship between the η_1 and η_2 latent variables (two straight lines with arrow heads from η_1 and η_2) as well as the relationship between the η variables and the ξ variables. The existence of the double arrows in the path diagram indicates that the partial correlation between η_1 and η_2, conditional on the latent variables ξ_1 and ξ_2, is not zero.

The ζs are random disturbance terms or structural errors. The β and γ are structural parameters that measure direct effects of the η variables on other η variables, and direct effects of the ξ variables on the η variables.

The simultaneous system of equations defined by the measurement and the structural model is also known as a LISREL model (Linear Structural Equation Models) or, more generally, as a covariance structure model.

The model parameters given on the path diagram are classified into three categories: fixed, constrained and free parameters. Fixed parameters are assigned a pre-specified value. For example, some of the factor loadings have been set to one for the purpose of defining the scale of the latent variables and others are set to zero when the latent variable does not affect the corresponding item. Constrained parameters are those that have unknown values but are related in a linear or non-linear way to the other model parameters. Often factor loadings in different groups or across time are constrained to be equal as a result of measurement invariance. Finally, the free parameters are both unknown and unconstrained.

We can easily see that the factor analysis model for continuous responses discussed in Chapter 7 is a special case of the SEM model if we drop the structural part of the model and allow only for exogenous latent variables.

The assumptions and properties of the structural equation model (including both the measurement model and the structural model) are given below:

1. $E(\delta) = E(\zeta) = E(\epsilon) = E(\xi) = E(\eta) = 0$ for all variables in the equations.

2. The error terms δ are assumed uncorrelated with the ξ variables.

3. The error terms ϵ are assumed uncorrelated with the η variables.

4. The error term ζ is assumed uncorrelated with the ξ variables.

5. The error term ζ is assumed uncorrelated with the measurement errors δ and ϵ.

6. The covariance of the ξ latent variables is denoted by $\boldsymbol{\Phi}$. In our path diagram the parameter ϕ_{21} defines the covariance between the exogenous latent variables ξ_1 and ξ_2. The variances of the two latent variables are also estimable parameters denoted by ϕ_{11} and ϕ_{22} respectively.

7. The covariance of the ζ error terms is denoted by $\boldsymbol{\Psi}$. In our path diagram the parameter ψ_{21} defines the covariance between the error terms ζ_1 and ζ_2. The variances are denoted by ψ_{11} and ψ_{22} for ζ_1 and ζ_2 respectively.

8. The covariance of the ϵ measurement error terms is denoted by Θ_ϵ. For our model specified in the path diagram, Figure 11.5, the covariance matrix of ϵ is diagonal.

9. The covariance of the δ measurement error terms is denoted with Θ_δ. That matrix is also diagonal for the specified model.

10. The covariance between ϵ and δ measurement error terms is denoted by $\Theta_{\delta\epsilon}$ and is equal to zero for the specified model.

The last three assumptions can be relaxed to allow the error terms to be correlated. If two error terms are correlated this implies that the latent variables in the model do not fully explain the correlations between two indicators. Typically covariances between error terms arise in longitudinal data where the same variable is measured at different time points.

Model specification using matrices

We now move on to express the general LISREL model in the notation of vectors and matrices. Vector and matrix notation becomes useful here because it conveys the structure of a large number of equations in a very compact form. The structural part may be written:

$$\boldsymbol{\eta} = \boldsymbol{\alpha} + \mathbf{B}\boldsymbol{\eta} + \boldsymbol{\Gamma}\boldsymbol{\xi} + \boldsymbol{\zeta}, \tag{11.1}$$

where $\boldsymbol{\eta}$ and $\boldsymbol{\xi}$ are vectors whose elements are the endogenous and the exogenous latent variables respectively and $\boldsymbol{\zeta}$ is the residual, or error, term. The $\boldsymbol{\alpha}$ vector contains the intercept terms and the matrices $\mathbf{B}$ and $\boldsymbol{\Gamma}$ contain factor coefficients.

Two sets of observable variables are needed for the measurement part of the model, one for the $\boldsymbol{\eta}$ (dependent) latent variables and another set for measuring the $\boldsymbol{\xi}$ (independent) latent variables. The measurement equations are thus:

$$\mathbf{x} = \boldsymbol{\alpha}_x + \mathbf{A}_x\boldsymbol{\xi} + \boldsymbol{\delta}, \tag{11.2}$$

and

$$\mathbf{y} = \boldsymbol{\alpha}_y + \mathbf{A}_y\boldsymbol{\eta} + \boldsymbol{\epsilon}, \tag{11.3}$$

where $\mathbf{x}, \mathbf{y}, \boldsymbol{\xi}$ and $\boldsymbol{\eta}$ are column vectors with elements $(x_1, \ldots, x_p)$, $(y_1, \ldots, y_q)$, $(\xi_1, \ldots, \xi_r)$ and $(\eta_1, \ldots, \eta_s)$ respectively. The matrices $\mathbf{A}_x$ and $\mathbf{A}_y$ are arrays that contain the factor loadings. The intercept vectors $(\boldsymbol{\alpha}, \boldsymbol{\alpha}_x, \boldsymbol{\alpha}_y)$ in all the above equations will become redundant when the variables are taken as deviations from theis means giving: $E(\mathbf{x}) = E(\mathbf{y}) = E(\boldsymbol{\xi}) = E(\boldsymbol{\eta}) = \mathbf{0}$.

More specifically, for the model given in Figure 11.5, the individual parameters involved in the vector form equations given in eqs. (11.2) and (11.3) are defined as follows:

$$
\mathbf{x} = \begin{pmatrix} x_1 \\ x_2 \\ x_3 \\ x_4 \\ x_5 \\ x_6 \\ x_7 \end{pmatrix}, \ \mathbf{y} = \begin{pmatrix} y_1 \\ y_2 \\ y_3 \\ y_4 \end{pmatrix}, \ \mathbf{A}_x = \begin{pmatrix} 1 & 0 \\ \alpha_{21}^{(x)} & 0 \\ \alpha_{31}^{(x)} & 0 \\ \alpha_{41}^{(x)} & \alpha_{42}^{(x)} \\ 0 & \alpha_{52}^{(x)} \\ 0 & \alpha_{62}^{(x)} \\ 0 & 1 \end{pmatrix}, \ \mathbf{A}_y = \begin{pmatrix} 1 & 0 \\ \alpha_{21}^{(y)} & 0 \\ 0 & \alpha_{32}^{(y)} \\ 0 & 1 \end{pmatrix}
$$

and

$$
\boldsymbol{\delta} = \begin{pmatrix} \delta_1 \\ \delta_2 \\ \delta_3 \\ \delta_4 \\ \delta_5 \\ \delta_6 \\ \delta_7 \end{pmatrix}, \ \boldsymbol{\epsilon} = \begin{pmatrix} \epsilon_1 \\ \epsilon_2 \\ \epsilon_3 \\ \epsilon_4 \end{pmatrix},
$$

where the parameters contained in eq. (11.1) can be written as:

$$
\boldsymbol{\Gamma} = \begin{pmatrix} \gamma_{11} & 0 \\ 0 & \gamma_{22} \end{pmatrix}, \ \mathbf{B} = \begin{pmatrix} 0 & \beta_{12} \\ \beta_{21} & 0 \end{pmatrix}, \ \boldsymbol{\zeta} = \begin{pmatrix} \zeta_1 \\ \zeta_2 \end{pmatrix}.
$$

Model interpretation

Let us take the structural model defined in Figure 11.5 and allow for the exogenous latent variable ξ_1 to affect η_2 and the exogenous latent variable ξ_2 to affect η_1:

$$
\begin{aligned}
\eta_1 &= \beta_{12}\eta_2 + \gamma_{11}\xi_1 + \gamma_{12}\xi_2 + \zeta_1 \\
\eta_2 &= \beta_{21}\eta_1 + \gamma_{21}\xi_1 + \gamma_{22}\xi_2 + \zeta_2.
\end{aligned}
$$

Interpreting the structural parameters of the model, we need to distinguish between total effects, direct effects and indirect effects, similarly to the path analysis for observed variables introduced in Section 6.15. More specifically, the coefficients β and γ measure direct effects on the dependent latent variables η. For example, the coefficient β_{12} measures the change that will be expected in η_1 for a unit change in η_2 assuming that all other variables in the model are held constant. However, we can see that a unit change in ξ_2 will not only bring a change in η_1 by an amount γ_{12} but also a change in η_2 by γ_{22} and therefore η_2 cannot be assumed to remain constant. Thus a change in ξ_2 has both a direct effect on η_1 and additional indirect effects when other variables cannot be assumed to be held constant. The total effect of ξ_2 on η_1 must include the direct effect (γ_{12}) and the indirect effect of ξ_2 that operates through η_2 by $(\gamma_{22} + \beta_{12})$. The indirect effects can be computed by subtracting the direct effects from the total effects.

Adequacy of a Structural Equation Model

All the test statistics and measures of fit discussed before in the chapter can be used to assess the suitability of a SEM. In addition, fit indices, model selection information criteria, and modification indices are used to help us decide on the best fitted model and on how to improve the fit of the model.

Fit Indices
Because of the limitation of the chi-squared statistic mentioned earlier in the chapter, alternative standardised forms of the chi-squared statistic have been proposed in the literature, which are less sensitive to the sample size and to the assumption that the model holds exactly in the population. Those statistics are known as fit indices. The best-known fit indices are discussed below. Fit indices are based on the sample and estimated (reproduced) covariance matrices. There is a bewildering array of such indices. We enumerate below those that are more robust.

   i) The root mean square error of approximation (RMSEA) takes into account the fact that the model might hold approximately rather than exactly in the population. Browne and Cudek (1993) suggested that values smaller than 0.05 indicate a close fit. A value greater than 0.1 indicates a poor fit. A p-value and a confidence interval can be computed.

   ii) The comparative fit index (CFI) takes values between 0 and 1. CFI values close to 1 indicate a very good fit. CFI computes the relative improvement in the overall fit of the fitted model compared to a null model; this is usually the independence model (where the observed variables are assumed to be uncorrelated).

   iii) Standardised root mean square residual (SRMR). This is a standardised summary of the average covariance residuals. Covariance residuals are discrepancies between the sample covariance matrix and the reproduced covariance matrix estimated under the model, assuming that the model is correct. The smaller the SRMR the better the fit. A value less than 0.08 indicates a good fit.

   iv) The Tucker-Lewis or non-normed fit index (NNFI). A value of 1 indicates a very good fit, where a value greater than 1 might indicate overfitting. Values smaller than 0.9 indicate a poor fit.

v) Bollen's incremental fit index (IFI). Again a value of 1 indicates a very good fit, where a value greater than 1 might indicate overfitting. Values smaller than 0.9 indicate poor fit. IFI is reported to be less variable than NNFI.

Indices CFI, NNFI and IFI are relative fit indices. They compare the chi-squared statistic of the fitted model with the chi-squared of a baseline model, usually the independence model (the degrees of freedom of each model are taken into account). Indices such as the RMSEA are calculated using the chi-squared statistics and the degrees of freedom of the fitted model.

Table 11.4 gives the fit indices for the two-factor confirmatory model applied to the subject marks data. All indices indicate a very good fit.

TABLE 11.4: Fit indices for the two-factor confirmatory model, p-values were available in brackets, subject marks data

Fit indices	Value
RMSEA	0.006
	(0.76)
CFI	1.00
SRMR	0.033
IFI	1.00
NNFI	1.00

Model selection criteria

In a SEM setting, one might want to compare models. Model comparison can be based on model selection criteria such as the Akaike Information Criterion (AIC), the Bayesian Information Criterion (BIC) or the consistent Akaike Information Criterion (CAIC). These criteria depend on the value of the chi-squared test (the minimised value of the fit function) and the degrees of freedom. The idea is to identify the most parsimonious model compatible with the data.

Modification indices

Although CFA is about testing the adequacy of a model representing a specified theory, it might be the case that the researcher is interested in improving the fit of the model by setting free more parameters to be estimated. In such cases, we would be hoping to modify our model so that it will fit the data better. Packages compute modification indices for each fixed parameter in the model or the path diagram. (By fixed parameters, we refer to those parameters that have not been estimated freely from the data.) The modification index computes the expected decrease in the chi-squared test statistic by the inclusion of that extra parameter (path in the diagram) in the model. Only large values of the modification indices should be taken into account in judging the substantive meaning of any parameter to be included in the model.

Identification rules

Identification conditions, in the general case, remain the same as those already discussed in Section 11.3 for the confirmatory factor analysis model. Checking for identifiability is not a straightforward procedure. As stated before, one hopes that software will give a warning message when a nonidentified model is proposed by the user.

Estimation methods

In path analysis, there are two covariance matrices: the sample covariance matrix of the observed variables and the theoretical covariance matrix derived from the model. The elements of the covariance matrix that correspond to the hypothesised model are functions of the model parameters such as intercepts, factor loadings, error variances, and covariance terms. The identification of the proposed model must be checked before proceeding with estimation.

Available software provides many different ways of estimating a confirmatory factor analysis model and structural equation models for continuous observed variables. These include unweighted least squares (ULS), generalised least squares (GLS) and maximum likelihood (ML) (see Chen et al. (2023) for a review of estimation methods for structural equation models). All these estimation methods are iterative which means that a number of cycles is required. One starts from initial values for the parameters and continues until some criterion of fit is satisfied. This criterion will involve minimising some measure of difference, or distance, between the theoretical covariance matrix (i.e. a function of the model parameters) and the sample covariance matrix. All such methods give consistent estimates (estimates close to the population values for large samples). Maximum likelihood estimation requires the assumption that the data are samples from a multivariate normal distribution with a covariance matrix specified by the model. ML estimation is found to be robust under deviations from normality at least as far as the estimation of the model parameters is concerned. Unfortunately, goodness-of-fit statistics tend to reject the model too readily, and t-statistics also tend to reject too often due to the underestimation of standard errors in the presence of non-normality. Satorra and Bentler (1994) have proposed ways of obtaining correct standard errors and test statistics under non-normality of the data. This applies to non-normal continuous data rather than ordinal data (see Savalei (2014) for a comprehensive discussion). Those suggestions have been incorporated in commercial software and appear as the Satorra and Bentler rescaled chi-squared test statistic and robust standard errors. ML estimation is recommended when data follow a multivariate normal distribution approximately, whereas robust ML is recommended for the non-normal case.

11.7   A worked example

The data analysed come from the Educational Testing Service in the United States and refer to a nationwide sample in 1963 and 1965 corresponding to seventh and ninth graders, respectively. The observed variables are scores from the verbal (SCATV) and quantitative (SCATQ) parts of the Scholastic Aptitude Test (SCAT) and achievement tests in mathematics (MATH), science (SCI), social studies (SS), listening (LIST), writing (WRIT) and reading (READ). The examinees were divided into four groups according to gender and whether or not they participated in an academic curriculum in Grade 12. We analyse here the group of academic girls in grades 7 and 9. The data have been analysed in Jöreskog and Sörbom (1999) and their source is Anderson and Maier (1963) and Hilton (1969). The analysis aims to study the progress of the students between the seventh and ninth grades so far as this is measured by the eight items described above. Items seem to distinguish between Verbal and Quantitative ability. Verbal and Quantitative ability at grade 7 will be considered as independent (exogenous) ξ-type latent variables, whereas the same latent variables measured at grade 9 will be considered as dependent (endogenous) η-type latent variables. Items measured at the 7th grade will be the x-type indicators of the independent latent variables (ξ) whereas the items measured at the 9th grade will be y-type indicators of the dependent latent variables η.

We started our analysis by first fitting a confirmatory factor analysis model with two factors for the items measured at the seventh grade. We found that Verbal ability was adequately measured by items SCI, READ and SCATV whereas Quantitative ability was adequately measured by items MATH, SCI and SCATQ. The same confirmatory model was fitted to the ninth graders. The Chi-squared test statistic was found to be 1.48 (p-value$= 0.69$) and 1.50 (p-value$= 0.68$) for the seventh and ninth graders respectively, each on 3 degrees of freedom and both indicating a very good fit. The correlation between the two latent variables (Verbal and Quantitative ability) was estimated to be 0.85 and 0.82 for the seventh and ninth graders. The latent variables, Verbal and Quantitative abilities, are scaled to have the same units of measurement as the observed variables SCATV and SCATQ, respectively. Figure 11.6 shows path coefficients set equal to one from Quant7 to SCATQ7, from Verbal7 to SCATV7 and similarly from Quant9 to SCATQ9 and from Verbal9 to SCATV9.

We then proceed to fit a structural equation model to all items measured at both time points. The pattern of factor loadings identified in the first step of our analysis is preserved here too. In addition, we allow Verbal ability at grade 9 to depend on Verbal ability at grade 7 and Quantitative ability at grade 9 to depend on Quantitative ability at grade 7. This means that there will be two zero entries in the matrix $\mathbf{\Gamma}$ of eq. (11.1). The error terms ζ_1 and ζ_2 are also taken to be uncorrelated which implies that Verbal ability and Quantitative ability at grade 9 have been accounted for by Verbal ability and Quantitative ability at grade 7, and whatever remains does not depend on anything else in the model. Finally, the error terms that appear in the measurement equations (δ and ϵ) are uncorrelated.

The maximum likelihood estimates of the above-specified model are given in the path diagram in Figure 11.6.

All parameter estimates were found to be statistically significant at a nominal 5% level, but the overall fit of the model was not adequate, judging from the chi-squared value of 151.85 on 30 degrees of freedom (p-value< 0.000) and the poor values of other fit indices.

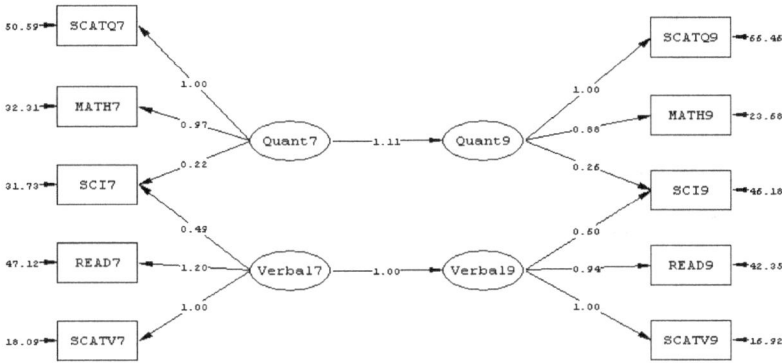

FIGURE 11.6: Path diagram, structural equation model, unstandardised solution, ability data

The poor fit of the fitted model suggests that some part of the model has been misspecified. We need to investigate whether the lack of fit is due to

1. a misspecified measurement model, or
2. a misspecified structural model.

Improving the fit of the model implies that certain parameters that were treated as fixed (set equal to zero here) should be freed and estimated. The measurement model has been tested separately for the samples of seventh and ninth graders, and we have already reported a good fit. Therefore, we should concentrate on the structural part of the model. The structural part of the model assumed a diagonal matrix for the parameter matrix $\mathbf{\Gamma}$ and all zero elements for the $\mathbf{B}$ matrix. Allowing the $\mathbf{\Gamma}$ matrix to be estimated freely does not improve things. The reduction in the chi-squared statistics was $(151.85 - 150.47) = 1.38$ on 2 degrees of freedom. The standardised residuals of the model in Figure 11.6 are given in Table 11.5. It is clear from the values of the standardised residuals that the model has not accounted for the correlation of the same items measured at the two time points. This implies that the error terms δ and ϵ should be set free to correlate between the two occasions.

The final model was the one that allowed the estimation of the covariances between the error terms of the measurement equations. Specifically, we allowed for the error term of item MATH7 to correlate with the error term of item MATH9, and the same for items SCI7 and SCI9, READ7 and READ9, SCATQ7 and SCATQ9, and for items SCATV7 and SCATV9. The fitted covariances are given in the matrix $\hat{\Theta}_{\delta\epsilon}$. In addition, the modification indices suggested only adding the path from Quantitative ability to Verbal ability at grade 9. The results of the final model are presented in Figure 11.7. The chi-squared statistic is now down to 23.08 on 24 degrees of freedom with a p-value= 0.52 showing a good fit. Also, the RMSEA=0.0 with a p-value= 0.99 for the hypothesis that its true value is smaller than 0.05. The smallest standardised residual is -1.78 and the largest 2.08, with the majority of them being close to zero.

TABLE 11.5: Standardised residuals for the structural equation model of Figure 11.6, ability data

	MATH9	SCI9	READ9	SCATV9	SCATQ9	MATH7	SCI7	READ7	SCATV7
MATH9	- -								
SCI9	1.41								
READ9	**3.22**	1.90	- -						
SCATV9	1.71	-0.10	-0.70	- -					
SCATQ9	-0.12	-1.29	1.87	-0.88	- -				
MATH7	-0.56	0.40	1.64	0.55	**-2.30**	- -			
SCI7	-1.04	**5.39**	-1.18	-0.90	**-3.00**	1.44	- -		
READ7	1.31	-0.94	**3.19**	**-5.29**	-1.70	**3.16**	0.99	- -	
SCATV7	-1.19	**-3.27**	**-3.82**	**5.00**	**-2.80**	0.57	-1.85	1.33	- -
SCATQ7	-0.51	**-2.29**	-0.38	**-2.60**	**4.80**	-0.61	0.00	-0.06	**-2.20**

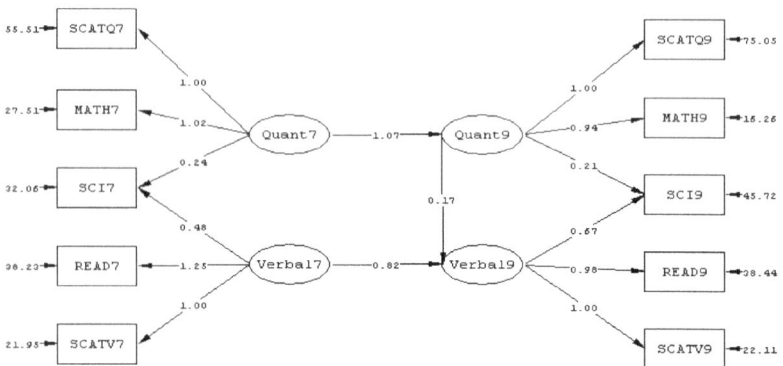

FIGURE 11.7: Path diagram, structural equation model, unstandardised solution, correlated error terms, ability data

Since the path diagram does not show all the results from the fitted model we give below the estimated parameters of the measurement part of the model in matrix notation. The rows of $\hat{\mathbf{A}}_x$ and $\hat{\mathbf{A}}_y$ correspond to the indicators in the order shown in Figure 11.7.

$$
\hat{\mathbf{A}}_x = \begin{pmatrix} 1 & 0 \\ 1.02 & 0 \\ 0.24 & 0.48 \\ 0 & 1.25 \\ 0 & 1 \end{pmatrix}, \quad \hat{\mathbf{A}}_y = \begin{pmatrix} 1 & 0 \\ 0.94 & 0 \\ 0.21 & 0.67 \\ 0 & 0.98 \\ 0 & 1 \end{pmatrix},
$$

where the matrices $\hat{\mathbf{A}}_x$ and $\hat{\mathbf{A}}_y$ contain the estimated factor loadings from the measurement equations.

The estimated variance-covariance matrices of the item-specific errors are given below. $\hat{\Theta}_\delta$ and $\hat{\Theta}_\epsilon$ contain the estimated variances of the item-specific errors associated with the x-type variables and the y-type variables, respectively. Since the δ errors are assumed to be uncorrelated and the ϵ errors are also assumed to be uncorrelated, their covariance matrix is diagonal (the off-diagonal elements are equal to zero). Matrix $\hat{\Theta}_{\delta\epsilon}$ contains the estimated covariances among the item-specific errors δ and ϵ for the same items measured at grade 7 and grade 9.

$$
diag\hat{\Theta}_\delta = \begin{pmatrix} 55.51 \\ 27.51 \\ 32.06 \\ 38.23 \\ 21.95 \end{pmatrix}, \quad diag\hat{\Theta}_\epsilon = \begin{pmatrix} 75.05 \\ 16.26 \\ 45.72 \\ 38.44 \\ 22.11 \end{pmatrix}
$$

and

$$
diag\hat{\Theta}_{\delta\epsilon} = \begin{pmatrix} 22.97 \\ -3.22 \\ 12.92 \\ 5.75 \\ 11.01 \end{pmatrix}.
$$

The notation $diag\hat{\Theta}_\delta$ denotes the vector formed by the diagonal elements of matrix $\hat{\Theta}_\delta$ and similarly for the matrices $\hat{\Theta}_\epsilon$ and $\hat{\Theta}_{\delta\epsilon}$.

The estimated parameters of the structural part of the model are:

$$
\hat{\Gamma} = \begin{pmatrix} 1.07 & 0 \\ 0 & 0.82 \end{pmatrix}, \quad \hat{\mathbf{B}} = \begin{pmatrix} 0 & 0 \\ 0 & 0.17 \end{pmatrix}, \quad \hat{\Psi} = \begin{pmatrix} 23.15 & 0 \\ 0 & 8.5 \end{pmatrix},
$$

$$
\hat{\Phi} = \begin{pmatrix} 98.74 & 88.98 \\ 88.98 & 112.85 \end{pmatrix},
$$

where the non-zero elements of matrix $\hat{\Gamma}$ denote the estimated effects of the exogenous latent variable Quant7 on Quant9 and of the exogenous latent variable Verbal7 on Verbal9 controlling for the endogenous latent variable Quant9. The non-zero element of the matrix $\hat{\mathbf{B}}$ denotes the estimated effect of the endogenous latent variable Quant9 on the endogenous latent variable Verbal9, controlling for the other exogenous latent variables in the model. The diagonal elements of the matrix $\hat{\boldsymbol{\Psi}}$ are the estimated variances of the error terms associated with the structural model. Those errors are uncorrelated in the specified model. Finally, matrix $\hat{\boldsymbol{\Phi}}$ contains the estimated variances and covariances of the exogenous latent variables (Quant7, Verbal7).

11.8   Structural equation model for categorical variables

To fit a SEM to categorical responses (binary and ordinal), we adopt the underlying variable approach discussed in Chapters 8 and 9 for binary and ordinal variables, respectively. We will focus here on ordinal variables, but the methods apply similarly to binary data. The measurement equations for the exogenous and endogenous categorical variables are:

$$\mathbf{x}^* = \boldsymbol{\alpha}_x^* + \mathbf{A}_x^* \boldsymbol{\xi} + \boldsymbol{\delta}^*, \qquad (11.4)$$

and

$$\mathbf{y}^* = \boldsymbol{\alpha}_y^* + \mathbf{A}_y^* \boldsymbol{\eta} + \boldsymbol{\epsilon}^*, \qquad (11.5)$$

where $\boldsymbol{\delta}^*$ and $\boldsymbol{\epsilon}^*$ are error terms and $\mathbf{x}^*$ and $\mathbf{y}^*$ are unobserved continuous variables underlying the categorical variables $\mathbf{x}$ and $\mathbf{y}$, respectively. The structural model of eq. (11.1) remains the same, as well as the model assumption given above. Polychoric correlations are estimated among all the categorical exogenous and endogenous variables. As with the factor model for ordinal data, there are three different sets of parameters to be estimated, namely the thresholds, the polychoric correlations between all the underlying variables, and, finally, the factor loadings of eqs. (11.4) and (11.5) and the structural parameters of eq. (11.1). Under the assumptions of the linear factor model that the latent variables are independent and normally distributed with mean 0 and variance 1, and that the residual terms are independent and normally distributed with mean 0 and diagonal covariance matrices $\boldsymbol{\Theta}_\delta^*$ and $\boldsymbol{\Theta}_\epsilon^*$, it follows that the underlying variables $x_1^*, \ldots, x_p^*, y_1^*, \ldots, y_q^*$ have a multivariate normal distribution with zero means, unit variances, and some correlation matrix $\mathbf{P} = \{\rho_{ij}\}$, where ρ_{ij} is a function of the factor loadings and structural parameters. Limited information estimation methods can be employed, such as the three-stage estimation method and pairwise likelihood estimation. In the application of the three-stage estimation method, thresholds are first estimated by maximising the univariate marginal likelihoods separately. Then, given the estimated thresholds, polychoric correlations are estimated by maximising the bivariate marginal likelihoods separately. In the third stage, the structural equation model given by eqs. (11.1), (11.4) and (11.5) is fitted to the estimated polychoric correlation matrix using a version of generalised least squares (GLS), such as unweighted least squares (ULS), diagonally weighted least squares (DWLS), and weighted least squares (WLS). In WLS, the weight matrix is an estimate of the inverse of the asymptotic covariance matrix of polychoric correlations, while DWLS involves only the diagonal elements of that weight matrix. Several studies have been conducted to compare these three least squares methods, and all have yielded similar conclusions. The WLS estimator converges very slowly to its asymptotic properties and therefore does not perform well in small sample sizes. DWLS and ULS are preferable to WLS, and they seem to perform similarly well in finite samples. However, to compute correct chi-squared measures of goodness-of-fit and standard errors for the parameter estimates, the full weight matrix is needed. The methods are then called robust. WLS, DWLS and ULS are implemented in all the standard SEM software, and PML is implemented in the R package lavaan. For example, lavaan uses diagonally weighted least squares (DWLS) to estimate model parameters, and the full weight matrix to compute robust standard errors, as well as a mean- and variance-adjusted chi-square test statistic. Within the maximum likelihood framework, the pairwise likelihood estimation method estimates thresholds, factor loadings, and structural parameters in a single step. Likelihood ratio type goodness-of-fit test statistics for overall fit and comparison of nested models, as well as model selection criteria, are also available and implemented in lavaan. Note that robust maximum likelihood methods developed for non-normal continuous variables are not appropriate for estimating models for categorical data, and the same is true for the corrected goodness-of-fit test statistics under non-normality.

There is a tendency for ordinal data to be treated as if it were numeric. However, that should be avoided, especially if the number of categories and the sample size are small. Using methods developed for categorical data produces more accurate estimates and inferences. Some key references for the treatment of categorical variables within SEM are Muthén (1984) and Jöreskog (1994). For a discussion of the limitations and practical difficulties that arise, see Jöreskog and Moustaki (2001).

Finally, the fit indices discussed above (RMSEA, CFI, SRMR, IFI, NNFI) for the continuous case do not directly generalise to models for categorical data. As reported in various research papers, the cut-off values used for fit indices, primarily developed for models with continuous data estimated using maximum likelihood, do not necessarily hold for models with ordinal data estimated using GLS estimation methods. Alternative fit indices have been developed for factor models for categorical data by Savalei (2021). Therefore, we encourage readers to be aware of the issues related to the fit indices and their behaviour in models with ordinal data estimated using limited information estimation methods, and to consult the relevant literature.

The Green consumer: a Greek study

This example is a subset of the variables analysed in Katsikatsou and Moustaki (2007). The sample consists of 330 university students in Greece. The study aimed to identify the factors that affect purchase behaviour for environmentally friendly products. We are interested in testing how factual knowledge about what causes environmental damage influences environmentally responsible purchasing behaviour, and how consumption habits mediate this relationship. The 330 students each provided answers for the 15 items listed below.

1. Poisonous metals (e.g. mercury, cadmium) are introduced into the food chain through groundwater [knen1]
2. Ozone near the ground may cause respiration problems [knen2]
3. Air pollution is mainly caused by cars [knen3]
4. Batteries cause severe soil pollution [knen4]
5. Mobile phones, when discarded, are very dangerous for the environment [knen5]
6. Chlorine is one of the most harmful substances for the water environment [knen6]
7. Washing powder and especially fabric softener contribute significantly to the degradation of maritime ecosystems [knen7]
8. I prefer to buy products in recyclable packaging [purc1]
9. I avoid buying products in aerosol containers (e.g. hair spray) [purc2]
10. I purchase light bulbs that are more expensive than ordinary ones but save energy [purc3]
11. I switch products for ecological reasons [purc4]
12. I try to use the least possible amount of water when I do household cleaning [cons1]
13. I try to cut down on electrical consumption in my household [cons2]
14. I try to use the least possible amount of paper [cons3]
15. I try to use the least possible amount of cleaning supplies [cons4]

Items 1 through 7 were scored on a five-point scale from 1 (agree strongly) to 5 (disagree strongly). Items 8 through 15 also used a five-point scale, ranging from 1 (Always) to 5 (Never). In designing the questionnaire, items 1 through 7 were designed to measure the construct "environmental knowledge"; items 8 through 11 the construct "environmental responsible behaviour when purchasing" (GPUR); items 12 through 15 the construct "environmental responsible behaviour when consuming" (GCON). Due to variables having very low frequencies in the first and last categories, we decided to group response categories one and two, and categories four and five, leaving all items with three categories.

The first step was to examine the measurement models by separately examining the indicators for each construct through an exploratory factor analysis. The first seven items on environmental knowledge were found not to be unidimensional. Two knowledge factors were identified: ENKN1, measured by items 1, 2, 3, 4, and 5, and ENKN2, measured by items 6 and 7. A one-factor model gives a good fit to items 8 to 11, which load on the GPUR factor, and similarly, items 12 to 15 load on a single factor GCON. We also fitted a confirmatory factor model to the 15 items with the above-described four latent variables freely correlated. The corrected chi-square test statistic (104.90 on 84 degrees of freedom, p-value=0.06) and the RMSEA (RMSEA=0.027, p-value=0.994) indicated a good fit. Examination of modification indices suggested that item 14 (cons3) is linked to factor (GPUR). We decided not to include that path because it was not theoretically justified. The chi-squared statistic reported is the mean- and variance-adjusted chi-squared statistic provided in lavaan, which is recommended for use when data are categorical.

We proceeded to fit the hypothesised structural equation model, now adapted to have two environmental knowledge factors. Note that the fit of the confirmatory factor analysis model fitted above is expected to be the same as the model fitted in Figure 11.8. The reason is that the structural model has the same number of parameters. The structural model in the confirmatory model has an unstructured factor correlation matrix with six free parameters. The structural equation model in Figure 11.8 has five parameters related to the five directed paths plus the correlation among the exogenous latent variables, ENKN1 and ENKN2.

Figure 11.8 gives the estimated path diagram (standardised solution) of the model that represents our theoretical framework. Looking at the estimated paths:

1. All the estimated factor loadings are statistically significant, showing a moderate to high correlation between the items and the factors they measure.

2. The estimated correlation between the two exogenous knowledge factors (ENKN1, ENKN2) is 0.41 and statistically significant.

3. The path coefficients from ENKN2 to GPUR and to GCON are not statistically significant. Given ENKN1, there is no evidence that ENKN2 would help in predicting GPUR or GCON, so that these potential links can be dropped.

4. The direct effect of ENKN1 on GPUR is not statistically significant. However, the indirect or mediating effect, defined by the paths from ENKN1 to GCON and from GCON to GPUR, is statistically significant. More specifically, the links from ENKN1 to GPUR

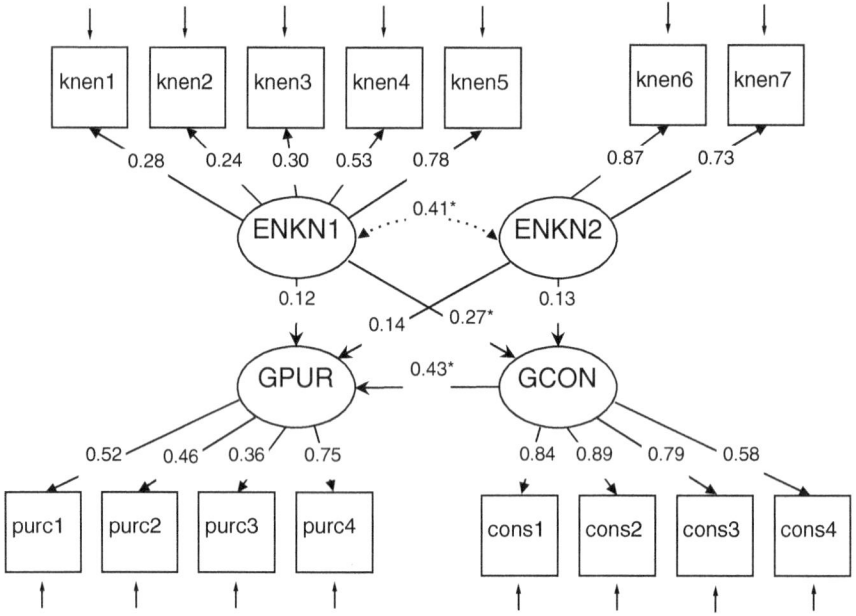

FIGURE 11.8: Path diagram, structural equation model, standardised solution, green consumer data. (*) statistically significant structural parameters

and GCON, suggest that reported knowledge (ENKN1) shapes how consumers behave (GCON), and this behaviour, in turn, drives reported purchasing (GPUR), underscoring a behavioural mediation pathway. For a comprehensive discussion of direct, indirect and total effects, see Kline (2023).

11.9   Concluding remarks and extensions

Structural equation modelling (SEM) is a method for estimating relationships within a hypothesised causal framework, including the direction and magnitude of effects—positive, negative, or null. It enables the estimation of both direct and indirect effects, accounts for measurement error in latent constructs, and has the ability to control for measured confounders.

Despite its strengths, SEM requires cautious interpretation. The estimated parameters do not establish causality, and fit indices cannot determine the direction of causal relationships. Model fit indices, discussed in this chapter, primarily reflect the adequacy of the measurement model. As noted by Tomarken and Waller (2003), limitations such as equivalent models and omitted variables constrain causal inference. Nonetheless, fit indices are a necessary first step in evaluating whether the model reasonably represents the data, serving as a starting point for interpreting coefficients, but not as the sole criterion for model evaluation.

The structural part of the model can be extended to include the effect of explanatory variables (covariates) on the endogenous latent variables (in a MIMIC model). Structural equation models can be adapted to fit data collected across time (longitudinal data), data collected from different sub-populations (multi-group analysis) as well as hierarchical data (multilevel analysis, see Section 13.10). Finally, non-linear effects such as interactions between latent variables and quadratic terms of latent variables can also be incorporated in the analysis.

11.10 Additional examples and further work

The test anxiety inventory

A brief description of the 20 items is given in Section 7.8. In that section, we presented an EFA of the Test Anxiety Inventory items. This inventory has been designed to provide a single scale of test anxiety, but the EFA suggested that two factors were needed.

This example is not presented to illustrate how CFA is used to test a theory but as an add-on to the results obtained from EFA to see whether a simpler structure (see Table 11.6) gives an adequate fit. As in Section 7.8, we proceed as if the data were multivariate normal. Gierl and Rogers (1996) found that the two-factor confirmatory model with correlated factors gave a better fit than the one-factor model. They derived one scale for "emotionality" and one for "worry". The indicators of factor ξ_1, named here as "emotionality", were found to be $(x_2, x_8, x_9, x_{10}, x_{11}, x_{15}, x_{16}, x_{18})$ while the indicators for the second factor ξ_2, named as "worry", were $(x_3, x_4, x_5, x_6, x_7, x_{14}, x_{17}, x_{20})$. In their analysis they excluded items $(x_1, x_{12}, x_{13}, x_{19})$ because they did not have large loadings in the two sub-scales. However, we did not come to the same conclusion when we analysed the 20 items with an exploratory two-factor model in Section 7.8. We found that those four items were strongly related to the "emotionality" factor. In our analysis, we used all twenty items. The estimated loadings for the two-factor confirmatory model are given in Table 11.6. The statistical significance of the individual factor loadings can be checked with the t-values obtained by dividing the estimated value by its standard error. For example, the t-value for the first loading is $0.63/0.05 = 12.6$. t-values for all items are greater than two suggesting that none of the factor loadings is zero in the population. The estimated correlation between the "emotionality" and the "worry" factor was found to be 0.85, with a standard error of 0.02, indicating a very strong correlation. The estimated correlation obtained in Section 7.8 from the oblique solution was 0.68.

TABLE 11.6: Parameter estimates and standard errors for the two-factor confirmatory model, test anxiety inventory items, males (parameters in blank cells are constrained to equal zero)

Item	$\hat{\alpha}_{i1}$	s.e.	$\hat{\alpha}_{i2}$	s.e.	R^2
x_1	0.63	0.05			0.40
x_2	0.64	0.05			0.40
x_8	0.71	0.05			0.51
x_9	0.68	0.05			0.46
x_{10}	0.51	0.05			0.26
x_{11}	0.76	0.05			0.58
x_{12}	0.66	0.05			0.44
x_{13}	0.60	0.05			0.36
x_{15}	0.80	0.05			0.64
x_{16}	0.78	0.05			0.60
x_{18}	0.68	0.05			0.46
x_{19}	0.66	0.05			0.43
x_3			0.58	0.05	0.34
x_4			0.68	0.05	0.46
x_5			0.59	0.05	0.34
x_6			0.54	0.05	0.29
x_7			0.72	0.05	0.52
x_{14}			0.68	0.05	0.47
x_{17}			0.68	0.05	0.47
x_{20}			0.68	0.05	0.46

In terms of exact fit, the chi-squared test rejected the model (see Table 11.7). However, the p-value for the RMSEA index shows that there is no evidence against the null hypothesis, suggesting an adequate fit when we relax the requirement for the model to hold exactly in the population. Furthermore, the fit indices reported in Table 11.7 show an adequate fit. Specifically, indices CFI, IFI and NNFI have values very close to one (0.94, 0.93, and 0.98, respectively), while the SRMR is close to zero, as expected when the model fits well.

An examination of the standardised residuals also suggests that the model is a good fit to the data. The smallest fitted residual (difference between a sample correlation and that estimated by the model) is −0.12 and the largest 0.15. Table 11.8 gives the pair of variables with the largest standardised residuals. Although the model provides an adequate fit, we also looked at the modification indices. Table 11.9 gives the decrease in the chi-squared statistic with the inclusion of each path separately. Although the software we used also produced modification indices for the correlation between the error terms, we do not include them here for reasons we have already explained. Adding item x_5 as an indicator for the "emotionality" latent variable will decrease the chi-squared test statistic by 16.6, and the estimated loading will be negative for that factor.

TABLE 11.7: Goodness-of-fit statistics and fit indices for the two-factor confirmatory model, (*p*-values where available in brackets), test anxiety inventory items, males

Statistics	Value
Chi-squared	355.50
	(0.00)
RMSEA	0.057
	(0.07)
CFI	0.94
SRMR	0.044
IFI	0.93
NNFI	0.98

TABLE 11.8: Pair of variables with largest standardised residuals, test anxiety inventory items, males

Pairs of variables	Standardised residual
x_1, x_{18}	−2.58
x_2, x_1	3.37
x_2, x_{19}	−2.72
x_3, x_7	3.86
x_3, x_{20}	−3.45
x_4, x_7	−3.04
x_4, x_{20}	2.56
x_5, x_9	−3.11
x_5, x_{16}	−3.43
x_5, x_{17}	4.53
x_5, x_{20}	−3.35
x_{11}, x_{15}	−3.04
x_{13}, x_{16}	−2.53
x_{13}, x_{18}	3.02
x_{12}, x_{20}	4.02
x_{14}, x_{17}	−2.52
x_{17}, x_{20}	−2.61

TABLE 11.9: Modification indices, test anxiety inventory items, males

Path	Decrease in Chi-Squared	New Estimate
$\xi_1 \rightarrow x_5$	16.6	−0.49
$\xi_2 \rightarrow x_{15}$	12.2	0.33
$\xi_1 \rightarrow x_{20}$	10.9	0.38

11.11 Software

There is a large choice of commercial software available for conducting confirmatory factor analysis and structural equation modelling. These include

LISREL (Jöreskog and Sörbom 1999), Mplus (Muthén and Muthén 2017), EQS (Bentler 2008), Amos (Arbuckle 2006) that runs under the environment of SPSS. The R Package lavaan (Rosseel 2012) can fit structural equation models for continuous and categorical data.

11.12 Further reading

Bollen, K. A. (1989). *Structural Equations with Latent Variables.* New York: John Wiley & Sons.

Chen, Y., Moustaki, I. and Zhang, S. (2023). On the estimation of structural equation models with latent variables in *Handbook of Structural Equation Modeling*, (2nd ed.). Rick H. Hoyle (Ed.) New York: Guilford Press.

Hoyle, R. H. (Ed.) (2022) *Handbook of Structural Equation Modeling* (2nd ed.). Guilford Press.

Kaplan, D. (2000). *Structural Equation Modeling.* Thousand Oaks, CA: Sage Publications.

Kline, R. B. (2023). *Principles and Practice of Structural Equation Modeling* (5th ed.). New York: Guilford Press.

Raykov, T. and Marcoulides, G. A. (2000). *A First Course in Structural Equation Modeling.* Mahwah, N.J.: Lawrence Erlbaum Associates.

Skrondal, A. and Rabe-Hesketh, S. (2004). *Generalized Latent Variable Modelling: Multilevel, Longitudinal, and Structural Equation Models.* Boca Raton, FL: Chapman and Hall/CRC.

12

Undirected Graphical Models

12.1 Introduction

Graphical models use diagrams to represent the relationships among manifest variables. The visualisation of these models usually allows for an intuitive understanding and interpretation of the underlying data structure. In this visual diagram, each manifest variable, which can be, for example, an individual, an organisational unit, or a questionnaire item, is represented by a node, and the relationships among the manifest variables are represented by the edges connecting the corresponding nodes. Depending on whether the edges are directed or undirected, graphical models can be classified into directed and undirected graphical models; see Figure 12.1 for examples of directed and undirected graphs. Sometimes, these graphs are also referred to as directed and undirected networks. Directed graphical models are closely related to path analysis in Chapter 6 and structural equation modelling in Chapter 11. These models are commonly used for causal learning, where the directed edges are used to represent cause-and-effect relationships among manifest variables. On the other hand, undirected graphical models do not impose any causal or directional relationships between variables. Instead, they focus on a notion of "neighbouring" relationships (e.g. between variables y_1 and y_2 as in Panels (b) and (c) of Figure 12.1), which have no inherent directionality, for the association between variables. These models are useful for explaining multivariate data structures and making predictions. This chapter provides an introduction to undirected graphical models.

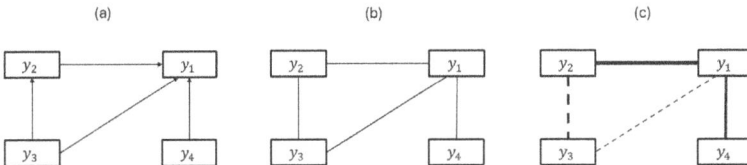

FIGURE 12.1: Examples of (a) a directed graph, (b) an unweighted undirected graph, and (c) a weighted undirected graph, where the solid and dashed lines in (c) represent certain positive and negative associations, respectively

DOI: 10.1201/9781003483342-12

An undirected graphical model imposes a joint distribution for the manifest variables without introducing latent variables. It is useful for modelling complex systems in social sciences, and is used in diverse fields such as psychology, political science, sociology, and finance. Below, we discuss some applications of undirected graphical models.

i) Modelling psychological constructs. Undirected graphical models are commonly used to model psychological constructs in a research field called network psychometrics. In this analysis, people's psychological constructs, such as mental disorder symptoms and personality facets or domains, are collected by questionnaires and modelled as variables in an undirected graphical model. A network of multiple psychological constructs can be learned from data that may be used to generate hypotheses about the underlying psychological phenomena and answer substantive research questions.

ii) Modelling voter behaviour in a legislative chamber. Consider a legislative chamber, such as a parliament, where its members vote on many legislative bills. An undirected graphical model may be applied to the voting data to understand the voters' political behaviours. In this analysis, each node corresponds to a voter (e.g. congressman, senator, or house representative), and a network of voters can be constructed. This analysis may provide insights into the partisan and ideological motivations behind voting behaviour.

iii) Modelling the credit risks of financial institutions. A Credit Default Swap (CDS) is a financial derivative contract in which the buyer of the protection periodically pays a small amount of money, known as "spread", to the seller of protection in exchange for the seller's payoff to the buyer if the reference financial institution defaults on its obligation. The spread depends on the creditworthiness of the reference financial institution and thus reflects its credit risk. An undirected graphical model may be applied to the daily CDS spread quotes of multiple financial institutions to understand the interconnectedness of their credit risks, which is closely related to the systemic risk of the market – the risk that many financial institutions fail together. In this analysis, each node corresponds to a financial institution, and a network of financial institutions can be learned from data.

An undirected graphical model visualises the dependence structure of the manifest variables, which facilitates people's understanding and interpretation of the data structure. Specifically, an unweighted graph shows whether an edge exists between two variables, where the absence of the edge indicates conditional independence between the corresponding two variables, given the rest of the variables. In most applications of undirected graphical models, including the ones mentioned above, the underlying graph is unknown and learned from data. In this process, the underlying graph is typically assumed to have a certain level of sparsity, i.e. a substantial proportion of edges are absent, as a fully connected graph lacks interpretability. An example of an unweighted graph is given in Panel (b) of Figure 12.1, according to which y_2 and y_4 are conditionally independent, and so are y_3 and y_4. Properties of the graph are then used to understand the dependence among the manifest variables. For example, a hub node refers to a node that has a large number of connections (edges) to other nodes in the graph. It often corresponds to a variable that is influential in the complex system of manifest variables. For example, in analysing mental disorder symptoms, the symptoms corresponding to the hub nodes are often the most important ones in the clinical sense. In the modelling of legislative voting, they tend to correspond to the most influential voters. The visualisation of an unweighted graph can be further enhanced by a weighted graph that conveys information about the direction and strength of the conditional associations. For example, Panel (c) of Figure 12.1 presents a weighted version of the unweighted graph in Panel (b). In this graph, solid and dashed lines represent positive and negative conditional associations, respectively, and the thickness of the edges represents the strength of the conditional associations. The weighted graph tends to be more informative than the unweighted one and thus is preferred in practice.

An undirected graphical model enables statistical inferences on the conditional associations between variables, such as testing the conditional independence of two variables, constructing confidence intervals for the conditional associations, and learning the undirected graph from data. These inference problems are closely related to those of regression analysis discussed in Chapter 6. In fact, an undirected graphical model for p manifest variables implies p regression models, each regressing one manifest variable on the rest of the variables. These regression models are multiple linear regression models under the Gaussian graphical model for continuous data in Section 12.2 and logistic regression models under the Ising graphical model for binary data in Section 12.3. The conditional association between two variables y_i and y_j in the graphical model is closely related to the regression coefficient for y_j of the model in which y_i serves as the response variable, which equals the regression coefficient for y_i of the model in which y_j serves as the response variable; see Sections 12.2 and 12.3 for more concrete cases. It is worth noting that, even though these two regression coefficients implied by a graphical model are equal at the population level, their estimates are generally different if separate regression models are fitted.

Drawing statistical inferences within an undirected graphical model has several advantages over performing p regression analyses independently. First, the regression analyses may not yield efficient and coherent inference results for each conditional association. That is, more parameters are estimated when performing p regression analyses independently, which leads to less efficient (i.e. less accurate) parameter estimates. In addition, the conclusions about the same conditional association may contradict each other, as one regression model may suggest two variables are conditionally independent, while the other may suggest the opposite. Second, an undirected graphical model allows us to draw statistical inferences about multiple conditional associations simultaneously, such as testing whether a pre-specified undirected graph can describe the conditional independence relationships among the manifest variables, which may not be feasible with multiple independent regression analyses.

An undirected graphical model can also be used to make predictions. Thanks to the conditional independence structure that an undirected graphical model implies, when predicting the value of one variable y_i based on the rest of the variables, only the values of the neighbouring variables of y_i in the undirected graph matter. For example, suppose that an undirected graphical model follows the graph in Panel (b) of Figure 12.1. When predicting the value of y_2 given the values of the other three variables, the value of y_4 is not needed, since y_2 and y_4 are conditionally independent given y_1. In an undirected graphical model with many variables and a sparse graph (i.e. a graph without many edges), this property often leads to a substantial reduction in computational complexity when making predictions.

Undirected graphical models are available for different data types, including continuous, binary, ordinal, and nominal variables, and even a mixture of them. In this chapter, we focus on the two most widely used graphical models, the Gaussian graphical model for continuous variables and the Ising model for binary variables. These two models and their statistical inference procedures can be naturally extended to other types of data.

We should also note that undirected graphical models may sometimes be confused with models for network data. In the latter, manifest data, instead of model parameters, are structured as a network. Examples of network data include friendship network of students, where an edge indicates the friendship between two students, email exchange network of employees, where an edge indicates two employees having email exchanges, and airline route network of cities, where an edge indicates the existence of direct flights between two cities. We refer readers to Kolaczyk (2009) for an introduction to models for network data.

12.2 Gaussian graphical model for continuous data

Model specification

The Gaussian graphical model is probably the most commonly used undirected graphical model when manifest variables are continuous. In this model, the manifest data $(y_1, ..., y_p)$ follow a multivariate normal distribution. However, instead of the standard parametrisation based on the mean vector and the covariance matrix, the Gaussian graphical model parametrises this distribution by $N(\boldsymbol{\mu}, K^{-1})$, where $\boldsymbol{\mu} = (\mu_1, ..., \mu_p)$ contains the mean values of the p variables, and $K = (\kappa_{ij})_{p \times p}$ is the inverse of the covariance matrix and referred to as the precision matrix. This parametrisation is essential because the precision matrix can fully characterise the conditional independence relationships among the manifest variables and thus determine the undirected graph. That is, under the multivariate normal assumption, $\kappa_{ij} = \kappa_{ji} = 0$ if and only if variables y_i and y_j are conditionally independent given the rest of the variables. In addition, the direction of the conditional association is determined by the corresponding precision parameter. Specifically, the conditional correlation between y_i and y_j is calculated as $-\kappa_{ij}/\sqrt{\kappa_{ii}\kappa_{jj}}$. Therefore, when κ_{ij} is positive, y_i and y_j have a negative conditional association, and vice versa. We note that the precision matrix K is positive definite, due to the positive definiteness of the covariance matrix.

Eq. (12.1) below gives an example of the precision matrix corresponding to the undirected graphs in Figure 12.1 and its inverse, i.e. the corresponding covariance matrix. We see that $\kappa_{24} = \kappa_{42} = \kappa_{34} = \kappa_{43} = 0$, which corresponds to the absence of the edges between y_2 and y_4 and between y_3 and y_4. We also see that the covariance matrix K^{-1} does not have zero entries and thus cannot directly reflect the conditional independence relationships among the variables.

$$K = \begin{pmatrix} 3.0 & -2.0 & 0.3 & -1.0 \\ -2.0 & 4.0 & 1.0 & 0.0 \\ 0.3 & 1.0 & 2.0 & 0.0 \\ -1.0 & 0.0 & 0.0 & 3.0 \end{pmatrix}, \quad K^{-1} = \begin{pmatrix} 0.77 & 0.47 & -0.35 & 0.26 \\ 0.47 & 0.58 & -0.36 & 0.16 \\ -0.35 & -0.36 & 0.73 & -0.12 \\ 0.26 & 0.16 & -0.12 & 0.42 \end{pmatrix}.$$
$$(12.1)$$

As previously mentioned, the Gaussian graphical model implies p multiple linear regression models. That is, for each manifest variable y_i, we can write a linear regression equation, with y_i being the response variable and the rest of the variables being the regressors. The parameters of this regression model are determined by $\boldsymbol{\mu}$ and K, and the regression coefficient for the regressor y_j is proportional to $-\kappa_{ij}$. Eq. (12.2) below shows the linear regression model for y_2 implied by a Gaussian graphical model whose mean vector $\boldsymbol{\mu} = (0, 0, 0, 0)$ and precision matrix is given in eq. (12.1).

$$y_2 = 0.5y_1 - 0.25y_3 + \epsilon_2, \tag{12.2}$$

where the error term ϵ_2 follows a normal distribution with mean zero and variance 0.25. In this regression model, y_4 does not appear due to $\kappa_{24} = 0$. As a result, if we were to predict the value of y_2 using the other three variables, only the values of y_1 and y_3 matter. In addition, the signs of nonzero regression coefficients are consistent with the corresponding conditional correlations, which are opposite to the signs of κ_{21} and κ_{23}.

Parameter estimation and model selection

Similar to the linear factor analysis described in Chapter 7, the statistical inference from an undirected graphical model starts with the sample covariance matrix of the manifest variables, or the sample correlation matrix when we prefer to work with variance-standardised data. We use S to denote this sample covariance or correlation matrix. A natural estimator of the precision matrix is $\hat{K} = (\hat{\kappa}_{ij})_{p \times p} = S^{-1}$, given the relationship between the precision and covariance matrices of a multivariate normal distribution. Using standard large-sample theory, the asymptotic standard errors for the estimated precision parameters can be calculated and used to test whether each $\kappa_{ij} = 0$ for all variable pairs. Based on the p-values of these tests, we can construct an unweighted graph as in Panel (b) of Figure 12.1, where an edge between two variables is formed only when the corresponding κ_{ij} is statistically significantly different from zero. As a total of $p(p-1)/2$ hypotheses are tested here, one needs to account for the multiple testing issue. More specifically, one may control for the familywise error rate, i.e. the probability of falsely rejecting one or more null hypotheses, which can be achieved, for example, by the Bonferroni correction that adjusts the significance level for each test based on the total number of hypotheses. For the edges that are found statistically significant, we estimate the corresponding conditional correlations by $-\hat{\kappa}_{ij}/\sqrt{\hat{\kappa}_{ii}\hat{\kappa}_{jj}}$, based on which a weighted graph can be constructed.

When the model involves many variables and the true precision matrix is believed to be sparse (i.e. conditional independence holds for many variable pairs), the previous estimator may not be optimal, as it does not utilise the sparsity assumption. In this setting, it is common to use a regularised estimator, known as the graphical LASSO (Friedman, Hastie, and Tibshirani 2008), to estimate the precision matrix and identify its nonzero entries. Recall that the zero pattern of the precision matrix determines the undirected graph of the variables' conditional associations. This estimator utilises the LASSO penalty mentioned in Section 6.12 of Chapter 6. This estimator minimises a loss function that combines a negative log-likelihood term for quantifying the distance between K^{-1} and S, and a LASSO penalty term for inducing sparsity on the off-diagonal elements of K. As a result, it tends to find $\hat{K}$ such that $\hat{K}$ is both sparse and close to S^{-1}, yielding a higher accuracy. The performance of the graphical LASSO estimator may be further improved by a refitting procedure (Epskamp and Fried 2018). That is, the parameter estimates from the graphical LASSO tend to be biased towards zero due to the regularisation effect, and therefore might imply underestimation of the conditional associations. To reduce the bias, one may first fit the model using graphical LASSO to obtain a graph structure (i.e. to determine which edges are present), and then refit a model with only those edges and without LASSO regularisation.

Example: Social mobility in the UK

We consider the same social mobility data as in Chapters 5 and 7 relating to the occupational and educational status of three generations of family members. A description of the ten variables for these data can be found in Table 5.13. We estimate the Gaussian graphical model based on the sample correlation matrix in Table 5.14.

We first consider the estimator $\hat{K} = S^{-1}$. Table 12.1 gives the estimated off-diagonal entries of the precision matrix using this estimator. The asymptotic standard errors of these parameters are calculated, based on which the p-values for testing each $\kappa_{ij} = 0$ are obtained. We use a Bonferroni correction to control the familywise error rate at 0.05, identifying 13 (out of 45) precision parameters significantly different from zero. These parameters are indicated by an "$*$" (asterisk) in Table 12.1. Table 12.2 gives the conditional correlations derived from $\hat{K}$. Figure 12.2 shows the unweighted and weighted graphs implied by the estimated precision matrix, where the weighted graph is constructed using the conditional correlations in Table 12.2. In both graphs, only the edges corresponding to the statistically significant precision parameters are presented.

We make several comments about the estimated model. First, the undirected graph constructed by hypothesis testing is relatively sparse, suggesting that many pairs of variables are conditionally independent. This visual representation provides insights into the interrelationships among these variables. For example, the graphs in Figure 12.2 suggest that the husband's qualification (H/Q) is conditionally independent of the rest of the variables, given his further education (H/FE), the wife's qualification (W/Q), and the firstborn's qualification (FB/Q). They also suggest that the firstborn's qualification (FB/Q) is conditionally independent of the rest of the variables, given the husband's qualification (H/Q), the firstborn's further education (FB/FE), and the firstborn's occupational status (FB/O). Second, according to Table 12.1, the significant precision parameters all have negative estimates. Based on the relationship between the conditional correlations and the precision matrix, all the corresponding conditional associations are positive, as confirmed by their values in Table 12.2. That means, for example, the husband's qualification (H/Q) tends to be higher if the wife's qualification is higher (W/Q), given the rest of the variables. Finally, the weighted graph in Panel (b) of Figure 12.2 highlights the strengths of the conditional associations. Specifically, we observe that the strongest conditional associations are between the further education and qualification for the husband, wife, and firstborn, which may suggest a significant effect of further education on one's qualification. Besides these, the conditional correlation between the husband's and firstborn's occupational statuses (H/O and FB/O) is also quite high, potentially implying the influence of fathers on their children's occupations.

TABLE 12.1: Estimated off-diagonal entries of the precision matrix (only the lower-triangular entries are shown) for social mobility variables. The "*" indicates the coefficients that are significant when controlling the familywise error rate with a Bonferroni correction

Variable	HF/O	WF/O	H/FE	H/Q	H/O	W/FE	W/Q	FB/FE	FB/Q	FB/O
HF/O	—									
WF/O	−0.30*	—								
H/FE	−0.10	−0.07	—							
H/Q	0.11	0.00	−0.67*	—						
H/O	−0.39*	−0.25*	−0.25*	−0.12	—					
W/FE	−0.05	−0.04	−0.14	0.03	−0.06	—				
W/Q	−0.06	0.02	−0.03	−0.38*	0.05	−0.58*	—			
FB/FE	−0.03	−0.05	−0.14	−0.09	0.01	−0.22*	0.02	—		
FB/Q	0.02	0.01	−0.01	−0.30*	0.13	0.10	−0.07	−0.58*	—	
FB/O	−0.11	−0.12	−0.06	0.06	−0.46*	−0.02	−0.03	−0.39*	−0.24*	—

TABLE 12.2: Estimated conditional correlations (only the lower-triangular entries are shown) for social mobility variables

Variable	HF/O	WF/O	H/FE	H/Q	H/O	W/FE	W/Q	FB/FE	FB/Q	FB/O
HF/O	—									
WF/O	0.23*	—								
H/FE	0.07	0.05	—							
H/Q	0.07	0.00	0.42*	—						
H/O	0.27*	0.18*	0.16*	0.07	—					
W/FE	0.04	0.03	0.10	−0.02	0.04	—				
W/Q	0.04	−0.02	0.02	0.25*	−0.03	0.41*	—			
FB/FE	0.02	0.03	0.09	0.05	−0.01	0.15*	−0.01	—		
FB/Q	−0.02	−0.01	0.01	0.20*	−0.09	−0.07	0.05	0.38*	—	
FB/O	0.07	0.08	0.04	−0.04	0.30*	0.01	0.02	0.25*	0.16*	—

(a)   (b)

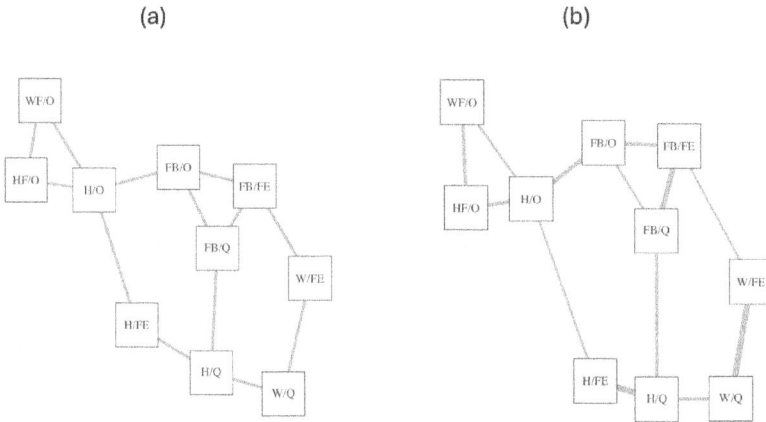

FIGURE 12.2: Panel (a): An unweighted undirected graph. Panel (b): A weighted undirected graph. The graphs are for social mobility variables and are constructed via hypothesis testing. In the weighted graph, the thickness of the edge indicates the strength of association

We next consider the graphical LASSO estimator. This estimator assumes that the underlying graph is sparse and uses the LASSO penalty to achieve a sparse estimate of the graph. Here, we apply the refitting strategy mentioned previously. This procedure identifies 14 (out of 45) nonzero precision parameters. Compared with the previous result from hypothesis testing, this estimator additionally selects the edge between the husband's further education (H/FE) and the firstborn's further education (FB/FE). Figure 12.3 presents the resulting weighted graph. We remark that the sample size in the current example is reasonably large compared to the number of variables, which leads

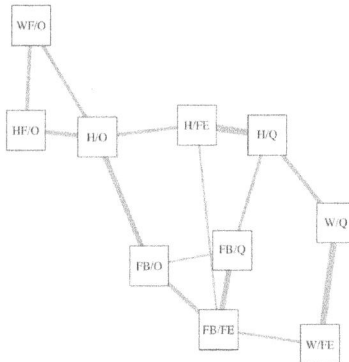

FIGURE 12.3: The weighted graph for social mobility variables constructed using a graphical LASSO estimator

to similar results from the unregularised and regularised estimators. For applications with a larger number of variables, the regularisation approach tends to outperform the unregularised one when the sparsity assumption holds.

Further topic: Gaussian graphical model with covariates

The conditional associations between manifest variables often depend on additional covariates, such as demographic characteristics of individuals. In recent years, the Gaussian graphical model has been extended to allow the precision matrix to vary with covariates (Kolar, Parikh, and Xing 2010; Ni, Stingo, and Baladandayuthapani 2022; Zhang and Li 2023). As a result, the strengths of conditional associations can vary across observation units. Methods for parameter estimation and model selection have also been developed for these extended models.

12.3   Undirected graphical model for binary data

Model specification

The Ising model (Ising 1925) is probably the most commonly used graphical model for multivariate binary data. Suppose we have p-variate binary data, $(y_1, y_2, ..., y_p)$, where each y_j takes a value of either 0 or 1. The parameters of the Ising model can be arranged into a $p \times p$ symmetric matrix, denoted by $K = (\kappa_{ij})_{p \times p}$. We intentionally use the same notation as the precision matrix in the Gaussian graphical model because the matrix K characterises the conditional associations among the manifest variables under the Ising model, mimicking the role that the precision matrix plays in the Gaussian graphical model. The Ising model assumes the following parametrisation for the manifest variables $(y_1, y_2, ..., y_p)$:

$$P(y_1 = c_1, y_2 = c_2, ..., y_p = c_p) = \frac{\exp\left(\sum_{i=1}^{p}\sum_{j=i+1}^{p}\kappa_{ij}c_ic_j + \sum_{i=1}^{p}\kappa_{ii}c_i\right)}{Z(K)},$$

(12.3)

where each c_i, $i = 1, ..., p$, takes a value of 0 or 1, and the denominator

$$Z(K) = \sum_{c_1=0}^{1}\sum_{c_2=0}^{1}\cdots\sum_{c_p=0}^{1}\exp\left(\sum_{i=1}^{p}\sum_{j=i+1}^{p}\kappa_{ij}c_ic_j + \sum_{i=1}^{p}\kappa_{ii}c_i\right)$$

(12.4)

is known as a normalisation term which ensures the probabilities in eq. (12.3) for all possible response patterns sum to one.

As mentioned previously, the Ising model implies p logistic regressions. That is, we can compute the conditional distribution of each y_i given the rest of the variables from the joint distribution (12.3), which takes the form

$$P(y_i = 1 | y_1, ..., y_{i-1}, y_{i+1}, ..., y_p) = \frac{\exp\left(\kappa_{ii} + \kappa_{i1}y_1 + \kappa_{i2}y_2 + \cdots + \kappa_{ip}y_p\right)}{1 + \exp\left(\kappa_{ii} + \kappa_{i1}y_1 + \kappa_{i2}y_2 + \cdots + \kappa_{ip}y_p\right)}.$$
(12.5)

From this equation, we see that the off-diagonal elements of K become the regression coefficients in the logistic regressions. Specifically, κ_{ij}, when $i \neq j$, is the regression coefficient for regressor y_j when y_i serves as the response variable and, due to the symmetry of K, is also the regression coefficient for regressor y_i when y_j serves as the response variable. Thus, κ_{ij} can be interpreted as the conditional log-odds-ratio between the two variables, given the rest of the variables. In particular, $\kappa_{ij} = 0$ if and only if y_i and y_j are conditionally independent. We also see that the diagonal elements of K, κ_{ii}, serve as the intercept parameters in the logistic regressions. Unlike the precision matrix in the Gaussian graphical model, the matrix K is not required to be a positive definite matrix. Therefore, the signs of the intercept parameters in the logistic regressions are not restricted.

Eq. (12.6) gives an example of a K matrix that is consistent with the undirected graphs in Figure 12.1, where the zero values in K correspond to the absence of an edge between y_2 and y_4 and between y_3 and y_4.

$$K = \begin{pmatrix} 1.0 & 2.0 & -0.3 & 1.0 \\ 2.0 & 0.5 & -1.0 & 0.0 \\ -0.3 & -1.0 & -1.0 & 0.0 \\ 1.0 & 0.0 & 0.0 & -2.0 \end{pmatrix}.$$
(12.6)

Under this Ising model, the implied logistic regression models with response variables y_1 and y_2 are, respectively,

$$P(y_1 = 1 | y_2, y_3, y_4) = \frac{\exp(1 + 2y_2 - 0.3y_3 + y_4)}{1 + \exp(1 + 2y_2 - 0.3y_3 + y_4)}$$
(12.7)

and

$$P(y_2 = 1 | y_1, y_3, y_4) = \frac{\exp(0.5 + 2y_1 - y_3)}{1 + \exp(0.5 + 2y_1 - y_3)}.$$
(12.8)

Specifically, we see that the coefficient of y_2 in eq. (12.7) and that of y_1 in eq. (12.8) are the same, which is due to the symmetry of K. In addition, y_4 does not appear on the right-hand side of eq. (12.8) due to $\kappa_{24} = 0$, which implies conditional independence between y_2 and y_4.

Parameter estimation and model selection

Unlike the Gaussian graphical model, there is no explicit relationship between the parameter matrix K and the population covariance matrix of manifest variables. Consequently, K cannot be easily estimated based on the sample covariance matrix. Instead, likelihood-based estimation methods are typically used to estimate K, which aim to find a K matrix such that the manifest and predicted distributions are as close as possible.

Two estimation methods are commonly used. The first is the maximum likelihood estimator, constructed from the joint distribution of eq. (12.3). This estimator aims to match the model-predicted joint distribution of $(y_1, y_2, ..., y_p)$ with the observed one. The statistical theory for the maximum likelihood estimator guarantees that it is asymptotically (i.e. when the sample size becomes large) almost unbiased and has virtually the smallest possible standard errors among all the estimators. In addition, its asymptotic sampling distribution can be approximated well by a normal distribution, based on which confidence intervals can be constructed for the model parameters. However, the maximum likelihood estimator for the Ising model has a drawback when the number of variables is large. The normalisation term in eq. (12.4) and its derivatives with respect to the unknown parameters need to be calculated many times when computing the maximum likelihood estimator. As the normalisation term involves a summation of 2^p response patterns, evaluating this function and its derivatives is computationally intensive when p is large. For example, in a setting with 20 variables, the number of terms in the summation exceeds one million. The computation of the maximum likelihood estimator on a typical personal computer quickly becomes infeasible when p exceeds 20.

The second estimation method is the pseudo-likelihood estimator, which is constructed using likelihood contributions based on the logistic regressions in eq. (12.5). This estimator aims to match the model-predicted conditional distributions of each y_i given the rest of the variables with the observed ones, instead of directly matching the respective joint distributions as in the maximum likelihood estimator. This estimator has statistical properties similar to those of the maximum likelihood estimator in terms of asymptotic consistency, unbiasedness and normality. However, its standard errors tend to be slightly larger than those of the maximum likelihood estimator, as there is information loss when focusing only on the conditional distributions implied by the joint model. The large-sample theory for this estimator follows the general theory for composite-likelihood methods (Varin, Reid, and Firth 2011). In terms of computation, this estimator is substantially more affordable than the maximum likelihood estimator, as it avoids computing the normalisation term.

Similar to the graphical LASSO estimator for the Gaussian graphical model, the LASSO penalty is also commonly combined with standard estimation methods, including those mentioned above, to estimate the K matrix and to simultaneously learn which elements are zero. In addition, a similar refitting strategy can be applied to the Ising model, as for Gaussian graphical models, to reduce the bias introduced by the LASSO regularisation.

Example: Workplace industrial relations data

We fit the Ising model to data from the 1990 Workplace Industrial Relations Survey (WIRS) regarding management/worker consultation in firms. The same data has been analysed using a factor model for binary data in Chapter 8 and a latent class model in Chapter 10. The six variables in this dataset include:

y_1. Informal discussion with individual workers.

y_2. Meetings with groups of workers.

y_3. Discussions in established joint consultative committee.

y_4. Discussions in specially constituted committee to consider the change.

y_5. Discussions with union representatives at the establishment.

y_6. Discussions with paid union officials from outside.

As the number of manifest variables is small, we apply the maximum likelihood estimator.

Table 12.3 presents the estimated off-diagonal entries of the K matrix in the Ising model. Similar to the analysis of social mobility data above, we test the hypotheses of $\kappa_{ij} = 0$ for all variable pairs, using a Bonferroni correction to control the familywise error rate at 0.05. This analysis identifies 9 (out of 15) nonzero parameters, which are indicated with an asterisk in the table. Figure 12.4 presents the resulting weighted graph, for which only the statistically significant edges are presented.

TABLE 12.3: Estimated off-diagonal entries of the precision matrix (only the lower-triangular entries are shown) for WIRS variables

Variable	y_1	y_2	y_3	y_4	y_5	y_6
y_1	–					
y_2	−1.43*	–				
y_3	0.32	0.09	–			
y_4	−0.51	0.72*	0.83*	–		
y_5	0.50*	0.49*	1.29*	0.67*	–	
y_6	0.50	0.22	0.80*	0.54	1.08*	–

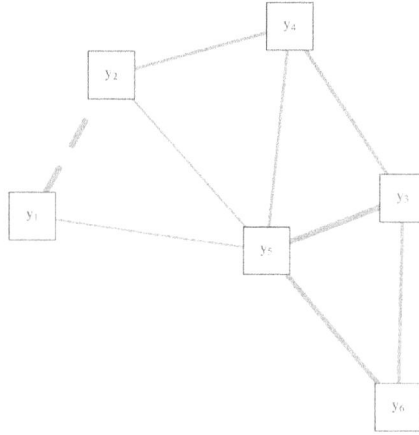

FIGURE 12.4: The weighted graph for WIRS variables, constructed via hypothesis testing. The solid and dashed lines represent positive and negative conditional associations, respectively.

The estimated Ising model implies some interesting relationships among the manifest variables, which were unavailable from the estimated factor models in Chapter 8. Unlike the social mobility example, where all the conditional associations are estimated to be positive, the estimated model finds both positive and negative conditional associations. More specifically, the conditional association between y_1 and y_2 is negative, and the rest are positive. The negative conditional association suggests that for individuals who give the same answers to the last four items, those who answer "yes" to the first item ("informal discussions with individual workers") tend to answer "no" to the second item ("meetings with groups of workers"), and vice versa. This result suggests that the two consultation methods tend to be mutually exclusive, i.e. firms tend to choose only one method from these two. On the other hand, y_1 has a significantly positive conditional association with y_5, meaning that these two consultation methods ("informal discussions with individual workers" and "discussions with union representatives at the establishment") tend to be implemented together, given the implementations of the other consultation methods.

In addition, the node representing y_5 can be seen as a hub node in the graph in Figure 12.4, as it is connected to all the other nodes. As the corresponding coefficients are all positive, it means that this consultation method ("discussions with union representatives at the establishment") tends to be co-implemented with all the others, which may highlight the importance of this consultation method in the workplace.

Finally, we may compare the fit of the Ising model and the factor models using the Bayesian information criterion (BIC). The one- and two-dimensional factor models, which contain 12 and 17 parameters, have BIC values of 6923.1

and 6800.6, respectively. The BIC value for the Ising model with the graph structure in Figure 12.4 is 6807.4, which is computed by refitting the Ising model with zero constraints implied by the graph. This comparison suggests that this Ising model fits the WIRS data substantially better than the one-dimensional factor model. At the same time, it is comparable to, but slightly worse than, the two-dimensional factor model. As such, the two-dimensional factor model and the Ising model seem to be preferred over the one-dimensional factor model, and the choice between the two-dimensional factor model and the Ising model may be made based on the purpose of the analysis (e.g. whether we aim to derive summary measures or understand the conditional associations among the variables) and the substantive interpretations of the estimated models.

Further topic: Ising model with covariates

Similar to Gaussian graphical models, the Ising model can also be extended to allow the strengths of conditional associations to depend on additional covariates. This is achieved by modelling the parameters in the Ising model as functions of the covariates; see Cheng et al. (2014).

12.4 Comparison with latent variable models

So far, we have introduced several models for modelling the joint distribution of multivariate data, including the undirected graphical models introduced above, the factor models in Chapters 7–9, the models with categorical latent variables in Chapter 10, and the structure equation models in Chapter 11. All these models involve latent variables, except for the undirected graphical models. Often, these models can be fitted to the same data. For example, with multiple continuous variables, we can fit the Gaussian graphical model, the linear factor model, and the Gaussian mixture model. If the data are binary, we also have the options of the factor model for binary data, the latent class model, and the Ising model. However, these models differ significantly in their underlying assumptions and the types of relationships they model.

The undirected graphical models and the latent variable models produce substantially different data dependence structures. We illustrate this point by comparing the Gaussian graphical model and the one-dimensional linear factor model, while pointing out that similar results hold for many other latent variable models and graphical models. Consider the one-dimensional factor model in eq. (7.2) in Chapter 7, with the joint normality assumption for all the random variables in the model. Then the conditional covariance between variables y_i and y_j is given by $\beta_i\beta_j/(1 + \sum_{l \neq i,j}(\beta_l^2/\sigma_l^2))$. Recall that β_i is the factor loading parameter and σ_i^2 is the unique variance for the ith variable. As the marginal covariance between the two variables is given by $\beta_i\beta_j$, the two variables are conditionally independent if and only if they are marginally independent. When all the loading parameters are nonzero, no pair of manifest variables is conditionally independent. When fitting a Gaussian graphical model to data generated from this one-dimensional factor model with nonzero loadings, the resulting undirected graph tends to be fully connected and, thus, the fitted model has low interpretability. On the other hand, the one-dimensional factor model cannot adequately fit data generated by a Gaussian graphical model if the population precision matrix is sparse while the covariance matrix is not (see eq. (12.1) for an example). As shown above, the one-factor model can only imply multivariate normal distributions for which the precision matrix has the same sparsity pattern as the covariance matrix. Although a satisfactory fit may eventually be achieved by fitting a model with many factors, the resulting factors may be hard to interpret.

In practice, we should choose the model based on the purpose of the analysis. Specifically, if the goal is to predict the values of new observations, we may compare the prediction performance of different models. In this regard, the cross-validation method discussed in Chapter 6 can be extended to evaluate the prediction performance of these models. If, instead, the goal is to understand the dependencies among the manifest variables, we may choose the model based on substantive theory or domain knowledge about the data. Specifically, if substantive theory or domain knowledge suggests the presence of one or multiple latent constructs, then a factor model may be suitable. Or, if they suggest hidden heterogeneity in the population, for example, individuals belonging to unobserved groups, then a model with categorical latent variables may be chosen. Otherwise, if they suggest that the dependence is due to the interaction between the variables, then an undirected graphical model is more suitable. Sometimes, multiple substantive theories may be suggested for the same data. For example, consider data on people's symptoms of a mental health disorder, for which multiple substantive theories may be available. One possibility is that there exists a latent psychological construct, for example, a depression factor, that drives the covariation of the symptoms. People at the higher end of the latent construct tend to have more symptoms, while those at the lower end tend to have fewer. Another possibility is that patients who suffer from the mental disorder are of multiple types, where different types of patients show different patterns of symptoms. Alternatively, an individual's symptoms may be explained by the interactions between symptoms over time (e.g. symptoms triggering each other). If there is no theory that is more plausible than the others, we may fit the models suggested by the theories and use model comparison methods, such as information criteria, to find the best-fitting model, which can lend credence to the theory that suggests this model.

12.5   Additional examples

Example: Post-traumatic stress disorder symptoms in U.S. military veterans

In this example, we analyse the conditional associations between symptoms of post-traumatic stress disorder (PTSD) for trauma-exposed U.S. military veterans, where PTSD is a mental health condition that can develop in individuals who have experienced or witnessed a terrifying event. The dataset in this example, which is from the first wave of the National Health and Resilience in Veterans Study (NHRVS, Fogle et al. 2020), is a subset of the data considered in von Stockert et al. (2018). This subset consists of 1,136 male trauma-exposed U.S. military veterans' responses to 20 items that assess PTSD symptoms defined in the Diagnostic and Statistical Manual of Mental Disorders, fifth edition (DSM-5; American Psychiatric Association 2013). DSV-5 classifies these 20 symptoms into four symptom clusters, including intrusion symptoms (IS), avoidance (AV), negative alterations in cognitions and mood (CM), and alterations in arousal and reactivity (AR). A short description of the symptoms and their classification is given in Table 12.4 below. The respondents were asked to report how much they had experienced each of the 20 symptoms in the past month on a scale from 0 (Not at all) to 4 (Extremely). Further details about this dataset can be found in von Stockert et al. (2018). We note that the data are ordinal but treated as continuous data in the current analysis.

TABLE 12.4: 20 items assessing PTSD symptoms

IS1: Intrusive thoughts	CM4: Negative trauma-related emotions
IS2: Nightmares	CM5: Loss of interest
IS3: Flashbacks	CM6: Detachment
IS4: Emotional cue reactivity	CM7: Restricted affect
IS5: Physical cue reactivity	AR1: Irritability/anger
AV1: Avoidance of thoughts	AR2: Self-destructive/reckless behaviour
AV2: Avoidance of reminders	AR3: Hypervigilance
CM1: Trauma-related amnesia	AR4: Exaggerated startle response
CM2: Negative beliefs	AR5: Difficulty concentrating
CM3: Blame of self or others	AR6: Sleep disturbance

The sample correlation matrix of the 20 variables is given in Table 12.5, based on which a Gaussian graphical model is fitted. More specifically, we use the graphical LASSO estimator to learn the graph and, further, a refitting procedure to estimate the free parameters given the learned graph. The learned graph is shown in Figure 12.5. In this graph, 57 out of 190 pairs of manifest variables are connected by edges, i.e. estimated to be conditionally dependent. Among these pairs, only three have negative conditional correlations, with values ranging between -0.13 and -0.04. The rest have positive conditional associations, with values between 0.04 and 0.48.

TABLE 12.5: Pairwise correlations between PTSD items

	IS1	IS2	IS3	IS4	IS5	AV1	AV2	CM1	CM2	CM3
IS1	1.00									
IS2	0.68	1.00								
IS3	0.66	0.71	1.00							
IS4	0.72	0.62	0.66	1.00						
IS5	0.64	0.65	0.68	0.70	1.00					
AV1	0.60	0.56	0.56	0.62	0.61	1.00				
AV2	0.60	0.60	0.57	0.65	0.65	0.69	1.00			
CM1	0.36	0.36	0.32	0.34	0.35	0.38	0.40	1.00		
CM2	0.51	0.43	0.43	0.53	0.45	0.44	0.52	0.37	1.00	
CM3	0.56	0.54	0.53	0.58	0.55	0.51	0.53	0.40	0.61	1.00
CM4	0.68	0.62	0.61	0.66	0.63	0.55	0.63	0.42	0.72	0.74
CM5	0.58	0.55	0.55	0.52	0.53	0.53	0.60	0.41	0.62	0.54
CM6	0.58	0.55	0.52	0.52	0.51	0.55	0.60	0.39	0.67	0.55
CM7	0.52	0.47	0.48	0.45	0.46	0.53	0.56	0.33	0.61	0.48
AR1	0.56	0.50	0.53	0.56	0.58	0.51	0.57	0.35	0.60	0.51
AR2	0.39	0.42	0.41	0.36	0.42	0.35	0.41	0.32	0.37	0.39
AR3	0.44	0.46	0.45	0.44	0.45	0.48	0.53	0.32	0.49	0.38
AR4	0.50	0.57	0.56	0.53	0.56	0.53	0.57	0.34	0.50	0.44
AR5	0.53	0.52	0.52	0.48	0.47	0.48	0.48	0.39	0.56	0.52
AR6	0.45	0.47	0.46	0.42	0.43	0.43	0.43	0.25	0.41	0.40

	CM4	CM5	CM6	CM7	AR1	AR2	AR3	AR4	AR5	AR6
IS1	0.68	0.58	0.58	0.52	0.56	0.39	0.44	0.50	0.53	0.45
IS2	0.62	0.55	0.55	0.47	0.50	0.42	0.46	0.57	0.52	0.47
IS3	0.61	0.55	0.52	0.48	0.53	0.41	0.45	0.56	0.52	0.46
IS4	0.66	0.52	0.52	0.45	0.56	0.36	0.44	0.53	0.48	0.42
IS5	0.63	0.53	0.51	0.46	0.58	0.42	0.45	0.56	0.47	0.43
AV1	0.55	0.53	0.55	0.53	0.51	0.35	0.48	0.53	0.48	0.43
AV2	0.63	0.60	0.60	0.56	0.57	0.41	0.53	0.57	0.48	0.43
CM1	0.42	0.41	0.39	0.33	0.35	0.32	0.32	0.34	0.39	0.25
CM2	0.72	0.62	0.67	0.61	0.60	0.37	0.49	0.50	0.56	0.41
CM3	0.74	0.54	0.55	0.48	0.51	0.39	0.38	0.44	0.52	0.40
CM4	1.00									
CM5	0.64	1.00								
CM6	0.67	0.72	1.00							
CM7	0.59	0.66	0.79	1.00						
AR1	0.67	0.57	0.63	0.62	1.00					
AR2	0.42	0.41	0.46	0.42	0.49	1.00				
AR3	0.49	0.49	0.55	0.49	0.48	0.39	1.00			
AR4	0.58	0.58	0.60	0.52	0.54	0.44	0.56	1.00		
AR5	0.60	0.69	0.69	0.60	0.57	0.38	0.44	0.58	1.00	
AR6	0.45	0.52	0.50	0.47	0.44	0.29	0.40	0.48	0.55	1.00

The highest conditional correlations are observed between symptoms in the same cluster. Specifically, the top five conditional correlations are between

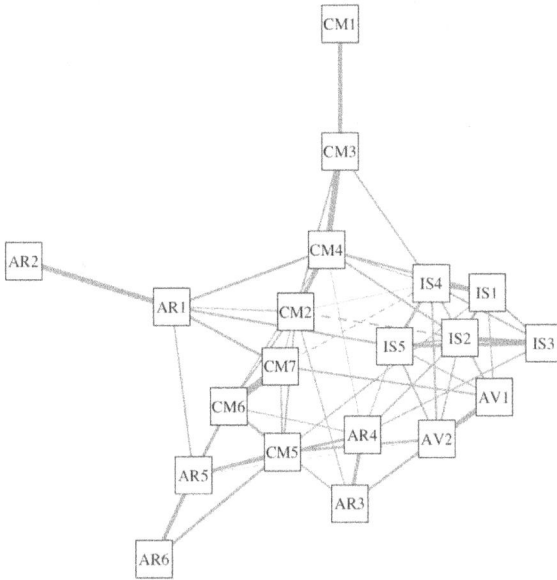

FIGURE 12.5: Undirected graph for PTSD symptoms learned by a graphical LASSO estimator

variables CM6 and CM7 (0.48), CM3 and CM4 (0.40), AR1 and AR2 (0.35), AV1 and AV2 (0.34), and IS2 and IS3 (0.32). In addition, symptoms in the same cluster tend to be more densely and positively connected. These patterns suggest a higher level of conditional associations within the symptom clusters, possibly because symptoms within the same cluster are more likely to trigger one another.

It is interesting to observe that AR symptoms are separated into two groups by the CM symptoms in the graph, with one group containing AR3 and AR4, and the other containing the remaining four AR symptoms. Any path connecting the two groups of symptoms has to pass at least one of CM2, CM4, CM5, CM6, and CM7. Statistically, it means that according to the estimated model, the corresponding two groups of variables are conditionally independent given variables CM2, CM4, CM5, CM6, and CM7. This result may suggest hypotheses about the underlying psychological process. Specifically, hypervigilance and exaggerated startle response indicate a hyperarousal state – a heightened sensory and physiological reactivity to the environment. The other symptoms may be consequences of this underlying state of hyperarousal, but are possibly mediated by the relevant CM symptoms.

We can further identify the hub nodes of the graph by calculating centrality measures like the degree of the nodes, which may suggest symptoms that potentially play a central role in the development of this mental disorder. For a weighted graph constructed by conditional correlations, the degree of a node may be computed as the sum of the absolute conditional correlations between the corresponding variable and the rest. Based on this criterion, the top five hub nodes of the learned graph are IS2, IS4, AV2, CM4, and CM6.

Example: Senate votes in the 108th Congress of U.S.

We consider the voting data of senators in the 108th Congress of the United States in 2003. This dataset involves 100 senators, who voted on 460 legislative bills. For each bill, a senator can vote in three ways, including "yea", "nay", and "not voting", indicating that they support, oppose, and are present but choose not to cast a vote, respectively. In the current analysis, we treat a "yea" response as value 1 and a "nay" or "not voting" response as value 0 and analyse the data with the Ising model. More specifically, we treat the senators as the variables, and the bills as the observation units. Information about the senators is given in Table 12.6, including the states they represent (two senators represent each state), names, and party memberships. In particular, the 100 senators are comprised of 51 Republicans, 48 Democrats, and one independent politician who caucused with the Democratic Party.

We use a LASSO-based estimator (van Borkulo et al. 2014) to learn the undirected graph of senators, and further, a refitting procedure based on the pseudo-likelihood estimator for estimating the unknown parameters given the learned graph. We note that the maximum likelihood estimator is computationally infeasible here, due to the large number of variables (senators) in the current analysis. The estimated graph is given in Figure 12.6. In this graph, the edges corresponding to positive and negative conditional associations are indicated by solid and dashed lines. In addition, the Democratic, Republican, and Independent senators are represented by nodes of square, circle, and rectangle shapes, respectively. The learned graph has 407 edges, so about 8% of the pairs of nodes are connected. Among these edges, 8 represent negative conditional associations, with estimated κ_{ij}s ranging between -1.85 and -0.15, and the remaining 399 represent positive conditional associations, with estimated κ_{ij}s ranging between 0.15 and 4.89.

The estimated graph shows several interesting patterns. First, the graph shows a clear two-cluster pattern, as the vertical dotted line indicates in Figure 12.6. The nodes within each cluster are densely and positively connected. In contrast, there are only a small number of edges connecting the two clusters, and most of these edges represent negative conditional associations. These two clusters of nodes are highly consistent with the two political parties of the U.S., the Democratic and Republican parties. This result confirms the partisan behaviour of the senators in the Congress, in the sense that the votes from senators of the same party tend to have strongly positive

TABLE 12.6: Information about the 100 senators in the 108th Congress of U.S., where "R", "D", and "I" in the column of Party represent the Republican, Democratic and Independent senators

	State	Surname	Party		State	Surname	Party
1	Alabama	Sessions	R	51	Montana	Baucus	D
2	Alabama	Shelby	R	52	Montana	Burns	R
3	Alaska	Murkowski	R	53	Nebraska	Hagel	R
4	Alaska	Stevens	R	54	Nebraska	Nelson	D
5	Arizona	Kyl	R	55	Nevada	Ensign	R
6	Arizona	McCain	R	56	Nevada	Reid	D
7	Arkansas	Pryor	D	57	New Hampshire	Gregg	R
8	Arkansas	Lincoln	D	58	New Hampshire	Sununu	R
9	California	Boxer	D	59	New Jersey	Corzine	D
10	California	Feinstein	D	60	New Jersey	Lautenberg	D
11	Colorado	Allard	R	61	New Mexico	Bingaman	D
12	Colorado	Campbell	R	62	New Mexico	Domenici	R
13	Connecticut	Dodd	D	63	New York	Clinton	D
14	Connecticut	Lieberman	D	64	New York	Schumer	D
15	Delaware	Biden	D	65	North Carolina	Edwards	D
16	Delaware	Carper	D	66	North Carolina	Dole	R
17	Florida	Graham	D	67	North Dakota	Conrad	D
18	Florida	Nelson	D	68	North Dakota	Dorgan	D
19	Georgia	Chambliss	R	69	Ohio	Dewine	R
20	Georgia	Miller	D	70	Ohio	Voinovich	R
21	Hawaii	Akaka	D	71	Oklahoma	Inhofe	R
22	Hawaii	Inouye	D	72	Oklahoma	Nickles	R
23	Idaho	Craig	R	73	Oregon	Smith	R
24	Idaho	Crapo	R	74	Oregon	Wyden	D
25	Illinois	Durbin	D	75	Pennsylvania	Santorum	R
26	Illinois	Fitzgerald	R	76	Pennsylvania	Specter	R
27	Indiana	Bayh	D	77	Rhode Island	Chafee	R
28	Indiana	Lugar	R	78	Rhode Island	Reed	D
29	Iowa	Grassley	R	79	South Carolina	Hollings	D
30	Iowa	Harkin	D	80	South Carolina	Graham	R
31	Kansas	Brownback	R	81	South Dakota	Daschle	D
32	Kansas	Roberts	R	82	South Dakota	Johnson	D
33	Kentucky	Bunning	R	83	Tennessee	Frist	R
34	Kentucky	Mcconnell	R	84	Tennessee	Alexander	R
35	Louisiana	Breaux	D	85	Texas	Cornyn	R
36	Louisiana	Landrieu	D	86	Texas	Hutchison	R
37	Maine	Collins	R	87	Utah	Bennett	R
38	Maine	Snowe	R	88	Utah	Hatch	R
39	Maryland	Mikulski	D	89	Vermont	Jeffords	I
40	Maryland	Sarbanes	D	90	Vermont	Leahy	D
41	Massachusetts	Kennedy	D	91	Virginia	Allen	R
42	Massachusetts	Kerry	D	92	Virginia	Warner	R
43	Michigan	Stabenow	D	93	Washington	Cantwell	D
44	Michigan	Levin	D	94	Washington	Murray	D
45	Minnesota	Dayton	D	95	West Virginia	Byrd	D
46	Minnesota	Coleman	R	96	West Virginia	Rockefeller	D
47	Mississippi	Cochran	R	97	Wisconsin	Feingold	D
48	Mississippi	Lott	R	98	Wisconsin	Kohl	D
49	Missouri	Talent	R	99	Wyoming	Enzi	R
50	Missouri	Bond	R	100	Wyoming	Thomas	R

conditional associations. Not surprisingly, the independent politician who caucused with the Democratic party, Senator Jeffords from Vermont (node 89), falls into the Democratic cluster. Interestingly, Senator Miller from Georgia

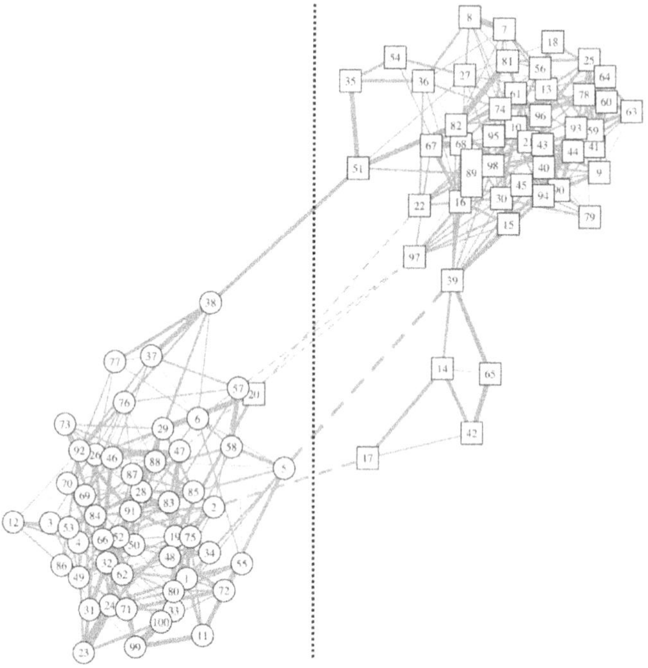

FIGURE 12.6: Undirected graph for senators of the 108th Congress of the U.S. learned by a LASSO-based estimator. The Democratic, Republican, and Independent senators are represented by nodes of square, circle, and rectangle shapes, respectively. The nodes are divided into two clusters by the vertical dotted line

(20), known as a conservative Democrat, falls into the cluster of Republican senators. It suggests that his votes were more similar to those of the Republican senators than to those of the Democrats. A subcluster of nodes shows up within the Democratic cluster, consisting of Senators Lieberman from Connecticut (14), Graham from Florida (17), Kerry from Massachusetts (42), and Edwards from North Carolina (65). These nodes are closely connected and deviate from the majority of the nodes of Democratic senators in the graph. This result aligns well with the political ideology of these senators. In fact, these senators are typically considered centrist democrats who tend to vote with Republicans on certain issues, particularly those related to national security, defense spending, and trade. This distinguishes their political ideology from the more consistently liberal wing of the party.

Second, senators representing the same state tend to show high conditional associations. In fact, among the pairs of senators with the top ten highest values of κ_{ij} estimates, nine are formed by senators representing the same states, including the senators from Arkansas (3 and 4), Idaho (23 and 24),

Maine (37 and 38), Michigan (43 and 44), New Jersey (59 and 60), North Dakota (67 and 68), Washington (93 and 94), and Wyoming (99 and 100). This is sensible, given that the senators who represent the same state respond to the same general electorate, economic conditions, and state-specific issues.

Finally, hub nodes may be identified by calculating centrality measures, such as the degree of the nodes, which may suggest the most influential senators. Specifically, based on the weighted graph in Figure 12.6, the top ten nodes with the highest degrees are Senators Chambliss from Georgia (19), Crapo from Idaho (24), Grassley from Iowa (29), Mikulski from Maryland (39), Levin from Michigan (44), Cochran from Mississippi (47), Burns from Montana (52), Nickles from Oklahoma (72), Reed from Rhode Island (78), Johnson from South Dakota (82), and Murray from Washington (94). These senators have high conditional associations with the senators within their parties. Further analysis is needed to check whether these senators are indeed more representative and influential than the other senators within the same parties.

12.6 Software

Several R packages are available for estimating undirected graphical models and learning the associated graphs. Specifically, the Gaussian graphical model can be estimated using the *glasso* (Friedman, Hastie, and Tibshirani 2019) and *qgraph* (Epskamp et al. 2012) packages. The Ising model can be estimated using the *IsingFit* (van Borkulo and Epskamp 2023) and *IsingSampler* (Epskamp 2025a) packages. Moreover, undirected graphical models for data consisting of both categorical and continuous variables can be fitted with the *mgm* package (Haslbeck and Waldorp 2020). The *psychonetrics* package (Epskamp 2025b) provides functions for estimating several graphical models, including the Gaussian graphical model and the Ising model. For the visualisation of the estimated graph, one can use the software *gephi* (Bastian, Heymann, and Jacomy 2009) and R packages *qgraph* (Epskamp et al. 2012) and *igraph* (Csárdi et al. 2025).

12.7 Further reading

Edwards, D. (2000) *Introduction to Graphical Modelling.* New York: Springer.

Højsgaard, S., Edwards, D. and Lauritzen, S. (2012) *Graphical Models with R.* New York: Springer.

Epskamp, S. and Fried, E. I. (2018). *Psychological Methods 23*, 617–634.

Isvoranu, A. M., Epskamp, S., Waldorp, L. J., and Borsboom, D. (Eds) (2022). *Network Psychometrics with R: A Guide for Behavioural and Social Scientists.* New York: Routledge.

13

Multilevel Modelling

13.1 Introduction

Most populations studied in the social sciences have a hierarchical structure. In educational research, for example, children's test scores may be clustered by class and school. Other individual outcomes, such as behavioural and attitudinal measures, may be clustered by household and geographical area. Both of these examples are of a three-level hierarchical structure with individuals at level 1 nested within classes (or households) at level 2 within schools (or areas) at level 3. When individuals form clusters or groups, we might expect that two randomly selected individuals from the same group will tend to be more alike than two individuals selected from different groups. For example, children learn in classes and features of their class, such as characteristics of the teacher and other children in the class, are likely to influence a child's educational attainment. Because of these class effects, we would expect test scores for children in the same class to be more alike than scores for children from different classes. Hierarchical structures can also arise from having multiple measurements on the same individual, either on different variables at one point in time (multivariate data) or the same variable at several time points (longitudinal or repeated measured data). In such cases, the multiple measures form the level 1 units and individuals are now at level 2. Measurements taken on the same individual will tend to be more highly correlated than two measurements from different individuals due to the presence of individual characteristics that affect all of his or her outcomes.

Multilevel models – also known as *hierarchical linear models*, *mixed models* and *random effects models* – have been developed to take into account and explore dependencies in hierarchical population structures. Ignoring these structures and analysing the level 1 units (e.g. children) as if they were independent would not be valid. Standard errors of regression coefficients would be biased. In particular, standard errors for coefficients of predictors that are defined at higher levels (e.g. school or area characteristics) would be underestimated, resulting in confidence intervals that would be too narrow and p-values too small, which may in turn lead to incorrect inferences. The effect of clustering, for example, of children by school, can be measured by the intra-cluster correlation. Unless this is zero the level 1 units within the same level 2 unit, group or cluster, will be correlated. If we are interested only in *adjusting* for clustering, rather than *exploring* it, there are methods other than multilevel modelling that can be used. Survey statisticians have developed methods to correct standard errors for design effects in the analysis of data from multistage sample designs (Kish 1965, pp.161-163). Another common approach, especially in longitudinal data analysis, is to use a marginal (or population average) model in which the dependency between observations is modelled explicitly (Diggle et al. 2002). Both methods treat clustering as a nuisance, rather than a feature of substantive interest.

Obtaining standard errors that are adjusted for clustering is just one reason for adopting a multilevel modelling approach. Multilevel models can be used to investigate the extent and nature of clustering, and the effects of higher-level characteristics on level 1 outcomes. For example, a multilevel analysis could assess whether the strength of the relationship between educational attainment at entry to secondary school and subsequent exam performance varies across schools. Schools in which prior attainment is found to be a weak predictor of subsequent attainment may be said to show greater equity because they have decreased differences in outcomes for children with different intake scores; hopefully this is achieved by raising the scores of children who performed poorly at intake, rather than reducing the scores of those who performed well. Where academic ability is assessed using a series of tests, multilevel factor analysis could be used to allow for the possibility that a child's score on each of these tests depends on his or her overall ability (represented by one or more child-level factor) and the ability of children in the same school (represented by school-level factors).

A note on the scope of this chapter

Multilevel modelling is not traditionally considered a multivariate technique, but it is closely related to factor analysis. Indeed multilevel models for two-level hierarchical structures can be framed as factor models with parameter constraints. In this chapter, following an introduction to multilevel data structures and a simple multilevel model, we explain the connection with factor analysis before discussing more general models for two-level structures. We then consider multilevel models for the analysis of multivariate data, including multilevel factor models for clustered multivariate data.

An important application of multilevel models is to longitudinal data, where a variable is measured repeatedly over time on the same individuals (or some other units such as organisations and areas). Repeated measures data can be viewed as a form of multivariate data, and widely used models for these data can be framed as multilevel models or factor models. As there are special considerations in the analysis of longitudinal data, in particular the temporal ordering of observations within individuals, this topic is covered separately in the next chapter.

13.2 Some potential applications

i) Suppose we wish to study school effects on students' academic progress. Students are tested on entry to secondary school at age 11 and take public examinations at age 16. The data have a two-level structure with students at level 1 and schools at level 2. By taking age 16 attainment as the outcome variable and age 11 attainment as an explanatory variable, we can estimate the variance between schools in student progress from ages 11 to 16. We can also estimate the effects of individual schools and confidence intervals for the "true" effects. Further analysis could be carried out to investigate whether some schools are more effective than others for certain types of student, for example according to their attainment on entry to secondary school. School characteristics can be included in the model as possible predictors of between-school variation.

ii) Epidemiologists are interested in the pathways through which income affects individual health outcomes (e.g. Wagstaff and van Doorslaer 2000; Wilkinson 1996.) Do people with a low income have poor health because their material living standards are lower than those with higher incomes? For example, having a low income may be associated with a low standard of housing or poor diet. Or does the relationship work at an area (macro) level? Is income inequality a stronger determinant of health than absolute income level, and do macro measures interact with individual income? An individual with a low income living in a society with large income inequalities might be expected to have poorer health than someone with the same income but living in a more egalitarian society. Multilevel modelling can be used to explore the effects of area-level income inequality and absolute (individual-level) income on individual health outcomes, where inequality could be measured by the within-area standard deviation of income or the Gini coefficient.

iii) Continuing the educational example in i), suppose we have data on exam scores in different subjects. We can treat the exam scores for each student as a multivariate response, with students nested in schools as before. We can jointly model the subject-specific scores as outcomes in a multilevel multivariate regression model, which allows the effects of explanatory variables and the between-school variance to differ across subjects. Are school effects stronger for some subjects than for others? A multivariate model also provides estimates of the between-subject correlations at both the student level and the school level. A particular advantage of a multilevel modelling approach is that students do not have to study the same subjects and the number of subjects studied may vary across students.

13.3   Comparing groups using multilevel modelling

A single-level model for the mean

Before introducing multilevel models, we consider the simplest possible model: a model for the mean of the outcome variable y with no explanatory variables. Such a null or empty model may be written

$$y_i = \beta_0 + e_i, \tag{13.1}$$

where y_i is the value of y for the ith individual $(i = 1, \ldots, n)$, β_0 is the mean of y in the population, and e_i is the "residual" for the ith individual, i.e. the difference between an individual's y value and the population mean. Figure 13.1 shows the residuals for four observations $(n = 4)$. We usually assume that the residuals are independent and follow a normal distribution with mean zero and variance σ^2, i.e. $e_i \sim N(0, \sigma^2)$. The variance summarises the variability around the mean; if this is zero all the points would have the same y-value and would therefore lie on the $y = \beta_0$ line. The larger the variance, the greater the departures about the mean.

Multilevel model for group means

Next consider the simplest form of multilevel model which allows for group differences in the mean of y. We now view the data as having a two-level structure with individuals at level 1 nested within groups at level 2. Throughout this chapter we use "group" as a generic term to describe clusters of individuals, for example schools, households or geographical areas. Group membership is indicated by a second subscript j so that y_{ij} is the value of the dependent variable for individual i in group j. Suppose there are J such groups with n_j individuals in group j and a total sample size of $n = \sum_{j=1}^{J} n_j$ individuals. In a two-level model the residual is split into two components corresponding to the two levels in the data structure. We denote the group-level residuals, also called group random effects or simply group effects, by u_j and the individual-level residuals by e_{ij}. The two-level extension of eq. (13.1) is given by:

$$y_{ij} = \beta_0 + u_j + e_{ij}, \tag{13.2}$$

where β_0 is the overall mean of y (across all groups). The mean of y for group j is $\beta_0 + u_j$, and so u_j is the difference between the mean for group j and the overall mean. The individual-level residual e_{ij} is the difference between the y-value for the ith individual and that individual's group mean, i.e. $e_{ij} = y_{ij} - (\beta_0 + u_j)$. Figure 13.2 shows y-values for eight individuals in two groups, with individuals in group 2 denoted by black circles and those in group 1 denoted by grey squares. The overall mean is represented by the solid line and the means for groups 1 and 2 are shown as dashed lines. Also shown are the group residuals and the individual residual for the 4th individual in the 2nd group (e_{42}). Group 1 has a below-average mean (negative u_j), while group 2 is above average (positive u_j).

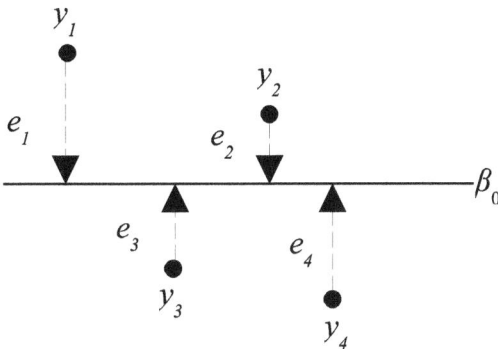

FIGURE 13.1: Residuals for four data points in a single-level model for the mean

We usually assume that the residuals at both levels follow independent normal distributions: $u_j \sim N(0, \sigma_u^2)$ and $e_{ij} \sim N(0, \sigma_e^2)$. The total variance is therefore partitioned into two components: the between-group variance σ_u^2, based on differences between the group means and the overall mean, and the within-group (between-individual) variance σ_e^2, based on differences between individual y-values and their group means. For this reason, eq. (13.2) is sometimes called a *variance components* model.

The variance partition coefficient (VPC) measures the proportion of total variance that is due to differences between groups:

$$\text{VPC} = \frac{\sigma_u^2}{\sigma_u^2 + \sigma_e^2}. \tag{13.3}$$

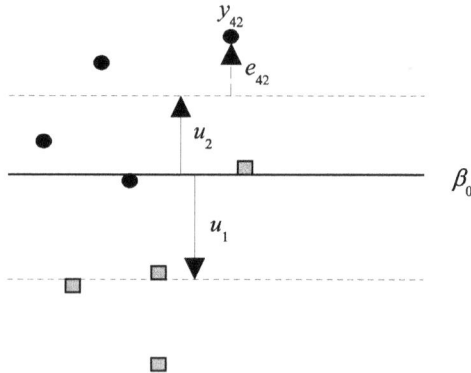

FIGURE 13.2: Individual and group residuals in a two-level model for the mean

The VPC ranges from 0 (no group differences, i.e. $\sigma_u^2 = 0$) to 1 (no within-group differences, i.e. $y_{ij} = y_j$ or $\sigma_e^2 = 0$).

For simple multilevel models, the VPC is equal to the intra-class correlation coefficient which is the correlation between the y-values of two randomly selected individuals from the same group.

Connection with factor analysis

The group random effects u_j in eq. (13.2) represent unobserved group-level characteristics affecting y_{ij}. As such u_j is a latent variable like f in factor analysis (Chapter 7) and in other latent variable models where f is assumed to be a continuous normally distributed random variable (Chapters 8, 9, and 11). It turns out that, with some changes in notation and parameter constraints, the linear single-factor model $y_i = \alpha_i + \beta_i f + e_i$ of eq. (7.2) can be transformed into the variance components model $y_{ij} = \beta_0 + u_j + e_{ij}$.

The first step is to view the p observed items $y_1, y_2, \ldots, y_p$ in factor analysis as a two-level hierarchical structure with the items at level 1 (indexed by subscript i) and individuals at level 2 (indexed by j). Adding a subscript for the individual to the factor model gives $y_{ij} = \alpha_i + \beta_i f_j + e_{ij}$. We next change f_j to u_j, a trivial change because both are assumed to be normally distributed.

Finally, we impose the constraints $\alpha_1 = \alpha_2 = \ldots = \alpha_p = \beta_0$ and $\beta_1 = \beta_2 = \ldots = \beta_p = 1$ in the factor model to obtain $y_{ij} = \beta_0 + u_j + e_{ij}$. For multivariate data (with items nested in individuals), we would not usually want to constrain the intercepts and factor loadings in this way because, in general, we wish to allow the means (α_i) and effects of the latent variable (β_i) to vary across items. In contrast, for a typical multilevel structure (with a single item measured on individuals nested in groups) we would not have enough degrees of freedom to estimate a different intercept and random effect loading for each individual, even if it made sense to do so.

The variance components model also resembles a Rasch model, introduced in Section 8.3 as a type of latent variable model for binary items which is widely used in educational testing. In a Rasch model, the factor loadings (referred to as discrimination parameters) are constrained to equal 1 for each item, while the intercepts may vary across items.

Testing for group effects

To test for group effects we test the null hypothesis that $\sigma_u^2 = 0$ (no group effects) against the alternative hypothesis that $\sigma_u^2 > 0$, where the alternative is one-sided because a variance cannot be negative. The null hypothesis implies that $u_j = 0$ for all j in eq. (13.2), leading to the simpler model $y_{ij} = \beta_0 + e_{ij}$ which is a standard single-level regression model including only an intercept term for the overall mean. The test involves a comparison of the variance components model of eq. (13.2) with the intercept-only regression model of eq. (13.1), and rejection of the null implies that the variance components model is preferred. As the two models are nested, we can compare them using a likelihood ratio test. The test statistic is calculated as

$$W = 2(\ell_A - \ell_0),$$

where ℓ_A and ℓ_0 are the log-likelihood values for the variance components model and the single-level model respectively. The test statistic is compared to a chi-squared distribution. Usually the number of degrees of freedom would be 1 when the null hypothesis imposes a restriction on a single parameter. However, this does not take into account that the parameter space for a variance parameter is restricted to positive values, as reflected in the one-sided alternative hypothesis. The value of zero under the null is said to be at the "boundary of the parameter space". Strictly, we should compare W with a 50 : 50 mixture of two chi-squared distributions, one on zero d.f. and one on 1 d.f, often written as $0.5\chi_0^2 + 0.5\chi_1^2$ and sometimes called a chi-bar-squared distribution. This reduces to $0.5\chi_1^2$ because $\Pr(X > x) = 0$ for any x when $X \sim \chi_0^2$. Therefore, the correct p-value is simply half that from a test that assumes 1 d.f. If we do not adjust the p-value, our test will be conservative, i.e. at a given level of significance we may incorrectly fail to reject the null and conclude that there are no cluster effects.

Example: Country differences in a European study of hedonism

We apply the multilevel model for group effects of eq. (13.2) in an analysis of between-country differences in human values, using data from the 2002/2003 European Social Survey (Jowell 2003). We focus on a measure of one of ten human values, hedonism, which is based on the extent to which respondents identify with a person with the following descriptions:

1. He seeks every chance he can to have fun. It is important to him to do things that give him pleasure.
2. Having a good time is important to him. He likes to "spoil" himself.

(The wording is adapted for female respondents.)

A respondent's own values are inferred from their self-reported similarity to a person with the above descriptions. Each of the two items is rated on a 6-point scale (from "very much like me" to "not like me at all"). The mean of these ratings is calculated for each individual. The mean of the two hedonism items is then adjusted for individual differences in scale use by subtracting the mean of all value items (a total of 21 are used to measure the ten values). These *centred* scores recognise that the ten values function as a system rather than independently. The centred hedonism score is interpreted as a measure of the *relative* importance of hedonism to an individual in their whole value system.

The scores on the hedonism variable range from -3.76 to 2.90, where higher positive scores indicate more hedonistic values. We analyse data from 20 countries with a combined sample size of 36,527. The data have a two-level hierarchical structure with individual respondents at level 1 and countries at level 2. The target of inference could be a wider population of countries from which those in the study can be considered a random sample. However, it is not clear which countries such a population would contain. In this case, it is more natural to think of the sample data as if they were a set of realisations from some underlying process, or "superpopulation", that could extend through time and possibly space. This process has driven the observations, but the statistics we compute from the observed data refer to a particular point in time and are subject to random fluctuations. We are interested in the underlying process that has generated the data we observe, and use the "sample" data to make inferences about this process.

Table 13.1 shows the estimates for model (13.2). The overall mean (taken across countries) is estimated as -0.203 and the estimate of the overall variance is $0.094 + 0.885 = 0.979$. This overall variance is made up of the variance between countries (estimated as 0.094) and the within-country (between-individual) variance which is 0.885. The VPC is $0.094/(0.094+0.885) = 0.096$.

Thus 9.6% of the total variance in hedonism scores can be attributed to between-country differences. We can also interpret 0.096 as the correlation between the hedonism scores for two randomly selected individuals from the same country. In hierarchies with individuals at level 1, it is usual for most of the variation to be at level 1. In longitudinal designs, however, where individuals form the level 2 units, most of the variation will be at level 2 because responses over time for the same individual will typically be highly correlated (leading to a relatively small amount of within-individual variation).

TABLE 13.1: Multilevel model with country effects fitted to the hedonism data

Parameter	Estimate	s.e.
β_0	−0.203	0.069
σ_u^2	0.094	0.030
σ_e^2	0.885	0.007

The −2 log-likelihood value for the multilevel model is 99,303, compared to 102,590 for the single-level model. The likelihood ratio test statistic W, calculated as the difference between these two values, is 3287 which in a conservative test is compared with a chi-squared distribution on 1 degree of freedom. We can clearly reject the null hypothesis of no country differences (the critical value for a test at the 5% level is 3.84) and conclude that there is evidence of between-country variation in hedonistic attitudes.

Random versus fixed effects

Another way of allowing for group effects would be to include as explanatory variables a set of dummy variables for groups (see Section 6.11). To recap, we can define J dummy variables $D_1, \ldots, D_J$ such that $D_j = 1$ for individuals in group j and $D_j = 0$ for individuals in any other group. If we denote the coefficient of D_j by β_j and the overall intercept by β_0 as before, the model with group dummy variables can be written:

$$y_{ij} = \beta_0 + \sum_{k=1}^{J} \beta_k D_k + e_{ij}. \tag{13.4}$$

With J groups, it is not possible to estimate β_0 and all J coefficients. The approach usually taken is to constrain one of the coefficients, β_J say, to equal zero so that β_j $(j = 1, \ldots, J-1)$ becomes the difference between the mean for group j and the mean for group J. Group J is called the reference or base category.

Note that, for any given individual, only one of $D_1, \ldots, D_J$ will equal 1 and the others will all equal zero. Thus $\sum_{k=1}^{J} \beta_k D_k = \beta_j$ for an individual in group j and eq. (13.4) simplifies to

$$y_{ij} = \beta_0 + \beta_j + e_{ij}$$

which has the same form as eq. (13.2). The crucial difference between the two models is that β_j are *fixed* rather than random effects. In the *fixed effects model* (13.4) the group effects are treated as parameters to be estimated together with β_0 and σ_e^2, while in the random effects model (13.2) they are assumed to follow a normal distribution summarised by the variance parameter σ_u^2.

The fixed effects approach to allowing for group differences is commonly called an analysis of variance (ANOVA), which was introduced in Sections 6.11 and 6.17. The fixed effects model can be used to compare any number of groups, but there are some potential drawbacks that should be considered. The fixed and random effects approaches are contrasted in more detail below.

i) When the number of groups is large, there will be many extra parameters to estimate. For example, with 20 groups there are 19 parameters for group effects compared to just one (the between-group variance σ_u^2) in the random effects model. In a more complex model which allows the effects of explanatory variables to differ across groups (Section 13.5), there are even more additional parameters to estimate.

ii) For groups with small sample sizes, the estimated group effects may be unreliable. The random effects approach recognises that there is little information for these groups by "shrinking" their residual estimates towards zero, and therefore pulling their mean towards the overall mean.

iii) The origins of the fixed effects approach lie in experimental design where there is typically a small number of groups to compare (e.g. several treatments and a control) and all groups of interest are sampled. Groups are treated as a *fixed* classification. More generally, we may have a sample of groups (e.g. a sample of schools or areas) and it is the population of groups from which our sample was drawn that is of interest. The fixed effects approach does not allow us to make inferences to groups outside our sample. The random effects approach, however, views the groups in our dataset as a random sample from some population and the estimate obtained for the between-group variance, $\hat{\sigma}_u^2$, is an estimate of the variance between groups in this population.

iv) It is straightforward to include group-level explanatory variables in a random effects model (see Section 13.6). For each extra level 2 variable that is included in a fixed effects model an extra constraint must be placed on the β_j in eq. (13.4), so that their interpretation becomes increasingly complicated. For this reason, group-level variables are rarely included in fixed effects models.

Estimating group effects in a multilevel model

In a multilevel model the group effects u_j (level 2 residuals) are random variables assumed to follow a normal distribution. Their distribution is therefore summarised by two parameters, the mean (fixed at zero) and variance σ_u^2. The variance is estimated together with the other parameters of model (13.2): the within-group variance σ_e^2 and the overall mean β_0. However, to make comparisons among groups we may wish to obtain an estimate or predicted value of u_j for each group. This is analogous to obtaining factor scores in factor analysis (see Section 7.7). The u_j are estimated after fitting the model and are based on the estimates of the model parameters $(\beta_0, \sigma_u^2, \sigma_e^2)$ and the data y_{ij}.

In a single-level model, we have a single set of residuals. For an individual sample unit, the residual is estimated as the difference between the observed value of y_{ij} and the value predicted by the fitted model, $\hat{y}_{ij}$. In a multilevel model, the *total residual* is $u_j + e_{ij}$ which is estimated as $r_{ij} = y_{ij} - \hat{y}_{ij} = y_{ij} - \hat{\beta}_0$. We need to split this into separate estimates of u_j and e_{ij}. A starting point for an estimate of u_j would be to take the mean of $y_{ij} - \hat{\beta}_0$ for group j. This is sometimes called the *mean raw residual*:

$$\bar{r}_j = \bar{y}_j - \hat{\beta}_0,$$

where $\bar{y}_j$ denotes the sample mean of y_{ij} in group j, and turns out to be the maximum likelihood estimator of u_j (Rabe-Hesketh and Skrondal 2022, Section 2.11.1).

To obtain an estimate of the residual for group j, we multiply the raw residual by a factor k called the *shrinkage factor*:

$$\hat{u}_j = k\bar{r}_j,$$

where

$$k = \frac{\hat{\sigma}_u^2}{\hat{\sigma}_u^2 + (\hat{\sigma}_e^2/n_j)}.$$

The estimated residuals $\hat{u}_j$ are called *shrunken residuals*. Other terms used in the literature include *empirical Bayes estimates* and *posterior estimates*. The shrinkage factor k is always less than or equal to 1 so that $\hat{u}_j$ will be less than or equal to the mean raw residuals $\bar{r}_j$. For large n_j the shrinkage factor will be close to 1 and therefore $\hat{u}_j$ will be close to $\bar{r}_j$. There will also be little shrinkage (k close to 1) when $\hat{\sigma}_e^2$ is small relative to $\hat{\sigma}_u^2$. The shrinkage factor will be noticeably less than 1 when n_j is small or $\hat{\sigma}_e^2$ is large relative to $\hat{\sigma}_u^2$ (high within-group variability). In either case we have relatively little information about the group; shrinkage pulls the raw residual towards zero with the result that the group mean $\hat{\beta}_0 + \bar{r}_j$ is pulled towards the overall mean $\hat{\beta}_0$. These shrinkage residuals are also called precision-weighted estimates because we have taken their reliability into account in their estimation. Unreliable estimates with small n_j will be shrunk towards the overall mean, while reliable estimates based on large n_j will remain close to their raw value.

As with any estimate based on sample data, presentation of the estimated level 2 residuals $\hat{u}_j$ should be accompanied by standard errors or confidence intervals to demonstrate the uncertainty in their estimation due to sampling variability.

Example: Estimated country effects on hedonism

Figure 13.3 shows the estimated country residuals $\hat{u}_j$ with 95% confidence intervals in what is sometimes called a caterpillar plot. The confidence intervals are narrow and of a similar width because the sample size is large in each country. Countries are ranked from lowest to highest according to their residual estimate. The residuals represent country departures from the overall mean, so a country whose confidence interval does not overlap the line at zero (representing the mean hedonism value across all countries) is said to differ significantly from the average at the 5% level. At the left-hand side of the plot, there is a cluster of countries whose mean hedonism is lower than average; the country with the lowest residual (-0.72) is Poland. At the other extreme, there is a cluster with above-average hedonism that includes Denmark, Belgium and Switzerland.

13.4   Random intercept model

An obvious way to extend the group effects model (13.2) is to add explanatory variables. Suppose we have a single continuous explanatory variable defined at level 1 and denoted by x_{ij}.

The simplest multilevel model with a single explanatory variable is

$$y_{ij} = \beta_0 + \beta_1 x_{ij} + u_j + e_{ij}. \tag{13.5}$$

In this model, the *overall* (cross-group) relationship between y and x is represented by a straight line with intercept β_0 and slope β_1. However, the intercept for a given group j is $\beta_0 + u_j$, i.e. it will be higher or lower than the overall

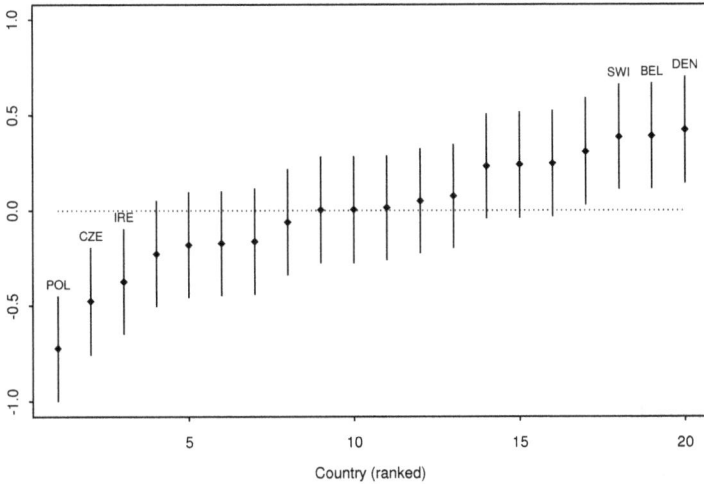

FIGURE 13.3: Caterpillar plot showing country residuals with 95% confidence intervals for hedonism

intercept β_0 by an amount u_j. As in eq. (13.2), u_j is a group effect or residual which is assumed to follow a normal distribution with a mean of zero and variance σ_u^2.

A multilevel model can be thought of as consisting of two components: a *fixed part* which specifies the relationship between the mean of y and explanatory variables, and a *random part* that contains the level 1 and 2 residuals. The fixed part of eq. (13.5) is $\beta_0 + \beta_1 x_{ij}$ with fixed part parameters β_0 and β_1, and the random part is $u_j + e_{ij}$ with random part parameters σ_u^2 and σ_e^2. The fixed part is extended by adding more predictors, while the random part is extended by allowing the effect of one or more predictors to vary across groups (see Section 13.5).

Model (13.5) is usually called a *random intercept model* because the intercept of the group regression lines is allowed to vary randomly across groups. This simply means that the intercept is allowed to take on different values from a distribution. To highlight the fact that the intercept is specific to a group, the model is sometimes written in the form of two equations as

$$y_{ij} = \beta_{0j} + \beta_1 x_{ij} + e_{ij},$$
$$\beta_{0j} = \beta_0 + u_j. \tag{13.6}$$

While the intercept may vary from group to group, the slope β_1 is assumed to be the same for each group. Thus a graph of the predicted regression lines for each group, $\hat{y}_{ij} = \hat{\beta}_0 + \hat{\beta}_1 x_{ij} + \hat{u}_j$ versus x_{ij}, will show a set of parallel lines.

Example: Relationship between hedonism and age (or birth cohort)

Table 13.2 shows the results from fitting the random intercept model (13.5) to the hedonism data with age as an explanatory variable; the results from the null model (13.2) are shown for comparison. Age has been centred at the sample mean of 46 years so that the intercept can be interpreted as the predicted hedonism score for respondents of mean age. Throughout this section we refer to the coefficient of age as an age effect, but it could also be picking up differences between cohorts. With cross-sectional rather than longitudinal data, an observed difference between the hedonism scores of two individuals aged 20 and 40 years may be due to their age difference, differences between their respective birth cohorts, or a combination of both. A pure age effect would imply that the 20 year old would have the same hedonism score as the 40 year old when he or she reaches that age, assuming all other characteristics associated with hedonism were the same. A pure cohort effect would imply that differences in hedonism are driven by changes in attitudes between birth cohorts that persist with age, so that a person, for whom other factors remained constant, would have the same hedonism score when they reached 40 as they had had at 20.

TABLE 13.2: Random intercept model with country and age effects fitted to the hedonism data

	Null model		Random intercept model	
	Estimate	s.e.	Estimate	s.e.
β_0	-0.203	0.069	-0.199	0.069
β_1 (age, centred at 46)	–	–	-0.017	0.0003
σ_u^2	0.094	0.030	0.094	0.030
σ_e^2	0.885	0.007	0.787	0.006

For any country, the effect of a one-year increase in age is to reduce the predicted hedonism score by 0.017 points. The intercept is estimated as -0.199 for the average regression line and $-0.199 + \hat{u}_j$ for country j. The between-country variance in the intercepts is 0.094. Based on the between-country standard deviation, we can calculate a 95% plausible values range for country intercepts in the population which is the range within which we would expect 95% of country intercepts to lie. The limits of this interval are estimated as $-0.199 \pm (1.96 \times \sqrt{0.094}) = -0.800$ and 0.402. Thus, for a typical respondent aged 46 years we would predict a hedonism score of -0.199. For respondents aged 46 in the bottom 2.5% of countries, however, we would predict a hedonism score less than -0.800; in the top 2.5% we would predict a score greater than 0.402.

As would be expected, the addition of age has led to a reduction in the level 1 variance (from 0.885 to 0.787). In contrast, the level 2 variance has not changed, thereby implying that the age distribution does not vary a great deal from country to country. The total variance in hedonism from the null model was $0.094 + 0.885 = 0.979$. The total residual, after accounting for age, is $0.094 + 0.787 = 0.881$. Thus, the proportion of the original variance explained by age is $(0.979 - 0.881)/0.979 = 0.100$, i.e. 10%. The VPC is now $0.094/(0.094 + 0.787) = 0.107$ so, after accounting for age, 10.7% of the unexplained variance in hedonism is due to differences between countries (compared to 9.6% of the total variance when age is ignored).

13.5 Random slope model

A random intercept model assumes that the relationship between y and x is the same for each group, i.e. the slope β_1 is fixed across groups. We can relax this constraint by allowing the slope to vary randomly across groups, leading to a *random slope model*:

$$y_{ij} = \beta_0 + \beta_1 x_{ij} + u_{0j} + u_{1j} x_{ij} + e_{ij}, \tag{13.7}$$

which can also be written as

$$\begin{aligned} y_{ij} &= \beta_{0j} + \beta_{1j} x_{ij} + e_{ij}, \\ \beta_{0j} &= \beta_0 + u_{0j}, \\ \beta_{1j} &= \beta_1 + u_{1j}. \end{aligned} \tag{13.8}$$

Comparing eq. (13.7) with the random intercept model (13.5), a new term $u_{1j} x_{ij}$ has been added and an additional "0" subscript has been added to u_j. The group-level random effects u_{0j} and u_{1j} are assumed to follow a bivariate normal distribution with zero mean, variances σ_{u0}^2 and σ_{u1}^2 respectively, and covariance σ_{u01}. Now the slope of the average regression line is β_1 and the slope of the line for group j is $\beta_1 + u_{1j}$. The covariance σ_{u01} is the covariance between the group intercept and slope residuals. A positive value of this covariance implies that groups with higher-than-average intercept residuals ($u_{0j} > 0$) tend to have higher-than-average slope residuals ($u_{1j} > 0$), and the lines will fan out as x increases. If the covariance, σ_{u01}, is negative then the lines will converge as x increases (and may eventually cross).

The term $u_{1j} x_{ij}$ can be thought of as an interaction between group and x. In a fixed effects model this interaction effect would be fitted by including as explanatory variables the products of the group dummy variables and x, which would lead to $J - 1$ additional parameters (see Section 6.11).

Example: Allowing the relationship between hedonism and age to differ across countries

Table 13.3 shows the results from fitting a random slope model to the relationship between hedonism and age, alongside the results from the random intercept model fitted earlier. There is little change in the parameters that are common to both models ($\beta_0, \beta_1, \sigma_{u0}^2$ and σ_e^2), apart from a slight reduction in the within-country variance (σ_e^2). However, two of these parameters, σ_{u0}^2 and β_1, have a different interpretation when a random slope is fitted to age. The intercept variance (σ_{u0}^2) is now interpreted as the country-level variance at age $= 46$, and $\hat{\beta}_1 = -0.018$ is now the estimated slope of the "average" country line, with the slope of the line for country j estimated as $-0.018 + \hat{u}_{1j}$. The limits of a 95% plausible values range for the coefficient of age are estimated as $-0.018 \pm (1.96 \times \sqrt{0.00002}) = -0.027$ and -0.009. Thus, we would expect a negative relationship between hedonism and age for the middle 95% of countries in the population (when ranked on their age effect).

TABLE 13.3: Random intercept and slope models with country and age effects fitted to the hedonism data

	Random intercept		Random slope	
	Estimate	s.e.	Estimate	s.e.
β_0	−0.199	0.069	−0.200	0.069
β_1 (age, centred at 46)	−0.017	0.0003	−0.018	0.001
Country-level random part				
σ_{u0}^2 (intercept variance)	0.094	0.030	0.095	0.030
σ_{u1}^2 (slope variance)	−	−	0.00002	0.00001
σ_{u01} (intercept-slope covariance)	−	−	0.001	0.0004
Individual-level random part				
σ_e^2	0.787	0.006	0.781	0.006

Compared to the random intercept model, the random slope model has two additional parameters: the between-country variance in the slope of age, σ_{u1}^2, and the covariance between the country intercepts and slopes, σ_{u01}. The likelihood ratio test statistic contrasting the random slope model with the simpler random intercept model is 232, which is compared to a chi-squared distribution on 2 degrees of freedom (because there are two new parameters). The p-value for the test is less than 0.001 so we conclude that there is evidence that the effect of age (or cohort) on hedonism varies across countries. (This is a conservative test because it does not adjust for the slope variance σ_{u1}^2 being non-negative.)

Figure 13.4 shows the country prediction lines obtained from the random slope model. All countries have a negative slope so, across all countries, it is the young who are the most hedonistic. The positive covariance between the intercept and slope residuals implies that countries with a large negative intercept ($u_{0j} < 0$, i.e. low mean hedonism score at the mean age) tend to have a steep slope ($u_{1j} < 0$, i.e. strong negative relationship between hedonism and age). The correlation between the intercepts and slopes of the country regression lines is

$$\rho_{u01} = \frac{\text{Cov}(u_{0j}, u_{1j})}{\sqrt{\text{Var}(u_{0j})\text{Var}(u_{1j})}} = \frac{\sigma_{u01}}{\sigma_{u0}\sigma_{u1}},$$

which is estimated as

$$\hat{\rho}_{u01} = \frac{0.001}{\sqrt{0.095 \times 0.00002}} = 0.725.$$

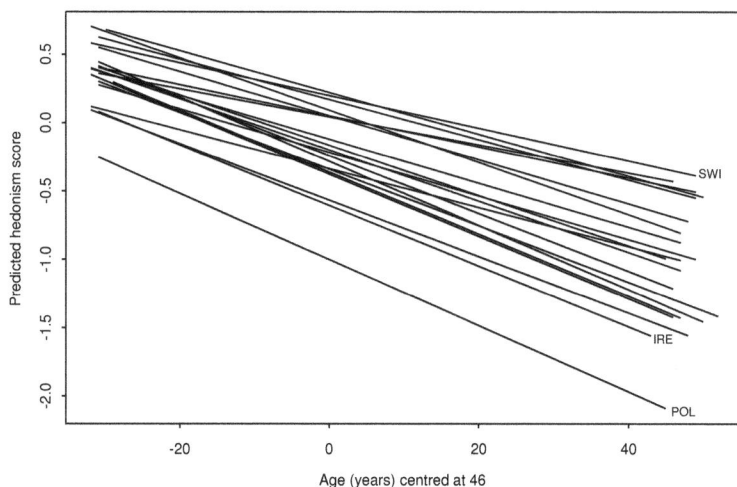

FIGURE 13.4: Predicted country lines from a random slope model fitted to the hedonism data

The covariance between intercept and slopes can also be seen in a plot of the intercept residuals $\hat{u}_{0j}$ versus the slope residuals $\hat{u}_{1j}$ (Figure 13.5). Countries around the intersection of $\hat{u}_{0j} = 0$ and $\hat{u}_{1j} = 0$, Austria and Portugal, will have prediction lines close to the overall average, i.e. with an average hedonism score at mean age (the intercept) and average effect of age (the slope). Bearing in mind that this average line has intercept -0.2 and slope -0.018, a country at the top right of the plot (with positive intercept and slope residuals) will have an above-average intercept but flatter-than-average negative slope for age; countries in this group include Hungary, the Netherlands and Switzerland. In contrast Poland, at the bottom left of the plot, has a below-average intercept and stronger-than-average negative effect of age.

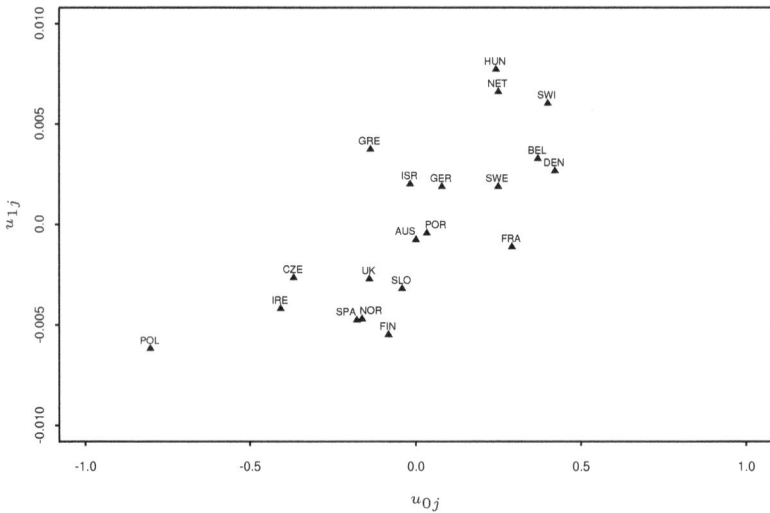

FIGURE 13.5: Estimated intercept and slope residuals for the relationship between hedonism and age

Summary of the analysis carried out so far

The first multilevel model fitted to the hedonism data included only country effects to allow for between-country variation in hedonism scores. We found that almost 10% of the variation in hedonism could be attributed to differences between countries. Estimates of the country (level 2) residuals were computed from this model to identify countries with especially low or high mean hedonism scores. We then introduced an individual-level explanatory variable, age, and began by assuming that the relationship between hedonism and age was the same in each country. Only the intercept of the regression model was permitted to vary across countries in a random intercept model. Next, this assumption was relaxed by fitting a random slope model in which the slope of the regression on age varies randomly across countries. Although there is evidence of between-country differences in the effect of age, a negative relationship was found in each country.

In the next section, we consider the addition of explanatory variables defined at the country level. To simplify the discussion, we return to a simple random intercept model and exclude age. In practice, having found significant age effects and between-country variance in the association between hedonism and age, we would wish to build on the random slope model.

13.6 Group-level explanatory variables

Thus far we have considered a model that includes an explanatory variable defined at the lowest level of the hierarchical structure. A particular advantage of multilevel modelling, however, is that it can be used to explore the effects of group-level variables while simultaneously allowing for the possibility that y may be influenced by unmeasured group factors. If the effects of level 2 variables are of interest, it is especially important to use a multilevel modelling approach because the standard errors of their coefficients may be severely underestimated when a single-level model is used.

Examples of research questions that involve level 2 explanatory variables include studies of the effects of teacher and peer group characteristics on student achievement where, for instance, a student's exam performance may be affected by the teacher's experience or the ability of other students in the class. In health research, there may be interest in the extent to which between-family variation in child health outcomes may be attributable to differences in family characteristics, shared by all children in a family, such as household socio-economic status, parental education and lifestyle factors.

A level 2 explanatory variable can be included in a multilevel model in exactly the same way as a level 1 variable. For example, if we have a level 1 variable x_{1ij} and a level 2 variable x_{2j} the random intercept model (13.5) becomes

$$y_{ij} = \beta_0 + \beta_1 x_{1ij} + \beta_2 x_{2j} + u_j + e_{ij}. \tag{13.9}$$

(Note that a level 2 variable does not have an i subscript because, by definition, its values do not vary from individual to individual within a level 2 unit.)

Level 2 variables may come from a number of sources. Data may be collected at level 2, for example community surveys in which key figures in the community are interviewed or data from geographical information systems (GIS) on the location of health facilities. Level 2 variables may also derive from aggregation of level 1 variables (e.g. means or, for categorical variables, proportions). Such data may come from an external source, e.g. a census, or the same source as the level 1 data.

When the level 2 mean of a level 1 variable is included in the model, eq. (13.9) becomes

$$y_{ij} = \beta_0 + \beta_1 x_{ij} + \beta_2 \bar{x}_j + u_j + e_{ij} \tag{13.10}$$

where $\bar{x}_j$ is the mean of x in group j.

In eq. (13.10) β_1 is called the *within-group effect* of x, β_2 is the *contextual effect* and $\beta_1 + \beta_2$ is the *between-group effect* of x. The within-group coefficient represents a comparison of the y values for two individuals in the same group whose x values differ by 1 unit. The contextual coefficient is the effect of a 1-unit increase in the group mean of x on an individual's y-value that is over and above the effect of their x-value on y. The between-group effect measures the relationship between x and y at the group level, i.e. the effect of a 1-unit increase in the group mean of x on the group mean of y.

So that the within-group and between-group effects can each be represented by a single parameter, eq. (13.10) can be conveniently re-expressed as

$$y_{ij} = \beta_0^* + \beta_1^*(x_{ij} - \bar{x}_j) + \beta_2^*\bar{x}_j + u_j + e_{ij}, \qquad (13.11)$$

where $\beta_1^* = \beta_1$ is the within-group effect, and $\beta_2^* = \beta_1 + \beta_2$ is the between-group effect. From eq. (13.11) we obtain an alternative interpretation of β_1 as the effect on y of a 1-unit increase in an individual's x-value relative to the mean of x in the group; this is equivalent to the interpretation given above for β_1 in eq. (13.10). Models (13.10) and (13.11) are equivalent, but (13.11) produces a direct estimate (and standard error) for the between-group effect of x.

Example: Within and between country relationships between hedonism and income

Suppose we are interested in the relationship between an individual's hedonism score and individual income and country income. Individual income (x_{ij}) is measured by monthly household income which was collected in 12 bands, and country income ($\bar{x}_j$) is the mean income band in a country. Table 13.4 gives the results from three random intercept models with the following income variables: (1) individual income only, (2A) individual income and the country mean of income, and (2B) the difference between individual and country income ($x_{ij} - \bar{x}_j$) and country income. As shown above, Models 2A and 2B are alternative parameterisations of the same model.

TABLE 13.4: Random intercept models with different specifications of the relationship between hedonism and income

Variable	Model 1 Estimate	s.e.	Model 2A Estimate	s.e.	Model 2B Estimate	s.e.
Fixed part						
Constant	−0.433	0.067	−0.787	0.245	−0.787	0.245
x_{ij}	0.038	0.003	0.038	0.003	–	–
$x_{ij} - \bar{x}_j$	–	–	–	–	0.038	0.003
$\bar{x}_j$	–	–	0.059	0.041	0.097	0.041
Random part						
σ_u^2	0.083	0.026	0.078	0.025	0.078	0.025
σ_e^2	0.880	0.007	0.880	0.007	0.880	0.007

From Model 1 of Table 13.4, we would conclude that there is a significant, positive relationship between an individual's income and their hedonism score. Each one-band increase in income is associated with an increase of 0.038 points on the hedonism measure. Note that we are careful not to interpret the relationship between hedonism and income in a causal way because it is highly likely that there are unobserved variables that determine both hedonism and income. It is also possible that hedonism affects income if people who place a high premium on having a good time are less inclined to choose a high-powered job with a high salary; without longitudinal data we cannot determine whether a change in income preceded a change in hedonism score, or vice versa. To reflect these concerns we use the word "association".

As noted above, Models 2A and 2B are equivalent. The coefficient of x_{ij} in Model 2A is equal to the coefficient of $x_{ij} - \bar{x}_j$ in Model 2B and, in both models, is interpreted as a measure of the within-country association between hedonism and income. The between-country association is represented by 0.097 (=0.038 + 0.059). We therefore conclude that, within a country, a higher income is associated with more hedonistic attitudes. However, the larger between-country association suggests that the relationship between income and hedonism is stronger at the country level. The contextual "effect" of income is the coefficient of $\bar{x}_j$ in Model 2A which is estimated as 0.059 but, with a standard error of 0.041, it is not significantly different from zero. Had it been significant we would have concluded that, over and above the individual-level association, there is a stronger contextual association: an individual in a given income band will, on average, be more hedonistic if he or she lives in a country with a higher income. The addition of country-level income reduces the variance at this level from 0.083 to 0.078, and so explains $(0.083 - 0.078)/0.083 = 6\%$ of the estimated between-country variance.

Cross-level interactions

As in multiple regression, we can allow for the possibility that the effect of one explanatory variable on y depends on the value of another. Such effects are called interaction effects and are represented in a model by including the product of the interacting variables as explanatory variables. Interactions can also be included in a multilevel model and these can be between any pair (or larger set) of variables, regardless of the level at which they are defined. An interaction between a level 1 variable and a level 2 variable is known as a *cross-level interaction*.

To give an example, suppose we wish to assess whether high ability children perform better in exams taken at the end of secondary school when taught with other high ability children or with a more mixed ability group. To investigate this question, we would include in the model a measure of the child's ability

(x_{1ij}), measured at the start of the school year, the mean ability in the class $(x_{2j} = \bar{x}_{1j})$, and their product $(x_{1ij}x_{2j})$. A random intercept model with a cross-level interaction is an extension of eq. (13.9):

$$y_{ij} = \beta_0 + \beta_1 x_{1ij} + \beta_2 x_{2j} + \beta_3 x_{1ij}x_{2j} + u_j + e_{ij}, \qquad (13.12)$$

which could also be written as

$$y_{ij} = (\beta_0 + \beta_2 x_{2j} + u_j) + (\beta_1 + \beta_3 x_{2j})x_{1ij} + e_{ij},$$

showing how the intercept and slope of the regression on x_1 depend on the value of x_2.

In eq. (13.12) β_1 is interpreted as the effect of a unit increase in x_1 when $x_2 = 0$. When $x_2 = 2$, for example, the effect of a unit increase in x_1 would be $\beta_1 + 2\beta_3$.

13.7   Multilevel multivariate regression

So far we have considered models for a two-level hierarchy with a single continuous response variable measured on individuals (level 1) within groups (level 2). Using a multilevel approach, we can allow for variation in the mean response that is due to unobserved characteristics of groups. Both the mean response and the between-group (level 2) variance may be modelled as a function of explanatory variables that can be defined at either levels of the hierarchy. These models can be extended in a number of directions, for example to handle categorical responses or to allow the within-group (level 1) variance to depend on explanatory variables (heteroskedasticity); some generalisations are discussed briefly in Section 13.10. The remainder of this chapter focuses on models for multivariate response data such as those considered in earlier chapters. Such data can be viewed as a two-level hierarchy with responses forming the lower level units and individuals the groups. For example, test scores in different subjects could be viewed as level 1 observations nested within individuals at level 2. In this section, we begin with a description of how a simple model for the means of the multivariate responses can be framed as a multilevel model and the advantages in doing so. We then discuss extensions to this basic model, including the addition of explanatory variables or further hierarchical levels. The use and interpretation of these models is illustrated in an analysis of test scores in science subjects.

A simple multivariate model

Denote by y_{ij} the response on variable i for individual j. A simple model which allows different means and variances for each response variable, and for correlation between an individual's responses, can be written

$$y_{ij} = \beta_{0i} + u_{ij}, \tag{13.13}$$

where β_{0i} is the mean for variable i and u_{ij} is a normally distributed residual term with mean zero and variance σ_{ui}^2. In addition, we allow for non-zero within-individual covariance between pairs of responses, i.e. $\text{Cov}(u_{ij}, u_{i'j}) = \sigma_{uii'}$. Suppose that there are m variables in total then we assume that the residuals $(u_{1j}, u_{2j}, \ldots, u_{mj})$ follow an m-dimensional multivariate normal distribution where the covariance matrix has diagonal elements σ_{ui}^2 and off-diagonal elements $\sigma_{uii'}$ (for $i \neq i'$).

There are several differences between eq. (13.13) and the standard two-level model $(y_{ij} = \beta_0 + u_j + e_{ij})$ given by eq. (13.2). First, now that level 1 units correspond to measurements or responses rather than individuals, we need to allow for a different mean for each level 1 unit so that β_0 is replaced by β_{0i}. Another difference between the models is that eq. (13.13) contains only one residual term. However, while u_{ij} might appear to be level 1 residuals they are more naturally viewed as individual-specific (level 2) residuals that have different variances for each variable i; after accounting for u_{ij}, there is no further source of variation to explain. For this reason, we will refer to eq. (13.13) as a simple multivariate model, and reserve the term multilevel for models with levels above the individual. A third distinction between eq. (13.13) and eq. (13.2) is that u_{ij} are permitted to be correlated across responses.

Estimation of eq. (13.13) proceeds by structuring the data so that for individual j there are n_j records, one for each observed response, where $n_j \leq m$. We then define a set of m response indicators or dummy variables $\{r_i; i = 1, \ldots, m\}$ where r_i is coded 1 for each observation on variable i and 0 otherwise. The r_i are included as explanatory variables in the model and their coefficients are the means or intercepts β_{0i}. The residuals u_{ij} are fitted by allowing the coefficients of r_i to vary randomly across individuals. Table 13.5 shows the data structure for two individuals, each with $m = 3$ responses.

TABLE 13.5: Multivariate data structure for two individuals, each with three responses

j	i	y_{ij}	r_1	r_2	r_3
1	1	y_{11}	1	0	0
1	2	y_{21}	0	1	0
1	3	y_{31}	0	0	1
2	1	y_{12}	1	0	0
2	2	y_{22}	0	1	0
2	3	y_{32}	0	0	1

An important advantage of the multilevel modelling approach to analysing multivariate data is that individuals with one or more missing response can be included straightforwardly. Suppose that there are m response variables. When viewed as a two-level structure, there is no requirement to have the same number of responses for each individual (just as the number of children in a class can vary across classes). Thus an individual j with $n_j < m$ responses can be included in the analysis and parameter estimates will be unbiased provided that responses can be assumed *missing at random*. (Values of a variable y are said to be missing at random if, possibly conditional on a set of predictors $x_1, \ldots, x_q$, the probability of being missing does not depend on the unobserved value of y.) Even if this assumption does not hold exactly, this method of handling missing data is better than excluding cases with missing values.

Multivariate regression and adding further levels

The model in eq. (13.13) could be used to obtain a correlation matrix when there are missing data, which could in turn be input for a principal components analysis or factor analysis. However, there may be interest in further modelling of individual responses, for example to investigate the effects of explanatory variables on each response or to allow for additional layers of clustering. When responses are correlated within individuals, it is preferable to model them jointly rather than fit a separate univariate regression model for each response. By analysing responses simultaneously, it is possible to test whether an explanatory variable has the same effect on each response variable and whether the extent of clustering is the same for each variable.

We begin by extending eq. (13.13) to include explanatory variables. Suppose we have a single predictor x_j. A model that allows for different effects of x_j on each variable is

$$y_{ij} = \beta_{0i} + \beta_{1i} x_j + u_{ij}, \tag{13.14}$$

where β_{0i} and β_{1i} are, respectively, response-specific intercepts and slopes.

In practice, the multivariate regression model is fitted to the expanded data structure shown in Table 13.5. Bearing in mind that the response indicator r_i is equal to 1 for response i and zero for all other responses, eq. (13.14) can be re-expressed as

$$y_{ij} = \beta_{0i} r_i + \beta_{1i} x_j r_i + u_{ij} r_i. \tag{13.15}$$

The main effect of r_i allows a different intercept to be fitted for each response, while the interactions between x_j and the m response indicators r_i allow the effect of x_j to differ across responses. Finally, response-specific residuals are fitted by permitting the intercept of r_i to vary randomly across individuals.

Special cases of eq. (13.14) can be derived by placing constraints on the β_{1i}. For example, β_{1i} can be fixed at zero for one or more response variable, or we might set them to be equal for a subset of responses. The adequacy of a constrained model can be assessed by comparing it with the unconstrained model (13.14) using a likelihood ratio test.

Another extension to the model is to allow for additional hierarchical levels, for example where individuals are nested within groups. Denote by y_{ijk} the response on variable i for individual j in group k. A model with group effects on each outcome can be written:

$$y_{ijk} = \beta_{0i} + \beta_{1i}x_{jk} + v_{ik} + u_{ijk}, \tag{13.16}$$

where x_{jk} is an individual-level predictor, $v_{ik} \sim N(0, \sigma_{vi}^2)$ are group residuals associated with response variable i and $u_{ijk} \sim N(0, \sigma_{ui}^2)$ are within-group individual residuals. Residuals at the same level can be correlated: $\text{Cov}(v_{ik}, v_{i'k}) = \sigma_{vii'}$ and $\text{Cov}(u_{ijk}, u_{i'jk}) = \sigma_{uii'}$. The group-level covariance between variables i and i' is the covariance between the true group means of these variables.

Model (13.16) includes only one predictor, x_{jk}, defined at the individual level but it is straightforward to add further predictors at either level.

Example: Multivariate regression analysis of science test scores

The multivariate models introduced above are now illustrated in an analysis of science scores for 2439 students in 99 Hungarian schools. The data consist of scores on four tests: a core test booklet – with components in earth science, physics and biology – plus an additional biology test (R) taken by a random subsample of 1222 students. Each test was scored out of ten, but we will analyse the standardised scores so that each score has mean zero and variance one. The data analysed here are a subset of the original dataset which contains scores on six tests: the three core tests, two further tests in biology and one further in physics. Each student responded to a maximum of five tests: the core tests plus a randomly selected pair of tests from the other three. See Goldstein (2010) and Steele (2005) for analyses of the full dataset.

Pairwise correlations between the four tests can be obtained from the multivariate model (13.13) as follows:

$$\text{Corr}(y_{ij}, y_{i'j}) = \text{Corr}(u_{ij}, u_{i'j}) = \frac{\text{Cov}(u_{ij}, u_{i'j})}{\sqrt{\text{Var}(u_{ij})\text{Var}(u_{i'j})}} = \frac{\sigma_{uii'}}{\sigma_{ui}\sigma_{ui'}}.$$

Table 13.6 shows the estimated correlation matrix. Because each student provides a score on all three core tests, the multivariate estimates of the correlations between these tests will equal the standard Pearson correlations. Pearson correlations with the biology R test, however, will be based on the 1222 students who took that test. In contrast, the multivariate approach uses data provided by *all* students, not just those for whom scores on the biology test are available. Where data can be assumed missing at random, as here, the multivariate approach will yield unbiased correlation estimates, while Pearson estimates from pairwise deletion would be biased. The multivariate estimates (as in Table 13.6) are maximum likelihood estimates and will be the same as those obtained using other techniques that assume data are missing at random, for example the E-M algorithm and multiple imputation.

TABLE 13.6: Multivariate estimates of pairwise correlations (and standard errors) between students' science test scores, ignoring differences between schools

	Earth science	Biology	Physics	Biology R
Earth science	1.000 (0.029)			
Biology	0.347 (0.021)	1.000 (0.029)		
Physics	0.313 (0.021)	0.525 (0.023)	1.000 (0.029)	
Biology R	0.160 (0.028)	0.194 (0.028)	0.180 (0.028)	1.000 (0.040)

Table 13.7 shows estimates from fitting a model of the form (13.14) with a single explanatory variable, gender (coded 1 for female and 0 for male). The intercept (coefficient of the "constant" variable) is the mean score for boys and the coefficient of the female dummy is the difference between the girls' and boys' means. By computing the ratio of the estimate of the female coefficients to their standard errors, and comparing the absolute value of the result with 1.96 (the critical value for a test at the 5% level) we find that girls do significantly worse than boys on physics. The residual covariance matrix in the lower part of Table 13.7 shows the variance of the scores in each test, and their covariances, after accounting for gender differences in the mean scores. Because the test scores have been standardised so that they all have unit variance, the amount of variance explained by gender can be assessed by a comparison of the residual variances in Table 13.7 with one. The correlation between exam performance and gender is strongest for physics and, as a result, this test has the smallest residual variance after including gender as an explanatory variable.

TABLE 13.7: Estimates from a multivariate model with gender effects, ignoring differences between schools, science test scores

Coefficients from a multivariate regression of test scores on gender				
	Earth science	Biology	Physics	Biology R
Constant	−0.005 (0.029)	0.020 (0.029)	0.151 (0.029)	−0.011 (0.041)
Female	0.010 (0.040)	−0.038 (0.040)	−0.295 (0.040)	0.033 (0.056)
Residual covariance matrix				
	Earth science	Biology	Physics	Biology R
Earth sci.	1.000 (0.029)			
Biology	0.347 (0.021)	0.999 (0.029)		
Physics	0.313 (0.021)	0.522 (0.023)	0.978 (0.028)	
Biology R	0.160 (0.028)	0.194 (0.028)	0.182 (0.028)	0.998 (0.040)

We next allow for clustering of students' test scores in schools by estimating a model of the form (13.16). Table 13.8 shows the gender effects and the school- and student-level covariance and correlation matrices obtained from this model. The gender effects now have a within-school interpretation; thus, for example, among students in a given school, we would expect a girl's score in the physics test to be 0.336 standard deviations lower than a boy's. Accounting for within-school correlation in student scores leads to an increase in the standard errors of the intercepts, but a decrease in the standard errors of the gender coefficients. The intercepts are the mean scores for boys (the reference category of gender). The standard errors of the boys' means are underestimated when within-school correlation is ignored. For the same reason, we would find that the standard errors of the girls' means would increase when school effects are included. (The simplest way to obtain standard errors for the girls' mean is to switch the coding of gender so that girls are the reference category.)

The gender coefficient estimates the population mean for girls minus the population mean for boys. In the multilevel model, if the proportion of girls, $\bar{x}_k$, was the same for every school, then the gender coefficient would be based entirely on within-school comparisons (involving only σ^2_{ui}). Conversely if all schools were single sex ($\bar{x}_k = 0$ for a boys' school and 1 for a girls' school), then the gender coefficient would be based entirely on between school comparisons (involving both σ^2_{vi} and σ^2_{ui}). Ignoring schools and analysing the data as if it were a simple random sample would lead to *over*estimating the standard error in the first case and to *under*estimating it in the second case. In intermediate cases, where the proportion of girls varies, the gender coefficient is based on both within-school and between-school information. Ignoring schools could lead to either underestimation or overestimation of the standard error of the gender coefficient, depending on the relative contributions of the within-school and between-school information. In our example, the proportion of girls varies from 0.25 to 0.81 across schools. The estimated standard errors for the gender coefficients ignoring schools (see Table 13.7) are slightly larger than those obtained by using the more correct multilevel model (see Table 13.8). The same two tables illustrate how ignoring schools leads to underestimating the standard errors of the intercepts (means for boys).

For each test score, the sum of the variances at the school and student levels will approximately equal the variance given in Table 13.7. Although the major source of variance is between students, there is also considerable between-school variation in performance. Table 12.9 shows the variance partition coefficients obtained by applying eq. (13.3) to each of the four test scores. We conclude that the proportion of residual variance that is attributable to differences between schools ranges from 7.4% for biology R to 24.9% for physics.

The school-level correlations in Table 13.8 are the correlations between the school mean scores for each pair of tests, after accounting for gender differences in the overall mean of each test. These are all positive and moderate to high in magnitude, suggesting that schools whose students perform well in one test tend also to score highly on other tests (and, conversely, schools with a low mean score on one test tend also to have low scores on other tests). Correlations at the student level are much lower, indicating a large amount of variation in individual students' score profiles across tests: there will be substantial prediction error if a student's score in one test is used to predict their performance in another test.

TABLE 13.8: Estimates from a multilevel multivariate model with gender effects, science test scores

Coefficients from a multivariate regression of test scores on gender				
	Earth science	Biology	Physics	Biology R
Constant	−0.009 (0.049)	0.021 (0.055)	0.149 (0.056)	−0.019 (0.048)
Female	−0.019 (0.038)	−0.081 (0.036)	−0.336 (0.035)	0.023 (0.055)
School-level residual covariance matrix (with correlations in *italics*)				
	Earth science	Biology	Physics	Biology R
Earth sci.	0.166 (0.029)			
	1.000			
Biology	0.133 (0.027)	0.228 (0.037)		
	0.683	*1.000*		
Physics	0.116 (0.027)	0.212 (0.035)	0.245 (0.039)	
	0.576	*0.897*	*1.000*	
Biology R	0.054 (0.018)	0.086 (0.022)	0.099 (0.023)	0.074 (0.021)
	0.493	*0.659*	*0.734*	*1.000*
Student-level residual covariance matrix (with correlations in *italics*)				
	Earth science	Biology	Physics	Biology R
Earth sci.	0.840 (0.025)			
	1.000			
Biology	0.222 (0.017)	0.780 (0.023)		
	0.274	*1.000*		
Physics	0.204 (0.017)	0.317 (0.017)	0.738 (0.023)	
	0.259	*0.418*	*1.000*	
Biology R	0.110 (0.026)	0.114 (0.025)	0.088 (0.024)	0.924 (0.039)
	0.125	*0.134*	*0.106*	*1.000*

TABLE 13.9: Proportion of residual variance (after accounting for gender differences) attributable to differences between school, science test scores

Earth science	Biology	Physics	Biology R
0.165	0.226	0.249	0.074

Application of multilevel multivariate regression models to multiple imputation for missing data

An important application of multilevel multivariate regression is to the imputation of missing values for clustered data. If the model of interest is multilevel, failure to account for clustering when imputing missing data will lead to understatement of between-group variances (and associated VPCs). Suppose there are missing values on m variables denoted by y_{ijk} where i indexes the variable, j the individual and k the group. These variables will typically be explanatory variables in the model of interest, but may also include the response. Suppose that there is one variable x_{jk} with fully observed data which is related to at least one of the y variables; such variables are often called *auxiliary variables*. The joint modelling approach to multiple imputation involves the following steps:

   i) Fit the multilevel multivariate model of eq. (13.16), i.e. $y_{ijk} = \beta_{0i} + \beta_{1i}x_{jk} + v_{ik} + u_{ijk}$, to the subset of observations with complete data on y_{ijk} for $i = 1, 2, \ldots, m$. This is referred to as the *imputation model*.

   ii) For observations with missing data on any y_{ijk}, use the parameter estimates from the model fitted in i) to obtain a predicted value $\hat{y}_{ijk}$. The observed and predicted values of y_{ijk} together form an imputed dataset.

   iii) Repeat step ii) M times to obtain M imputed datasets.

   iv) Fit the model of interest to each imputed dataset (with the addition of the fully observed variables) and apply simple formulae (known as "Rubin's rules") to combine estimates across the M datasets to obtain a single estimate of each parameter and its standard error.

Further details can be found in Carpenter and Kenward (2013) and Enders, Hayes, and Du (2018), including generalisations to handle variables defined at different levels, categorical variables and more complex random effects models. The above approach, using Bayesian estimation methods, has been implemented in the jomo R package (Quartagno and Carpenter 2023).

13.8 Multilevel factor analysis

As discussed in Chapter 7, the purpose of collecting responses on a set of items may be to provide a measure of an underlying construct that is not directly observable. For example, scores on multiple tests may be used to measure academic ability. In such cases, factor analysis can be used to explore the dimensionality of the latent construct and factor scores may be computed to obtain a measure for each dimension. The multivariate models of the previous section can be extended so that the correlation between the response variables is assumed to be explained by their shared dependency on one or more common factor.

Simple factor model with one factor

The one-factor model for single-level data was introduced in Section 7.2. Here we give the identical model but use different notation. For individual j, the values of the *observed* response variables are denoted by y_{1j} to y_{mj}, the value of the factor or latent variable by f_j and the values of the response-specific residuals by u_{1j} to u_{mj}, for $j = 1, \ldots, J$, giving

$$y_{ij} = \alpha_{0i} + \alpha_{1i} f_j + u_{ij}, \tag{13.17}$$

where $f_j \sim N(0, \sigma_f^2)$ and $u_{ij} \sim N(0, \sigma_{ui}^2)$ independently for $i = 1, \ldots, m$ and $j = 1, \ldots, J$. The intercept and factor loading for response i are α_{0i} and α_{1i} respectively.

Eq. (13.17) is an extension of the multivariate model given in eq. (13.13), with β_{0i} replaced by α_{0i}. The introduction of the factor, f_j, is assumed to account for all the correlations between y_{1j} to y_{mj}. Hence, whereas in eq. (13.13) the response-specific residuals u_{ij} are correlated, in eq. (13.17) they are assumed independent so that the y_{ij} are conditionally independent given f_j.

The factor model (13.17) can also be viewed as a variation on the multivariate regression given in eq. (13.14), with β_{0i} and β_{1i} replaced by α_{0i} and α_{1i}. The two models differ crucially in that the explanatory variable x_{ij} in eq. (13.14) was observable and was not assumed to account for the correlations between y_{1j} to y_{mj}, whereas in eq. (13.17) the factor, f_j, is an unobservable (latent) variable and is assumed to account for *all* the correlations between y_{1j} to y_{mj}.

To fix the scale of the latent variable f_j, some constraint must be placed either on one of the factor loadings or on the factor variance σ_f^2. Common choices are to set one of the loadings, α_{11} say, to one so that the factor has the same scale as the first response, as is commonly done in structural equation modelling (see Chapter 11), or to fix $\sigma_f^2 = 1$ as we did in Chapter 7.

The u_{ij} are sometimes called specific or unique factors because they are unique residuals to a particular response, and σ_{ui}^2 are the residual or specific variances. If σ_f^2 has been set equal to one and if the ys have also been standardised to unit variance, then the proportion of variance in variable i that is explained by f_j, i.e. its communality, is α_{1i}^2 which can also be calculated as the relative change in the residual variance between the multivariate model (13.13) and the factor model (13.17).

Multilevel factor model with one factor at each level

If individuals are nested within groups, we might expect the covariance between responses to depend not only on an individual-level common factor but also on a factor defined at the group level. For example, the correlation between scores on multiple tests may be due to a shared dependency on a student's underlying academic ability (represented by an individual-level factor) and the ability of other students in the same school (a school-level factor). A *multilevel* factor model includes one or more factors at each level of the hierarchical structure. Furthermore, there may be unique factors or residuals at each level to allow for within-group and between-group variance in a given response variable that is not explained by the common factors.

A two-level version of eq. (13.17) with one factor at each level can be written as

$$y_{ijk} = \alpha_{0i} + \alpha_{1i} f_{1jk} + \alpha_{2i} f_{2k} + u_{ijk} + v_{ik}, \quad (13.18)$$

where f_{1jk} is the latent score of individual j in group k on the level 1 factor, f_{2k} is the value for group k on the level 2 factor, and we assume $f_{1jk} \sim N(0, \sigma_{f1}^2)$ and $f_{2k} \sim N(0, \sigma_{f2}^2)$. We will often choose to set $\sigma_{f1}^2 = \sigma_{f2}^2 = 1$. The level 1 and 2 factor loadings for variable i are denoted by α_{1i} and α_{2i}. Communalities can also be defined for each level. Provided $\sigma_{f1}^2 = \sigma_{f2}^2 = 1$, the proportion of within-group variance that is explained by the level 1 factor is $\alpha_{1i}^2/(\alpha_{1i}^2 + \alpha_{ui}^2)$ and the proportion of between-group variance explained by the level 2 factor is $\alpha_{2i}^2/(\alpha_{2i}^2 + \alpha_{vi}^2)$. The response-specific residuals at each level are u_{ijk} and v_{ik}, which are assumed to be independently and normally distributed with mean zero and variances σ_{ui}^2 and σ_{vi}^2.

Example: Multilevel factor analysis of science test scores

The single-level and multilevel factor models given by eq. (13.17) and eq. (13.18) are fitted to the Hungarian science test data. For each model, the factor variances are constrained to one. Thus, because the test scores have also been standardised, the factor loadings at a given level can be interpreted as correlations between each variable and the factor.

Table 13.10 shows the results from the single-level factor model. The factor loadings are all positive but there is a large amount of variation between tests in their correlation with the underlying "ability" measure. Scores for biology R have the weakest correlation with the factor; only 7% of the variance in scores on this test is explained by a single common factor.

TABLE 13.10: Estimates from a single-level factor model, science test scores

Variable i	$\hat{\alpha}_{1i}$ (s.e.)	$\hat{\sigma}_{ui}^2$ (s.e.)	$\hat{\alpha}_{1i}^2$
Earth science	0.458 (0.023)	0.791 (0.026)	0.209
Biology	0.760 (0.026)	0.425 (0.033)	0.578
Physics	0.690 (0.025)	0.525 (0.029)	0.476
Biology R	0.266 (0.033)	0.930 (0.039)	0.071

The results from the multilevel factor model are given in Table 13.11. The communalities are now partitioned into within-school (between-student) and between-school components, given respectively by $\hat{\alpha}_{1i}^2$ and $\hat{\alpha}_{2i}^2$. For variable i, the sum of the communalities across levels is approximately equal to $\hat{\alpha}_{1i}^2$ in Table 13.10. For all three core tests, the student-level factor is dominant but there is nonetheless an important contribution of school "ability" to individual performance. Correlations between the test scores and the school-level factor range from 0.205 to 0.482. The residual variance is also partitioned into within-school and between-school components, so that the sum of $\hat{\sigma}_{ui}^2$ and $\hat{\sigma}_{vi}^2$ is approximately equal to $\hat{\sigma}_{ui}^2$ in Table 13.10. For all tests, most of the variation that is not explained by the common factors is at the student level.

TABLE 13.11: Estimates from a multilevel factor model, science test scores

Student level			
Variable i	$\hat{\alpha}_{1i}$ (s.e.)	$\hat{\sigma}_{ui}^2$ (s.e.)	$\hat{\alpha}_{1i}^2$
Earth science	0.379 (0.024)	0.699 (0.023)	0.144
Biology	0.597 (0.029)	0.427 (0.031)	0.356
Physics	0.544 (0.027)	0.473 (0.027)	0.296
Biology R	0.192 (0.035)	0.895 (0.039)	0.037
School level			
Variable i	$\hat{\alpha}_{2i}$ (s.e.)	$\hat{\sigma}_{vi}^2$ (s.e.)	$\hat{\alpha}_{2i}^2$
Earth science	0.285 (0.045)	0.094 (0.020)	0.081
Biology	0.482 (0.045)	0.013 (0.012)	0.232
Physics	0.466 (0.047)	0.040 (0.015)	0.217
Biology R	0.205 (0.041)	0.034 (0.017)	0.042

13.9   Additional examples and further work

School effects on students' academic progress

The ILEA (Inner London Education Authority) dataset contains the results of public examinations taken at age 16 for 2114 children from 114 London secondary schools. The data refer to a single cohort of children who took the examinations in 1987. Together with the age 16 exam scores (the response variable in our analyses), we have information on the students' performance at age 11 when they entered secondary school, their gender, and two school measures (whether the school was mixed or single sex, and its religious denomination). A list of variables contained in the data file is given in Table 13.12.

TABLE 13.12: Variables in the ILEA dataset

Variable name	Description and codes
student	Anonymised student identifer
school	Anonymised school identifer
exam16	Average score in exams taken at age 16 (ranges from 1 to 70)
vrband11	Verbal reasoning band, a London-wide measure of performance at age 11. There are three bands; band 1 contains the top 25% of children, band 2 the middle 50%, and band 3 the lowest 25%.
female	Sex of student (1=female, 0=male)
schgend	School gender (1=mixed school, 2 = boys' school, 3 = girls' school)
schdenom	School denomination (1=county, i.e. non-denominational, 2=Church of England, 3=Roman Catholic)

Nuttall et al. (1989) carried out extensive multilevel analyses of three cohorts of ILEA children to explore between-school variation in children's academic progress between ages 11 and 16. Table 13.13 shows the results from fitting a simple random intercept model to data from the 1987 cohort. The age 11 verbal reasoning band is included in the model as a measure of prior attainment, so that the residual between-school variance can be interpreted as the variance in school effects on children's *progress* since entering secondary school. The variance partition coefficient is estimated as $21.42/(21.42 + 109.49) = 0.164$, which implies that 16.4% of the variation in students' progress between ages 11 and 16 can be attributed to differences between schools.

TABLE 13.13: Random intercept model fitted to the ILEA data

Parameter	Estimate	Standard error
Constant	31.37	0.634
Verbal reasoning (base=high)		
Middle	−10.82	0.55
Low	−19.71	0.73
Between-school variance	21.42	3.73
Within-school variance	109.49	3.46

We suggest that you carry out further analyses of these data to explore gender differences in academic progress, and the extent to which the between-school variance can be explained by differences between mixed and single sex schools and between schools of different religious denominations.

Pupil ratings of school managers

School managers from 96 schools were rated by 854 pupils on their management style (Krüger 1994). There were six questions about the school manager. The dataset also contains the pupils' and school managers' gender. These data could be analysed using multivariate regression or multilevel factor analysis. Hox (2002) fits multilevel multivariate models to the data.

Suggested exercises

1. Fit a multilevel multivariate model to the six management items (with no covariates). Obtain the between- and within-school correlation matrices and, from these, calculate the variance partition coefficient for each item.

2. Use multilevel multivariate regression to answer the following questions: (i) Are male and female managers rated differently by pupils? (ii) Do male and female pupils rate managers differently? (iii) Does the difference in ratings for male and female managers depend on pupils' gender?

3. Fit a single-level factor model to the six management items (excluding co-variates). You could try one and two factor models.

4. Fit a multilevel factor model with one factor at each level. What is the evidence for an underlying school-level latent variable representing a school's management style?

13.10   Further topics

Categorical responses and mixed response types

In this chapter we have restricted our discussion to models for continuous responses. These models may be generalised to handle different types of response, including binary, ordered and unordered categorical, count and duration data. For example, a random intercept logit model for binary responses may be written

$$\log \left(\frac{\pi_{ij}}{1 - \pi_{ij}} \right) = \beta_0 + \beta_1 x_{ij} + u_j$$

where $\pi_{ij} = \Pr(y_{ij} = 1)$. It is also possible to handle mixtures of different response types in a multivariate model. For example, Goldstein (2010, Section 4.7) describes an application where a binary response indicating whether an individual smokes, is modelled jointly with a continuous response (defined only for smokers) for the number of cigarettes smoked per day.

Multilevel structural equation modelling

In Section 13.8, we saw how factor models can be extended to analyse grouped data, leading to factors and residuals defined at both levels of a two-level hierarchy. Structural equation models (Chapter 11) can be extended in a similar way. For example, a two-level version of the Multiple Indicator Multiple Causes (MIMIC) model would consist of two components: the measurement model of eq. (13.18) and a structural model in which the factors can depend on other factors and possibly covariates. For example, a level 1 factor may depend on other level 1 factors and level 2 factors, and a level 2 factor may depend on other level 2 factors. Detailed accounts of multilevel SEM include Muthén (1994) and Skrondal and Rabe-Hesketh (2004). A summary is given by Steele (2005).

More complex data structures

We have considered models for two-level hierarchical data, but all models can be generalised to handle non-hierarchical structures with more than two levels or classifications. There are many applications where the data structures are more complex. Suppose a student is classified as belonging sequentially to a particular combination of primary school and secondary school and we have followed a sample of such students through each school and wish to relate measurements made at the end of secondary school to those made earlier in the primary schools. The students will be identified by a *cross classification* of primary schools and secondary schools. Another example is where students are simultaneously classified by the school they attend and the area where they live, both classifications affecting the value of a response variable. A further complication occurs when we cannot assign a lower-level unit to a single higher-level unit. For example, students may move between schools in which case they can be regarded as *multiple members* of schools. If our response is, say, a test score at the end of secondary school, then for such a student we will need to share out the school "effect" among all the schools attended, using a suitable weighting function. Models for cross-classified and multiple-membership structures are discussed in Browne, Goldstein, and Rasbash (2001) and Goldstein (2010, Chapters 12 and 13).

13.11   Estimation procedures and software

The two most common procedures for estimating multilevel models for continuous responses are the E-M algorithm and the iterative generalised least squares (IGLS) or the related Fisher scoring algorithm. Goldstein (2010, Chapter 2 appendices) gives details of these. These methods are iterative and implemented in major statistics packages including SAS (Stroup et al. 2018), `lme4` in R (Bates et al. 2015), SPSS, and Stata (Rabe-Hesketh and Skrondal 2022) as well as specialist multilevel modelling packages such as HLM (Raudenbush and Congdon 2021) and MLwiN (Charlton et al. 2024). For non-hierarchically structured continuous responses, cross-classified models can also be fitted using these packages while multiple membership models are implemented in MLwiN.

Multilevel factor analysis is implemented in Mplus (Muthén and Muthén 2017; using two-stage weighted least squares), Stata (adaptive quadrature), and MLwiN (using Markov chain Monte Carlo estimation; see Browne 2023). Mplus and Stata (StataCorp 2023) can also be used to fit more general structural equation models to any mixture of continuous and categorical responses.

13.12   Further reading

We recommend the following textbooks on multilevel modelling.

> Goldstein, H. (2010). *Multilevel Statistical Models* (4th ed.). London: Wiley.

> Rabe-Hesketh, S. and Skrondal, A. (2022). *Multilevel and Longitudinal Modelling using Stata, Volume I* (4th ed.). Stata Press.

> Raudenbush, S.W. and Bryk, A.S. (2002). *Hierarchical Linear Models* (2nd ed.). Newbury Park: Sage.

> Snijders, T.A.B. and Bosker, R.J. (2012). *Multilevel Analysis: An Introduction to Basic and Advanced Multilevel Modelling* (2nd ed.).

London: Sage.

Skrondal, A. and Rabe-Hesketh, S. (2004). *Generalized Latent Variable Modelling: Multilevel, Longitudinal, and Structural Equation Models*. Boca Raton, FL: Chapman and Hall/CRC.

Readers may also find useful the LEMMA online training materials that have been developed by the Centre for Multilevel Modelling. These can be freely downloaded from https://www.bristol.ac.uk/cmm/.

14

Longitudinal Data Analysis

14.1   Introduction

An important type of clustered data arises in longitudinal studies where repeated measurements are made on individuals over time. In Chapter 13 on multilevel modelling, it was noted that repeated measures of a response variable y have a two-level multilevel structure with measurements at level 1 nested within individuals at level 2 (where individuals can be replaced by other units such as organisations or countries). Another way to view repeated measures is as a multivariate response, but where we have measurements on the same variable at multiple time points rather than measurements on multiple variables at a cross-section. This duality in how we view longitudinal data means that they can be analysed using the multilevel models of Chapter 13 or the latent variable models of earlier chapters, in particular factor models (Chapter 7) and structural equation models (Chapter 11) for continuous responses. In this chapter, we focus on *growth curve models* that can be framed as either multilevel or factor models, but some kinds of models (including certain generalisations of growth curve models) fit more naturally in one framework than the other. For example, we also consider *mixture models* that are related to latent class analysis (Chapter 10).

Whatever modelling approach we take, there are several important considerations unique to longitudinal data analysis. These include the temporal ordering of observations within individuals. The ordering is important because we expect that the correlation between a pair of responses on the same individual (often called *autocorrelation*) will depend on the length of time between their measurement, with higher correlation between measurements taken closer in time. For this reason, models fitted to longitudinal data should capture the within-person correlation structure. Related considerations are the choice of time metric (e.g. calendar year or age at measurement) and the specification of time effects in the model.

There are a variety of ways to analyse longitudinal data and several excellent books on the topic (some of which are listed in Section 14.11). We restrict attention to random effects models for "growth" processes where interest lies in describing individual response trajectories over time. The random effects (or latent variables) may be continuous (as in standard growth curve models), categorical (as in latent class growth analysis) or a combination of the two (as in growth mixture models). Throughout the chapter, we discuss connections with the methods discussed earlier in the book.

14.2 Some potential applications

i) Suppose we have repeated measurements on children's height taken at several ages. Longitudinal models can be used to study the nature and predictors of child growth, allowing initial height and change in height (the growth rate) to vary across children. As well as studying the effects of predictors such as sex and social class on the *mean* growth rate, we can assess the extent to which the *variance* in growth rates between children depends on their individual characteristics.

ii) In studies of ageing there is interest in the nature of physical and cognitive decline with age. Research questions include: What is the average rate of decline in older people? How does a person's physical and cognitive "trajectory" depend on earlier lifestyle factors (such as amount of exercise and diet) and their current circumstances?

iii) In i) and ii) repeated measurements were made on individuals. More generally, longitudinal data can be collected on entities such as geographical areas, schools and organisations. Suppose, for example, that data are available on the annual profits of a sample of companies. Questions of interest could include: How do profit trajectories vary across companies? To what extent can differences between companies in their profit trajectories be explained by company characteristics such as industry sector and external economic factors?

14.3 Longitudinal data structures

Wide and long form data

To fix ideas, data structures and models will be described for a setting where individuals are the units of analysis. Using the notation of Chapter 13, we denote by y_{ij} the response at occasion i of individual j. We will also refer to measurement occasions as "waves", a term widely used in survey research to describe rounds of data collection. Suppose that we have T_j observations on individual j so that $i = 1, 2, \ldots, T_j$. (Note the change in notation from n_j for the size of group j in Chapter 13 to T_j in the longitudinal case.)

Longitudinal data are often in the form of one record per individual with each response (and any time-varying covariates) stored as a set of variables, one for each measurement occasion (see Table 14.1 for an example). This data structure is referred to as *wide form* and is consistent with treating the ys for a given individual as a multivariate response. The data need to be in wide form when analysed using factor models.

TABLE 14.1: An extract of wide form data for two individuals showing responses y_{ij} at $T_j = 4$ occasions

	Occasion i			
Individual j	1	2	3	4
1	y_{11}	y_{21}	y_{31}	y_{41}
2	y_{12}	y_{22}	y_{32}	y_{42}

An alternative way of structuring the data is to have one record per measurement occasion, leading to multiple records per individual. The wide form data of Table 14.1 has been reshaped to *long form* in Table 14.2. Long form data follows a two-level hierarchical structure where individual "clusters" are indicated by an individual identifier. The data need to be in long form when analysed using multilevel models.

TABLE 14.2: An extract of long form data for two individuals showing responses y_{ij} at $T_j = 4$ occasions

Individual j	Occasion i	Response y_{ij}
1	1	y_{11}
1	2	y_{21}
1	3	y_{31}
1	4	y_{41}
2	1	y_{12}
2	2	y_{22}
2	3	y_{32}
2	4	y_{42}

Example: Physical health functioning

In later sections we will analyse data on British civil servants from the Whitehall II study (Marmot and Shipley 2005). Health functioning was assessed by the SF-36, a 36-item instrument that comprises eight subscales covering

physical, psychological and social functioning. These eight scales can be summarised into physical and mental health components. These are scaled to have
mean values of 50 and low scores imply poor functioning. We will study change
in physical health functioning which was measured on up to six occasions for
each respondent. We analyse data on 4350 individuals who together contribute
21,104 observations (a mean of 4.85 observations per person). Further analyses
of these data can be found in Steele (2014).

Table 14.3 shows the long form data structure for three individuals in
the dataset. As in most longitudinal studies, there is some missing data for
individuals 1 and 3. Individual 1 missed the final measurement, while individual 3 was present at only the first two occasions. Both individuals display
a *dropout* (or *monotone*) missing data pattern because they do not return
after leaving the study. For the 4350 individuals in the full dataset, 34.4 percent have missing data for at least one occasion. For our analysis, records
with missing data will be excluded (e.g. leaving 13 observations for the three
individuals in Table 14.3). Excluding these records makes a *missing at random*
assumption which was defined in Section 6.16 and will be further discussed
for longitudinal data in Section 14.8.

TABLE 14.3: An extract of the physical functioning dataset in long form

Individual j	Occasion i	age_{ij}	phf_{ij}
1	1	55.0	39.6
1	2	57.0	38.6
1	3	60.4	39.9
1	4	63.6	21.9
1	5	66.6	25.7
1	6	–	–
2	1	56.0	29.3
2	2	60.0	22.7
2	3	62.1	24.2
2	4	65.6	21.4
2	5	67.6	17.2
2	6	71.0	23.0
3	1	53.0	46.7
3	2	57.0	41.8
3	3	–	–
3	4	–	–
3	5	–	–
3	6	–	–

Apart from the variables indexing the individual and occasion, there are
two time-varying variables: phf_{ij} is the physical health functioning score at
occasion i for individual j and age_{ij} is their age in years at the same occasion.
At any given occasion, there is variation between individuals in their age. We
can also see that the length of time between measurements varies between
individuals, for example the gap between occasions 1 and 2 is 2 years for
individual 1 and 4 years for individuals 2 and 3. It will be important to allow

for variations in the timing of measurements in our analysis because we would expect health functioning to depend strongly on age. We return to this issue in the next section.

14.4   Introduction to growth curve models

Growth curves

The aim of growth curve modelling (GCM) is to fit a curve or trajectory through each individual's repeated measurements. The name "growth" comes from early applications in child development studies where curves were fitted to measurements of children's height or weight. More generally, GCMs can be fitted to other developmental or ageing processes, for example cognitive development in children (usually measured by reading or verbal reasoning tests) or cognitive and physical decline among older people. (See questions (i) and (ii) in Section 14.2.)

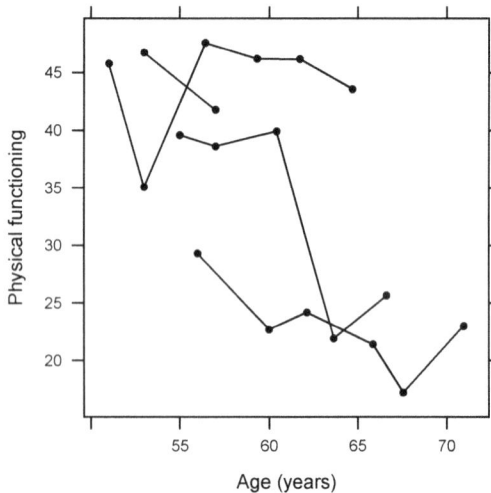

FIGURE 14.1: Physical functioning trajectories for four individuals

Figure 14.1 shows the observed physical health functioning trajectories for four individuals. We can observe several features of the data and trajectories from this small sample. First, due to between-individual variation in the age at each measurement, the curves start at different ages. Second, the curves are of different lengths, due mainly to missing data but also differences in the length of time between measurements. Third, there is substantial variation in individuals' initial functioning scores and their rate of change with age. Finally, although there is a general tendency for functioning to decline with age, there are large within-individual fluctuations and the trajectories are nonlinear. The objective of GCMs is to fit curves that approximate the observed trajectories, allowing for between-individual variation and nonlinearity with as few parameters as possible.

Formulating a growth curve model as a random slope model

A GCM can be fitted as a type of random slope model (described in Section 13.5). Before setting out the details of the model specification, we first consider the idea behind this approach. A starting point for modelling how a response variable depends on time is to fit an overall curve representing the average growth across the population. The simplest formulation would assume a linear relationship between the response and time, treating time as a continuous explanatory variable. For example, to fit an overall trajectory for physical functioning as a linear function of age, we could fit a linear regression of phf_{ij} on age_{ij} to the long-form data (illustrated in Table 14.3).

For a more realistic model, however, we need to allow for individual variation about this overall trajectory. This is where multilevel modelling comes in. We can allow each individual to have their own trajectory by including individual-level random intercepts and random slopes: a random intercept allows for individual variation in the intercept of the trajectory, while a random slope on age allows for individual variation in the rate of change with age. As before, these random intercepts and slopes are assumed to come from a (bivariate) normal distribution. As the shape of a person's curve is determined by these random effects, which are continuous latent variables, GCMs are also known as *latent trajectory models* and *latent curve models*.

In general, a random slope model allows:

i) The relationship between y and time to vary across individuals;

ii) The between-individual variance in y to depend on time.

In addition, and of particular importance for longitudinal data, a random slope for time allows the correlation between two responses to depend on the time that each was measured.

Choice and coding of the time variable

Among the important decisions in the specification of a GCM are the choice of the time variable and its coding. Denote by t_{ij} the timing of measurement i for subject j. When subjects are all observed at the same fixed occasions, such that there is no between-individual variation in the relevant time variable at a given occasion i, we can write $t_{ij} = t_i$. However, we will use the more general notation t_{ij} throughout.

Choice of time variable

Consider the following extract from a typical annual panel study. Note that individual 2 entered at wave 2.

Individual j	Occasion i	Wave	Year	Age
1	1	1	1990	30
1	2	2	1991	31
1	3	3	1992	32
2	1	2	1991	25
2	2	3	1992	26

Occasion is a within-person observation index and is rarely of interest. Wave and Year are equivalent (Wave=Year-1989). The question is then which of Age and Year/Wave to use? For developmental processes such as physical growth or cognitive outcomes, or ageing processes, Age is a natural time metric. In such cases, the outcome depends directly on a person's age rather than the calendar year or month at which the measurement was taken. For other kinds of process, Year (or equivalently Wave) may be a suitable choice. For example, a person's employment status may vary by year due to economic factors affecting their local area or the whole country. Such time effects are often referred to as *period effects*. Apart from age and period effects, there could be an effect of when a person was born. For example, a person's attitude towards nuclear deterrents is likely to differ for someone who grew up during the Cold War compared to someone born later, regardless of their age when their attitudes were measured. This type of time effect is known as a *cohort effect*.

Although age, period and cohort effects are conceptually distinct, in practice they cannot be disentangled without strong assumptions (see Bell (2021) for a discussion). If two of age, period and cohort are known, the other can be deduced. For example, in the example above, we can deduce from Year and Age that individual 1 was born in 1960 and individual 2 in 1966. This implies that, if age, period and cohort effects are all linear, we can identify only two of them due to their linear dependence.

Is individual variation in timing of measurements important?

In most large-scale studies there will be variation between individuals in when the measurement at occasion i was taken. The importance of individual variation in timing depends on the time metric and whether differences in timing are likely to influence the response y. Consider the following examples.

i) In a study of reading, it is intended that children are tested at ages 5, 7 and 9. However, the ages of children in the same school year may vary by up to 1 year. As reading scores depend strongly on age, it would be preferable to use a child's age in months at occasion i.

ii) In a large household panel study, the length of the fieldwork period at a given wave may be long. Suppose y_{ij} is the attitude towards abortion of individual j at wave i. Although there may be some within-individual fluctuations in a person's attitude over time, on average, we would not expect an impact of a few months difference in when the interview was conducted, in which case it would be sufficient to use wave (or, equivalently, calendar year).

Coding of the time variable

There are two issues to consider in the coding of t_{ij}:

i) *Coding should reflect the spacing of measurements.* It is common for measurements to be unequally spaced. In an annual panel study, for example, our response variable of interest may not be measured at every wave. Suppose that the total number of measurements per person is 4, with 2 years between measurement occasions 1 and 2, and 1 year between occasions 2 and 3 and between occasions 3 and 4. A suitable coding for t_{ij} is $(1, 3, 4, 5)$ or any coding that preserves the relative gaps between measurements.

ii) *Choice of origin.* As discussed in the next section, the coding of an explanatory variable with a random slope affects estimates of parameters associated with the intercept, including the intercept variance. In a GCM, t_{ij} will have a random slope so $t_{ij} = 0$ should be meaningful. If age is the time variable of interest, it is common to centre age at a suitable value, e.g. the mean age at occasion 1. For fixed measurement occasions (with $t_{ij} = t_i$), it is common to code $t_1 = 0$.

14.5   Growth curve models

Linear random intercept model

The simplest GCM we consider is a linear random intercept model which has the same form as eq. (13.5) in Section 13.4, replacing x_{ij} with t_{ij}:

$$y_{ij} = \beta_0 + \beta_1 t_{ij} + u_{0j} + e_{ij}, \tag{14.1}$$

where β_0 is the intercept, β_1 is the linear growth rate, $u_{0j} \sim N(0, \sigma_{u0}^2)$ is the individual-level random effect and $e_{ij} \sim N(0, \sigma_e^2)$ the time-varying residual. (Although the "0" subscript on u_j is unnecessary when there is only one random effect, we include it in anticipation of the extension to a random slope model.) Under this model, the intercept for individual j is $\beta_{0j} = \beta_0 + u_{0j}$ but the growth rate β_1 is constant across individuals.

In the longitudinal setting, we can refer to the intra-cluster correlation, $\text{ICC} = \sigma_{u0}^2/(\sigma_{u0}^2 + \sigma_e^2)$ as the intra-*individual* correlation, because it is now interpreted as the correlation between the y-values at two randomly selected measurement occasions for the same individual.

Linear random slope model

We now consider a more realistic growth model which relaxes the strong assumption of eq. (14.1) that the growth rate β_1 is constant across individuals. A linear random slope model, with a random coefficient on t_{ij}, allows for between-individual variation in the linear effect of t_{ij}. This model has the same form as the random slope model (13.7) in Section 13.5, again replacing x_{ij} with t_{ij} to obtain

$$y_{ij} = \beta_0 + \beta_1 t_{ij} + u_{0j} + u_{1j} t_{ij} + e_{ij}, \tag{14.2}$$

where individual intercept and slope random effects, u_{0j} and u_{1j}, are assumed to follow a bivariate normal distribution with zero mean, variances σ_{u0}^2 and σ_{u1}^2 respectively, and covariance σ_{u01}.

Under this model, the growth rate for individual j is given by $\beta_{1j} = \beta_1 + u_{1j}$ and the between-individual variance in growth is σ_{u1}^2. The intercept variance σ_{u0}^2 now represents the between-individual variance at $t_{ij} = 0$. The interpretation of the intercept-slope covariance σ_{u01} given in Section 13.5 translates to the longitudinal case. Briefly, for $\beta_1 > 0$ a positive intercept-slope covariance indicates that individuals with above-average intercepts $(u_{0j} > 0)$ tend also to have steeper-than-average slopes $(u_{1j} > 0)$. Similarly, individuals with below-average intercepts tend also to have flatter-than-average slopes. For $\beta_1 < 0$, a positive slope residual $(u_{1j} > 0)$ implies a flatter-than-average negative slope or a positive slope (if $u_{1j} > \beta_1$).

While the random intercept model implies a constant between-individual variance $(\text{var}(u_{0j}) = \sigma_{u0}^2)$, the random slope model implies that the between-individual variance depends on t_{ij}. Taking the individual-level random component of eq. (14.2):

$$\begin{aligned} \text{var}(y_{ij}|t_{ij}) = \text{var}(u_{0j} + u_{1j}t_{ij}) &= \text{var}(u_{0j}) + 2\text{cov}(u_{0j}, u_{1j})t_{ij} + \text{var}(u_{1j})t_{ij}^2 \\ &= \sigma_{u0}^2 + 2\sigma_{u01}t_{ij} + \sigma_{u1}^2 t_{ij}^2, \end{aligned} \tag{14.3}$$

which is a quadratic function of t_{ij} with the random effect variances and covariance as coefficients. In applications of GCMs to developmental processes such as reading or physical growth, for example, it is common for between-individual differences to become more apparent at older ages, so that the between-individual variance increases with t_{ij}.

Moreover, the within-individual covariance is no longer constant when the growth rate varies across individuals. Under the assumption that $\text{cov}(e_{ij}, e_{i'j}) = 0$ for $i \neq i'$, the covariance between measures i and i' on individual j (conditional on t_{ij}) is

$$
\begin{aligned}
\text{cov}(y_{ij}, y_{i'j}|t_{ij}, t_{i'j}) &= \text{cov}(u_{0j} + u_{1j}t_{ij}, u_{0j} + u_{1j}t_{i'j}) \\
&= \sigma_{u0}^2 + \sigma_{u01}(t_{ij} + t_{i'j}) + \sigma_{u1}^2 t_{ij}t_{i'j}.
\end{aligned}
\tag{14.4}
$$

Using the expressions for the covariance and variance given by eq.s (14.4) and (14.3) we obtain the intra-individual correlation for occasions i and i':

$$
\begin{aligned}
\text{cor}(y_{ij}, y_{i'j}|t_{ij}, t_{i'j}) &= \frac{\text{cov}(y_{ij}, y_{i'j}|t_{ij}, t_{i'j})}{\sqrt{\text{var}(y_{ij}|t_{ij})\text{var}(y_{i'j}|t_{i'j})}} \\[2ex]
&= \frac{\sigma_{u0}^2 + \sigma_{u01}(t_{ij} + t_{i'j}) + \sigma_{u1}^2 t_{ij}t_{i'j}}{\sqrt{(\sigma_{u0}^2 + 2\sigma_{u01}t_{ij} + \sigma_{u1}^2 t_{ij}^2 + \sigma_e^2)(\sigma_{u0}^2 + 2\sigma_{u01}t_{i'j} + \sigma_{u1}^2 t_{i'j}^2 + \sigma_e^2)}}.
\end{aligned}
\tag{14.5}
$$

This is a complex expression, but the main point to note is that, in contrast to the equal ICC assumption of the random intercept model, the ICC implied by the random slope model depends on the timing of both occasions t_{ij} and $t_{i'j}$. Eq. (14.5) can accommodate complex within-individual correlation structures, including the common case where the correlation between the responses at occasions i and i' decays with $|t_{ij} - t_{i'j}|$.

Example: Linear growth models for physical health functioning

We illustrate the application of linear GCMs in analyses of the physical health functioning data, taking age as the time variable. There is variation between individuals in their age at a given occasion. The mean age is 50 years at occasion 1 and 64 years at occasion 6, with an age range of 39 to 76 across the study period. In the following analyses, age has been centred at 50 years. The length of time between measurements also varies, though on average they are approximately three years apart.

The results from fitting the linear random intercept model (14.1) are shown in columns 2 and 3 of Table 14.4. The $-$log-likelihood value for this model is 70,514 compared to 74,555 for the single-level model (without individual effects). The likelihood ratio test statistic for a test of the null hypothesis that $\sigma_{u0}^2 = 0$ is calculated as 8082 which is highly significant when compared with a chi-squared distribution on 1 degree of freedom so we select the random intercept model. As expected, physical functioning declines with age. The between-individual variance is estimated as 38.68 and the intra-individual correlation (conditional on age) as $38.68/(38.68 + 31.74) = 0.55$, which implies that 55% of the total residual variance in physical functioning is due to differences between individuals.

We next consider the random slope model (14.2) which allows the change in health functioning with age to vary across individuals and relaxes the unrealistic assumption of the random intercept model that the correlation between any pair of health functioning scores on the same person is equal, regardless of the length of time between them. From the $-$log-likelihood values in Table 14.4, the likelihood ratio test statistic for a comparison of the random slope and random intercept models is 657 on 2 degrees of freedom (in a conservative test that does not adjust for the slope variance being non-negative). We therefore conclude that there is between-individual variation in the effect of age on health functioning and we select the random slope model. Turning to the parameter estimates for the random slope model in columns 4 and 5 of Table 14.4, the coefficient of age is now interpreted as the "average" age effect or, equivalently, the effect of age for individuals at the mean of the slope distribution (with $u_{1j} = 0$). Under the assumption that the random effects are normally distributed, the 95% plausible values range for the individual slopes is calculated as $-0.25 \pm (1.96 \times \sqrt{0.09}) = (-0.84, 0.34)$, so there is substantial variation between individuals. The correlation between the intercept and slope effects is estimated as $0.44/\sqrt{29.3 \times 0.09} = 0.27$. Taken together with the negative coefficient for age, a positive random effects correlation implies that individuals with high functioning at age 50 ($u_{0j} > 0$) tend to have a larger positive slope effects ($u_{1j} > 0$) which translates to a slower decline or an increase in functioning ($\beta_1 + u_{1j}$) according to the magnitude of u_{1j} relative to β_1.

TABLE 14.4: Linear random intercept and random slope growth models for physical health functioning

	Random intercept		Random slope	
	Estimate	s.e.	Estimate	s.e.
Fixed part				
β_0	51.82	0.11	51.90	0.008
β_1 (age, centred at 50)	−0.25	0.01	−0.25	0.01
Random part				
σ_{u0}^2	38.68	1.01	29.30	0.95
σ_{u1}^2	−	−	0.09	0.007
σ_{u01}	−	−	0.44	0.06
σ_e^2	31.74	0.35	28.60	0.36
$-$log-likelihood	70,514		70,185	

Figure 14.2 shows the fitted health trajectories for the four individuals whose observed trajectories were shown in Figure 14.1. The impact of allowing for between-individual variation in the slope of age is most apparent for the individuals in the bottom row, both of whom experienced steeper-than-average decline in functioning with age.

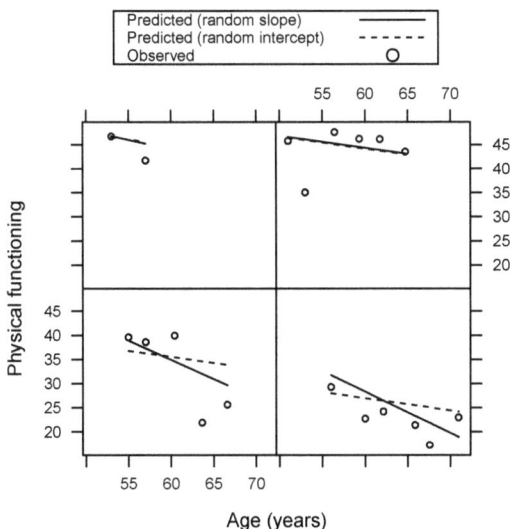

FIGURE 14.2: Fitted physical health trajectories for four individuals from random slope and random intercept models

The random slope model implies that the between-individual variance is a quadratic function in age. The variance function plotted in Figure 14.3 is obtained by substituting the estimates of the individual-level random effect variances and covariance from the random slope model in Table 14.4 in eq. (14.3). There is greater individual heterogeneity in physical functioning at older ages, which is consistent with a "fanning out" pattern in individual trajectories (see Section 13.5). The estimate of the between-individual variance at a given age can be used to obtain a 95% plausible values range for the individual trajectories, calculated as $\hat{\beta}_0 + \hat{\beta}_1 \text{age} \pm 1.96\sqrt{\hat{\sigma}_{u0}^2 + 2\hat{\sigma}_{u01}\text{age} + \hat{\sigma}_{u1}^2\text{age}^2}$ and plotted in Figure 14.4.

Finally, the random slope model allows the correlation between a pair of measurements on the same person to depend on their respective ages, as in eq. (14.5). To illustrate how the correlations depend on age, Table 14.5 shows the within-person correlations in physical health functioning for two individuals, derived from the estimates of the random part parameters given in Table 14.4. Both individuals were observed at four occasions, but the first was observed at relatively young ages over a 9-year period (ages 42 to 51.2

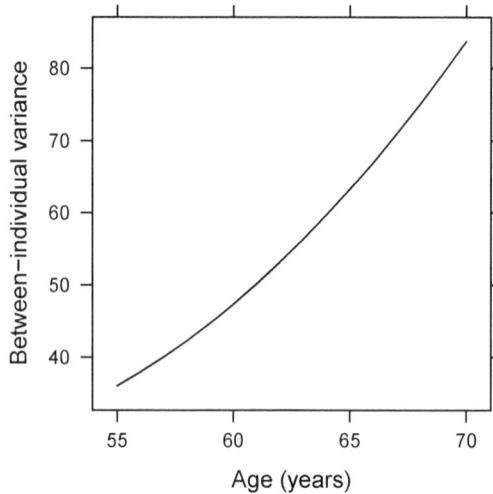

FIGURE 14.3: Estimated between-individual variance in physical health functioning from random slope model

years) while the second was observed at older ages over a 15-year period (ages 61 to 76.1 years). The correlations are lower for the first individual and decrease as the gap between measurements increases; all correlations are lower than the estimate of 0.55 from the random intercept model in Table 14.4. The estimated correlations are substantially higher for the second individual.

TABLE 14.5: Estimates of the within-person correlations implied by the random slope model of Table 14.4 for two individuals observed at four different ages (t_{ij}), physical functioning

Age (years)	Age (years)			
	42	45	48.2	51.2
42	1			
45	0.484	1		
48.2	0.464	0.483	1	
51.2	0.439	0.472	0.498	1
	Age (years)			
	61	67.4	72.8	76.1
61	1			
67.4	0.666	1		
72.8	0.676	0.740	1	
76.1	0.678	0.749	0.786	1

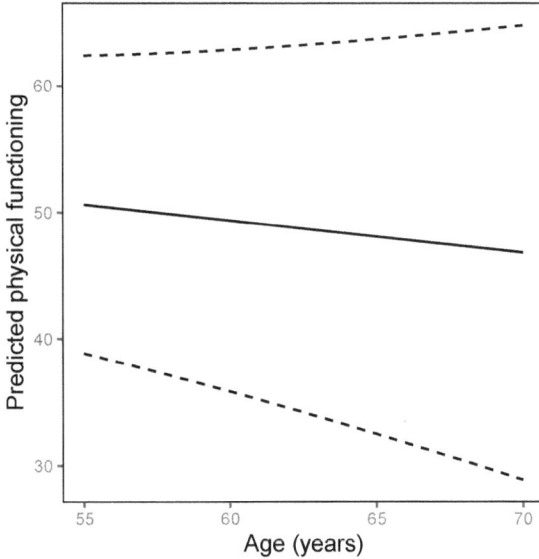

FIGURE 14.4: Mean physical health trajectory with 95% plausible values range

Nonlinear growth

In many applications, the assumption that the response y_{ij} changes linearly with time will be overly restrictive. One way to allow for nonlinearity is to fit a polynomial in time. For example, a quadratic function would be suitable when growth increases or decreases monotonically before flattening off at higher values of t_{ij}. A quadratic growth model is specified by including t_{ij}^2 as an additional predictor, leading to the following extension of the linear random slopes model

$$y_{ij} = \beta_0 + \beta_1 t_{ij} + \beta_2 t_{ij}^2 + u_{0j} + u_{1j} t_{ij} + u_{2j} t_{ij}^2 + e_{ij} \qquad (14.6)$$

which can also be expressed in multi-equation form as

$$\begin{aligned}
y_{ij} &= \beta_{0j} + \beta_{1j} t_{ij} + \beta_{2j} t_{ij}^2 + e_{ij} \\
\beta_{0j} &= \beta_0 + u_{0j} \\
\beta_{1j} &= \beta_1 + u_{1j} \\
\beta_{2j} &= \beta_2 + u_{2j}.
\end{aligned} \qquad (14.7)$$

The individual random effects, u_{0j}, u_{1j} and u_{2j}, are assumed to follow a trivariate normal distribution, so there are three additional random effect parameters compared to the linear growth model (the variance of u_{2j} and its covariances with u_{0j} and u_{1j}). In some applications it may be possible to simplify the model by specifying a fixed effect for the quadratic term such that $\beta_{2j} = \beta_2$.

Returning to the physical functioning example, the linear random slope model is first extended to a quadratic in age with a fixed coefficient for t_{ij}^2. The likelihood ratio test statistic for a test of the null hypothesis that $\beta_2 = 0$ is 81.4 which is compared to a chi-squared distribution on 1 degree of freedom. We conclude that β_2 is non-zero and further extend the model to allow for individual-specific coefficients for t_{ij}^2 as in eq. (14.7). The test statistic for a comparison of this model with the previous model (with $\beta_{2j} = \beta_2$) is 313.9 on 3 degrees of freedom (in a conservative test that does not adjust for testing on the boundary of the parameter space for $\text{var}(u_{2j}) = \sigma_{u2}^2$). We therefore conclude that the coefficient of t_{ij}^2 varies between individuals.

The fitted equation of the average quadratic trajectory, representing the relationship between physical functioning and age (centred at 50 years) for an individual at the mean of the distribution for each random effect ($u_{0j} = u_{1j} = u_{2j} = 0$), is

$$\hat{y}_{ij} = 51.9 - 0.19\,\text{age}50_{ij} - 0.005\,\text{age}50_{ij}^2$$

which implies that the decline in physical functioning with age accelerates at older ages.

Higher-order polynomials, and other transformations of time such as the log and exponential functions, allow a variety of nonlinear growth patterns to be represented by a relatively simple function. For greater flexibility, a piecewise polynomial function (also known as a spline) can be fitted. This involves splitting the observation period into intervals defined by "knot points" and specifying a polynomial for each interval such that they join together to form a continuous function. A piecewise-linear function often gives sufficient flexibility, while retaining interpretability. Further details of the use of splines in GCMs, including the choice of knot points, can be found in Howe et al. (2016) and Pan and Goldstein (1998).

Autocorrelated residuals

So far we have assumed that the occasion-specific residuals e_{ij} are uncorrelated, which implies that responses y_{ij} on the same individual j are conditionally independent given the individual-specific random effects. We can relax this assumption to better account for within-individual correlation, leading to what is sometimes referred to as a multilevel time series model (Goldstein, Healy, and Rasbash 1994).

For simplicity, suppose that we have T measurements for each individual. Denote by $\mathbf{\Omega}_e$ the covariance matrix for the level 1 residuals, so that $\mathbf{e}_j = (e_{1j}, \ldots, e_{Tj})' \sim N(\mathbf{0}, \mathbf{\Omega}_e)$. In the models considered thus far $\mathbf{\Omega}_e = \sigma_e^2 \mathbf{I}_T$

where $\mathbf{I}_T$ is the $T \times T$ identity matrix. There are many possible alternative forms for $\boldsymbol{\Omega}_e$ that allow non-zero covariance between the residuals for any pair of observations on the same individual. A common choice is the first-order autoregressive, AR(1), structure where the correlation between a pair of residuals is given by

$$\mathrm{cor}(e_{ij}, e_{i'j}) = \rho^{|i-i'|},$$

where ρ is the AR(1) parameter. The correlation decreases as $|i - i'|$ increases and $\boldsymbol{\Omega}_e$ has the form

$$\boldsymbol{\Omega}_e = \sigma_e^2 \begin{pmatrix} 1 & \rho & \rho^2 & \cdots & \rho^{T-1} \\ \rho & 1 & \rho & \cdots & \rho^{T-2} \\ \vdots & \vdots & \vdots & \ddots & \vdots \\ \rho^{T-1} & \rho^{T-2} & \rho^{T-3} & \cdots & 1 \end{pmatrix}.$$

A likelihood ratio test can be used to test for autocorrelation. In the case of the above AR(1) structure, for example, a comparison of the above model with a model assuming $\mathrm{cor}(e_{ij}, e_{i'j}) = 0$ provides a test of the null hypothesis that $\rho = 0$.

It is important to note that under an AR(1) model the correlation depends on the distance between measurement occasions $|i - i'|$, not the distance between the timing of these occasions for a given individual $|t_{ij} - t_{i'j}|$. For this reason, an AR(1) structure is only appropriate for fixed measurement occasions, i.e. when $t_{ij} = t_i$. Unequal spacing between occasions due to missing data can be accommodated, but this must be reflected in the coding of the time variable i used to specify the AR(1) structure.

Allowing for autocorrelated occasion-level residuals may be necessary in a simple random intercept model which otherwise assumes equal correlation between any pair of responses on the same individual. However, a random slope model with independent residuals may be adequate to capture the within-individual correlation structure because this already allows $\mathrm{cor}(y_{ij}, y_{i'j})$ to depend on t_{ij} and $t_{i'j}$. Including additional individual-specific random effects, such as a quadratic in t_{ij} with a random coefficient for t_{ij}^2, allows for a more flexible correlation structure because the within-individual correlation function given by eq. (14.5) will be extended to depend on an additional three random effect parameters. If an individual's pattern of responses over time is adequately captured by the individual-specific trajectory implied by the model, it may be reasonable to assume that e_{ij}, the deviations in y_{ij} about this trajectory, are independent random perturbations. There is a trade-off between specifying a highly complex function for the growth curve and assuming independent e_{ij} and fitting a simpler growth curve with autocorrelated e_{ij}. A random intercept model with AR(1) residuals is useful in situations when we want to approximate the within-individual correlation structure, but we are not interested in individual trajectories (or when the response of interest does not follow a growth process). An example of the application of GCMs with AR(1) residuals is given in Section 14.9.

14.6   Specifying growth curve models as factor models

As described in Section 13.3, multilevel models for 2-level hierarchical structures can be framed as factor models with parameter constraints. The connection between the two modelling frameworks is especially important for longitudinal data analysis where repeated measures can be viewed as a multivariate response, represented as a wide form data structure with the responses at each occasion stored as separate variables (see Section 14.3). In this section, we show how the linear random intercept and random slope growth models of Section 14.5 can be specified as one- and two-factor models respectively before considering the relative advantages of each framework for extensions to these models.

Random intercept model as a one-factor model

For simplicity, we consider the special case of fixed measurement occasions $(t_{ij} = t_i)$ and T observations per individual. We begin by reformulating a single-factor model as a linear random intercept growth model. With the addition of a j subscript to index individuals, model (7.2) becomes

$$y_{ij} = \alpha_i + \beta_i f_j + e_{ij}, \tag{14.8}$$

where i indexes the items forming the multivariate response, α_i are item-specific intercepts, f_j is a factor (latent variable) with item-specific loadings β_i and e_{ij} is a residual. We assume that f_j and e_{ij} are normally distributed and independent.

In the longitudinal setting, item i corresponds to the measurement of y at occasion i. Model (14.8) can be transformed into the linear random intercept growth model (14.1) by setting $\alpha_i = \beta_1 t_i$ and $\beta_i = 1$ for $i = 1, \ldots, T$ and $f_j = \beta_{0j} = \beta_0 + u_{0j}$. The equality constraint on the factor loadings β_i assumes that unmeasured time-invariant characteristics, represented by the latent variable, have the same effect on the responses at each occasion. The additional constraint that the loadings are equal to 1 is necessary for identification (equivalently, we could set $\text{var}(u_{0j}) = 1$).

Random slope model as a two-factor model

To specify a linear random intercept growth model as a factor model, we start by adding a second factor to eq. (14.8) to obtain

$$y_{ij} = \alpha_i + \beta_{1i} f_{1j} + \beta_{2i} f_{2j} + e_{ij} \tag{14.9}$$

where the latent variables f_{1j} and f_{2j} are assumed to follow a bivariate normal distribution. This model can be specified as the linear random slope growth model (14.2) by setting $\alpha_i = 0$, $\beta_{1i} = 1$ and $\beta_{2i} = t_i$ for $i = 1, \ldots, T$, $f_{0j} = \beta_{0j} = \beta_0 + u_{0j}$ and $f_{1j} = \beta_{1j} = \beta_1 + u_{1j}$. A particular feature of formulating the random slope model as a factor model is the specification of the linear time effect through the constraints on the loadings for the second factor. This idea can be generalised to specify polynomial growth models. For example, the quadratic model (14.7) can be formulated as a three-factor model by adding a term $\beta_{3i}f_{3j}$ to eq. (14.9), where f_{3j} is a factor that is correlated with the other two factors, and applying the constraints $\beta_{3i} = t_i^2$.

Choice between multilevel modelling and factor analysis approaches

While the simple growth models considered above can be specified as either a multilevel model or a factor model, there are some extensions that are only possible, or at least more straightforward, in one framework. The choice of approach should therefore be guided by the data and research questions. This will in turn guide the choice of software, although the distinctions have become increasingly blurred as some SEM and multilevel modelling software, or functions within mainstream packages, can be used to fit both SEMs and multilevel models.

One advantage of the multilevel modelling framework is that the long-form data structure and treatment of time as a predictor variable make it easier to accommodate between-individual variation in the timing of measurement occasions (i.e. $t_{ij} \neq t_i$) and flexible nonlinear functions of time such as splines. The two-level models considered so far can also be extended to allow for further levels of clustering which can be hierarchical or non-hierarchical (see Section 13.10 for a brief discussion of cross-classified and multiple-membership structures, for example). On the other hand, the SEM framework includes models with latent responses or predictors measured by multiple indicators where the latent variables may be continuous or categorical. Suppose, for example, that we specify physical functioning as a latent variable measured by an individual's responses to a set of items rather than the sum of those items. We could then fit an SEM comprising 1) a measurement model for the relationship between the observed functioning items and a latent variable for functioning, and 2) a structural model which takes the form of a GCM with the manifest response y_{ij} replaced by latent functioning. The SEM framework also includes mixture models where individuals are grouped into latent classes based on their response patterns over time; these models are the subject of the next section. For further details on the correspondence between multilevel and structural equation modelling approaches to growth curve analysis, see Curran (2003).

14.7 Mixture (latent class) models

Mixture models combine the GCMs described above with the latent class analysis (LCA) approach of Chapter 10. As with LCA, our objective is to identify clusters of individuals with similar response patterns where, in the longitudinal setting, individuals in the same cluster should exhibit similar growth trajectories (patterns in their responses over time $y_{1j}, \ldots, y_{Tj}$). We consider two types of mixture model: latent class growth analysis and generalisations known as growth mixture models. Both approaches stem from finite mixture models (Heckman and Singer 1984) to approximate the distribution of unobserved individual differences in growth curves by a finite number of latent sub-populations. In the case of continuous responses, these models are extensions of the Gaussian mixture models of Section 10.2, but they can be generalised to other types of response (e.g. binary, ordinal and counts).

We assume that individuals are classified on a single latent categorical variable and that each individual belongs to one of C latent classes. In Chapter 10 the number of classes was denoted by J and j was a class index. We change notation here in order to retain j as an individual index. As in LCA, we denote by η_c the prior probability of belonging to class c for $c = 1, \ldots, C$.

Latent class growth analysis

The simplest type of mixture model combines LCA with a special case of the GCMs considered above that excludes individual random effects. Latent class growth analysis (LCGA), also known as group-based trajectory modelling, was first proposed by Nagin and colleagues for the study of criminal careers (e.g. Nagin and Land 1993). Assuming fixed measurement occasions, $t_{ij} = t_i$, and linear growth, the trajectory for an individual in class c is given by

$$y_{ij} \mid c = \beta_{0c} + \beta_{1c} t_i + \epsilon_{ij}, \tag{14.10}$$

where $\epsilon_{ij} \sim$ iid $N(0, \sigma_\epsilon^2)$. More generally, the variance of the residuals ϵ_{ij} may also be class-specific. Based on estimates of the model parameters, β_{0c}, β_{1c}, σ_ϵ^2 and η_c, we can obtain estimates of the posterior class membership probabilities

$$\Pr(\text{individual } j \text{ is in class } c \mid y_{1j}, \ldots, y_{Tj}),$$

which can be used to allocate individuals to classes. For example, modal class allocation assigns an individual to the class with the highest posterior probability.

The lack of random effects implies that individuals in the same class have the same intercept and growth rate: all between-subject variation in the intercept and slope of t_i is captured by the latent classes.

Growth mixture models

A growth mixture model (GMM) (Muthén and Shedden 1999) extends model (14.10) to include individual random effects, allowing for between-individual variance about an average class-specific growth curve. A linear random slopes model has the form

$$y_{ij} \mid c = \beta_{0c} + \beta_{1c} t_i + u_{0c,j} + u_{1c,j} t_i + e_{ij} \tag{14.11}$$

where in class c the individual intercept and slope random effects, $u_{0c,j}$ and $u_{1c,j}$, are assumed to follow a bivariate normal distribution with zero mean, variances σ_{u0c}^2 and σ_{u1c}^2 respectively, and covariance σ_{u01c}. The residuals e_{ij} are also assumed to be normally distributed, possibly with class-specific variances.

While the linear random slope GCM (14.2) assumes that individuals are drawn from a single population with a common covariance structure, the more general model (14.11) allows the between-individual variances at each occasion, and therefore within-individual correlations, to vary across classes. A GCM is a special case of the GMM with $C = 1$, while a LCGA with C classes is a special case of a GMM with the same number of classes and the random effect variances constrained to zero.

Considerations in the application of mixture models

As for the Gaussian mixture model for multivariate responses (Section 10.2), the likelihood function of a LCGA model or GMM could have several local maxima, and the fitting algorithm may find one of these rather than the global maximum. It is therefore advisable to run the algorithm multiple times using different starting values and, if different solutions are obtained, choose the solution that gives the largest value of the likelihood function. There is a greater risk of finding a local maximum when the number of latent classes is large. This is one of the factors to consider when choosing the number of classes; other factors are goodness of fit (using measures such as the BIC and AIC, as illustrated in the example below), class size (avoiding small classes), theoretical justification and interpretability. It is important to bear in mind that the number and interpretation of classes will often depend on the number of repeated measurements available and the length of time between occasions.

In practice, it is advisable to compare the results from applying GCM, GMM and LCGA. In many applications, a GCM with normally distributed random effects for the intercept and coefficient of time (and possibly nonlinear functions of time) will adequately capture heterogeneity in individual trajectories, but the existence of latent sub-populations can be explored using LCGA and GMM. Jung and Wickrama (2008) recommend using LCGA to identify latent classes prior to introducing class-specific random effects in a GMM. If the research objective is to investigate whether class-specific trajectories are consistent with substantive theory about the number and nature of classes, it is recommended to compare the classes obtained using both approaches. GMM will typically require fewer classes to capture individual variation in trajectories, but it is desirable that the two approaches give broadly consistent results if the classes are to be used for interpretation. While GMM relaxes the assumption of LCGA that individuals in each class follow the same trajectory, this is at the cost of stronger parametric assumptions. Moreover, Bauer and Curran (2003) show that non-normality of the repeated measures can lead to a two-class GMM solution even when data are generated from a single population (i.e. a GCM).

The use of mixture models is not restricted to the discovery of distinct, interpretable trajectory classes. In common with finite mixture modelling, the original goal of LCGA was to approximate the distribution of unobserved individual differences using a categorical latent variable rather than continuous (usually normally distributed) latent variable(s). Mixture models are especially useful in the analysis of binary repeated measures where the population contains latent sub-groups of individuals whose response will never change, leading to all-zero and/or all-one response patterns; such individuals are often called "stayers", while those who have a non-zero probability of change are called "movers". Singer and Spilerman (1976) proposed the use of a "spiked" distribution to accommodate stayers, combining a parameteric distribution with "spikes" (or endpoints) at plus and/or minus infinity (depending on whether one or both endpoints are needed). Berridge and Crouchley (2011, Chapter 13) discuss a mover-stayer model for binary responses which can be viewed as a three-class hybrid of LCGA and GMM. Two of the classes capture stayers, with no between-individual variation as in a LCGA, while the third class captures movers with between-individual heterogeneity represented by a normally distributed random effect. See Marquart and Hayes (2019) for an overview of methods for the analysis of longitudinal binary data with latent sub-populations.

Further discussion of LCGA and GMM for longitudinal data, with applications and a comparison with GCMs, can be found in Herle et al. (2020). See also Nagin and Odgers (2010) for a non-technical overview of LCGA and GMM, including their different origins and an annotated bibliography.

Example: cigarette smoking

We illustrate the application of LCGA and GMM using annual panel data on cigarette smoking in Britain 1991–2002. We analyse data on a subset of 1112 individuals who reported smoking in 1991 and participated in the study at all 11 subsequent waves. The response variable is the number of cigarettes usually smoked per day which ranges from 0 to 70 with a mean of 13.7 and standard deviation of 9.9, and an intra-individual correlation of 0.645.

We begin with LCGA and consider a generalisation of the linear model (14.10) with quadratic calendar year effects (where year is coded $0, 1, \ldots, 11$). Table 14.6 shows goodness-of-fit statistics and the prior class membership probabilities for 1–6 class quadratic models. The AIC and BIC are smallest for the 6-class model, suggesting that more classes are needed to capture between-individual variation in smoking trajectories. However, class 2 of the 6-class model is very small with an estimated 12 individuals. We therefore focus on the interpretation of the 5-class model where the smallest class (class 4) has an estimated 91 individuals. Figure 14.5 shows the predicted trajectories from the 5-class model. Classes 1, 3 and 5 (accounting for an estimated 63% of individuals) differ in their intercepts, with mean consumptions of approximately 27, 19 and 12 cigarettes per day in 1991, but show little change over time. Class 2 (26%) has the lowest initial consumption of approximately 9 cigarettes per day which decreases slowly over time, while class 4 (8%) has a similar initial consumption as class 2 but decreases rapidly to zero by the end of the study period.

TABLE 14.6: Results from latent class growth analysis of smoking with quadratic effects of calendar time, number of classes $C = 1$ to 6

				$\hat{\eta}_c = \hat{\Pr}(\text{class} = c)$ for class c					
C	$-\log L$	AIC	BIC	1	2	3	4	5	6
1	49468.4	98944.8	98974.8	1.00	–	–	–	–	–
2	46078.5	92173.0	92233.0	0.54	0.46	–	–	–	–
3	44904.2	89832.4	89922.4	0.59	0.16	0.35	–	–	–
4	44354.6	88741.2	88861.2	0.30	0.24	0.34	0.12	–	–
5	44020.2	88080.5	88230.4	0.11	0.26	0.29	0.08	0.26	–
6	43754.9	87557.8	87737.8	0.25	0.01	0.30	0.24	0.08	0.12

We next consider GMMs which extend LCGA models to allow for individual variation in smoking trajectories within latent classes. As in LCGA, we specify a quadratic trajectory for each class, but now include individual random effects on the intercept and linear slope of year. The fitted models are a generalisation of the linear model (14.11) with quadratic calendar year effect assumed to be fixed across individuals. The residual (within-individual)

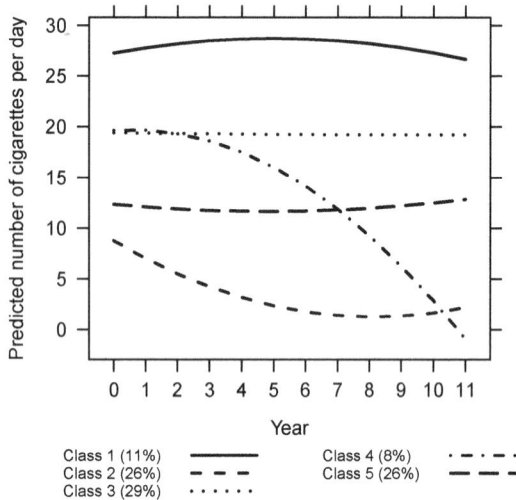

FIGURE 14.5: Predicted smoking trajectories from 5-class quadratic LCGA model

variance σ_e^2 is assumed equal across classes. Convergence problems were encountered when a random coefficient was added to the age-squared term. Table 14.7 shows goodness-of-fit statistics and the prior class membership probabilities for quadratic random slope models with $C = 1, \ldots, 4$ latent classes. Random intercept models were also considered but for each C the effect of year was found to vary significantly between individuals. The 1-class model is a standard growth model where the individual random effects are assumed to be drawn from a single population, but both the AIC and BIC statistics indicate that individuals can be grouped into latent sub-populations with different random effects variances and covariances. Moreover, from a comparison of the goodness of fit statistics with those for the LCGA models (Table 14.6), we find that the addition of class-specific random effects leads to a substantial improvement in model fit. For example, the BIC for the selected 5-class LCGA model is 88,230, compared with 86,981 for the 1-class GMM. Among the models considered, the 4-class model provides the best fit. A 5-class model was also fitted, but one of the classes was found to be very small. We therefore illustrate interpretation of a GMM using the 4-class model.

Figure 14.6 shows the predicted trajectories from the 4-class model. These represent the average trajectory for each class, based on the fixed part parameter estimates in Table 14.8.

There is variation about the intercept and linear slope of year of these average trajectories, quantified by the estimates of the between-individual

TABLE 14.7: Results from random slope growth mixture models of smoking with quadratic effects of calendar time, number of classes $C = 1$ to 4

				$\hat{\eta}_c = \hat{\Pr}(\text{class} = c)$ for class c			
C	$-\log L$	AIC	BIC	1	2	3	4
1	43457.3	86928.6	86981.1	1.00	–	–	–
2	43190.8	86409.6	86514.5	0.14	0.86	–	–
3	43023.3	86088.5	86246.0	0.13	0.14	0.74	–
4	42896.0	85848.0	86058.0	0.11	0.10	0.14	0.65

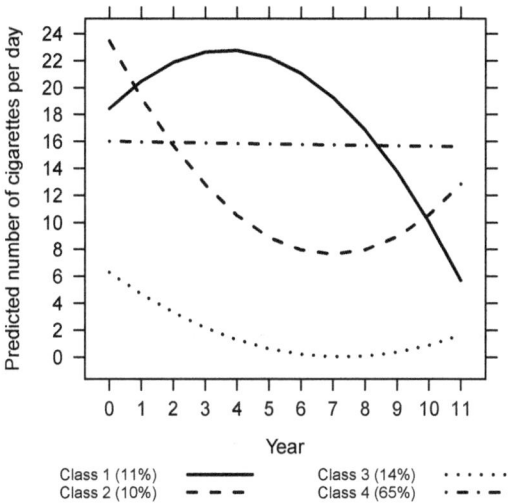

FIGURE 14.6: Predicted average smoking trajectories from 4-class quadratic random slope growth mixture model

intercept and slope variances in Table 14.8. In class 1 the average trajectory shows an increase in consumption from 1991 to 1995, followed by a sharp decline. However, there is substantial variation between individuals about the average. The 95% plausible values range is calculated as $18.44 \pm (1.96 \times \sqrt{65.69}) = (2.6, 34.3)$ for the intercept (consumption in 1991) and $2.36 \pm (1.96 \times \sqrt{1.24}) = (0.18, 4.54)$ for the linear slope of age. Class 2 has the highest average consumption in 1991, which declines and then increases after 1998, but again there is considerable individual variation about this trend. Individuals in class 3 have the lowest average consumption in each year, with little between-individual variation in either the intercept or change over time. The average trajectory in this class is similar to that of class 2 from the 5-class LCGA model

TABLE 14.8: Parameter estimates from 4-class growth mixture model of smoking with quadratic effect of calendar time

		Estimates for class c			
Parameter		1	2	3	4
Fixed part					
Intercept	β_0	18.44	23.48	6.27	16.01
Year slope	β_1	2.36	-4.54	-1.72	-0.04
Year^2 slope	β_2	-0.32	0.32	0.12	0.001
Random part					
Intercept	σ^2_{u0c}	65.69	71.30	1.02	37.10
Year slope	σ^2_{u1c}	1.24	1.10	0.01	0.22
Intercept-slope covariance	σ_{u01c}	-3.90	-3.05	-0.10	-0.29
Residual*	σ^2_e	23.38	23.38	23.38	23.38

*The residual (within-individual) variance is constrained to be equal across classes

(Figure 14.5). In class 4, the largest group, average consumption is constant over time at a value between that of classes 3 and 5 in the LCGA model, but there is again a large amount of between-individual variation.

14.8   Missing data

Missing data arises in most studies and Section 6.16 of the Regression Analysis chapter gave a brief overview of the assumptions about missing data mechanisms underlying different analysis approaches. As missing data is a particular challenge in longitudinal studies, we now discuss the topic in the longitudinal setting. The most common form of nonresponse in longitudinal studies is drop-out (or attrition) where an individual does not participate after a certain measurement occasion, but nonresponse can also be intermittent. In this section, we give a brief overview of different assumptions about the relationships between the probability of nonresponse and the (observed and missing) responses and covariates, and how these relate to different approaches to handling missing data in longitudinal data analysis. The analysis of incomplete data due to nonresponse is an active area of methodological research and there are many excellent texts and tutorial articles where readers can find details of the methods, applications and practical guidance. A small selection of these that are specific to the longitudinal setting are Hedeker and Gibbons (2006, Chapter 14), Hogan, Roy, and Korkontzelou (2004) and Verbeke and Molenberghs (2000, Chapters 14–21).

While nonresponse leads to a reduction in sample size, of greater concern is the potential for bias due to systematic differences between individuals who drop out and those who remain in the study. When deciding how to handle missing data, it is important to understand the assumptions about the underlying nonresponse process made by different approaches. These assumptions are usually framed in terms of the taxonomy of missing data mechanisms proposed by Rubin in the 1970s (Little and Rubin 2019) which has since been adapted to longitudinal studies (see, for example, Hedeker and Gibbons 2006, Chapter 14). Suppose that for each individual we partition their planned or complete response vector $\mathbf{Y}$ into its observed and missing components denoted by $\mathbf{Y}_{obs}$ and $\mathbf{Y}_{mis}$, and denote by $\mathbf{R}$ a vector of binary indicators of whether an individual responds at each occasion, and $\mathbf{X}$ a matrix of covariates which, for simplicity, we assume to be fully observed. The strongest assumption, referred to as "missing completely at random" (MCAR), is that $\mathbf{R}$ does not depend on $\mathbf{Y}$ or covariates $\mathbf{X}$, which implies that $\mathbf{Y}_{mis}$ is a random sample of $\mathbf{Y}$. A more plausible variant of MCAR, called covariate-dependent MCAR, allows $\mathbf{R}$ to depend on $\mathbf{X}$. A weaker assumption, referred to "missing at random" (MAR), is that $\mathbf{R}$ may depend on the observed values of the response variable $\mathbf{Y}_{obs}$ and $\mathbf{X}$ but, conditional on $(\mathbf{Y}_{obs}, \mathbf{X})$, not the missing responses $\mathbf{Y}_{mis}$. Suppose, for example, that $\mathbf{Y}$ are repeated measures of test scores. Under MAR, the probability that a student sat the test at occasion i may depend on their observed scores on tests taken at other occasions, but not the score they would have achieved at i had they taken the test; for instance, a student with a history of low scores (captured by $\mathbf{Y}_{obs}$) may be less likely to sit the test at a later occasion. In sequential MAR, a version of MAR specific to the longitudinal setting, the probability of response at occasion i may depend on observed values of Y and X up to i (e.g. Hogan, Roy, and Korkontzelou 2004). However, the most likely nonresponse mechanism, termed "missing not at random" (MNAR), is that $\mathbf{R}$ depends on $\mathbf{Y}_{mis}$, even after conditioning on $(\mathbf{Y}_{obs}, \mathbf{X})$. For example, a student may fail to take the test at occasion i for unmeasured reasons such as lack of preparation which are highly likely to be related to the unobserved test score at i.

One treatment of missing data is to restrict analysis to individuals with complete data, i.e. with $\mathbf{Y} = \mathbf{Y}_{obs}$, but this is valid only under MCAR (or covariate-dependent MCAR if the model for $\mathbf{Y}_{obs}$ conditions on $\mathbf{X}$). A more reasonable strategy, which uses all available data, is to retain the observed responses for individuals with missing data, as in our analysis of physical functioning above. This approach is valid under a MAR assumption and in multilevel modelling, where the data are in long form, is implemented by simply deleting records for occasions with missing data. In structural equation modelling of longitudinal data, where data are in wide form with one record per individual, missing data are usually handled using 'full-information' maximum likelihood estimation which also makes a MAR assumption (this approach was referred to as simply 'maximum likelihood' in Section 7.11). An alternative approach, also valid under MAR, is weighting. An advantage of weighting is that weights may be constructed to adjust not only for drop-out but for nonresponse at the first occasion and for unequal selection probabilities at any occasion, for example due to over-sampling of rare sub-groups. Longitudinal weights are often available in longitudinal survey datasets and typically incorporate auxiliary information available only to the data provider.

A variety of approaches have been proposed for the most likely, and more challenging, MNAR scenario. These include selection models, pattern mixture models and shared parameter (frailty) models, and involve specifying a joint model for $\mathbf{Y}$ and $\mathbf{R}$. We give a brief outline of shared parameter models because they can be fitted in a multilevel or structural equation modelling framework. Suppose that the model of interest for $\mathbf{Y}$ is the linear random slope model (14.2). Let $R_{ij} = 1$ if Y_{ij} is observed and $R_{ij} = 0$ if Y_{ij} is missing, and define $\pi_{ij} = \Pr(R_{ij} = 1)$. A model for π_{ij} can be specified as

$$\log\left(\frac{\pi_{ij}}{1 - \pi_{ij}}\right) = \alpha_0 + \alpha_1 u_{0j} + \alpha_2 u_{1j}, \tag{14.12}$$

where, as before, u_{0j} and u_{1j} are the random intercept and random slope of t_{ij}. Eq.s (14.2) and (14.12) must be estimated jointly because of their shared dependence on u_{0j} and u_{1j}. This model assumes that the association between $\mathbf{R}$ and $\mathbf{Y}$, including $\mathbf{Y}_{mis}$, is due to unmeasured time-invariant individual characteristics that influence both $\mathbf{Y}$ and $\mathbf{R}$. Under the model assumptions, a test of $H_0 : \alpha_1 = \alpha_2 = 0$ provides a test of MNAR. However, this and other joint models depend on strong, untestable parametric assumptions. It is therefore recommended to compare results from different approaches and to conduct sensitivity analysis (e.g. Verbeke and Molenberghs 2000, Chapters 19 and 20).

14.9   Additional examples and further work

Cigarette consumption in US states

So far, we have considered examples where the subjects that are repeatedly measured are individuals. We now illustrate the application of growth curve models and mixture models in an example where the subjects are geographical areas (US states). The data (an example from Baum (2006)) are annual statistics on the demand for cigarettes in 48 states for 1985–95. Demand is measured by the average number of packs of cigarettes (**packpc**) consumed per capita which has a mean of 106.5 and standard deviation of 23.1. For each state, data were available for all 11 years.

We begin by fitting growth curve models to these data, including year as the time variable t_i (coded as $0, 1, \ldots, 10$). Table 14.9 shows the results from a sequence of likelihood ratio tests, gradually building from a random intercept model (M1) to the quadratic random slope model (M4) of (14.7), with random coefficients on t_i and t_i^2. The test statistics comparing each model with the previous (nested) model are all statistically significant and we select M4 as our final model.

TABLE 14.9: Results from likelihood ratio tests in building a growth curve model of cigarette consumption in US states

	$-\log$-like	$2\Delta\log$-like	d.f.
M1: Linear random intercept	1782.4	–	–
M2: Linear random slope	1700.6	163.6	2
M3: Quadratic r.s. (fixed effect for t^2)	1667.6	65.8	1
M4: Quadratic r.s. (random effect for t^2)	1639.9	55.4	3

The fitted equation of the average quadratic trajectory, representing the expected cigarette consumption in year i for a state at the mean of the distribution for each random effect $(u_{0j} = u_{1j} = u_{2j} = 0)$, is

$$\widehat{\text{packpc}}_{ij} = 123.6 - 4.57\,t_i + 0.16\,t_i^2 \qquad (14.13)$$

which implies that consumption declined after 1985, but the decline slowed down over time. There is substantial variation between states about this average curve. The estimates of the between-state variances in the intercept, coefficient of t and coefficient of t^2 are respectively 448.7, 4.6 and 0.04. These estimates can be used to calculate 95% plausible values ranges for the intercept and coefficients of t and t^2. For example, in 95% of states we would expect the consumption in 1985 (the intercept) to lie in the interval $123.6 \pm 1.96\sqrt{448.7} = (82.1, 165.1)$ and the linear change in consumption thereafter to lie in $-4.57 \pm 1.96\sqrt{4.6} = (-8.8, -0.4)$. The expression for the between-subject variance implied by a random slope model (14.3) can be extended to models with more than two random effects. For the quadratic growth model, the between-subject variance is given by $\text{var}(y_{ij}|t_{ij}) = \text{var}(u_{0j} + u_{1j}t_{ij} + u_{2j}t_{ij}^2)$ which turns out to be quartic function in t_{ij}, where the coefficients are the random effect variances and covariances. Figure 14.7 plots the estimated between-state variance in cigarette consumption by year from model M4. The variance decreases from 1985 to 1989 ($t_i = 0$ to $t_i = 4$) and then increases sharply thereafter. Although average consumption has declined over the period, the variation between states has increased.

We next analyse these data using LCGA. We start with a two-class model which allocates 26 states to class 1 and the remaining 22 states to class 2. A three-class model gives a very small class with only two states, so we focus on the two-class solution. The predicted consumption trajectories in each class are given by:

$$\text{class 1:} \quad \widehat{\text{packpc}}_{ij} = 111.2 - 5.11\,t_i + 0.20\,t_i^2,$$
$$\text{class 2:} \quad \widehat{\text{packpc}}_{ij} = 138.4 - 3.92\,t_i + 0.13\,t_i^2.$$

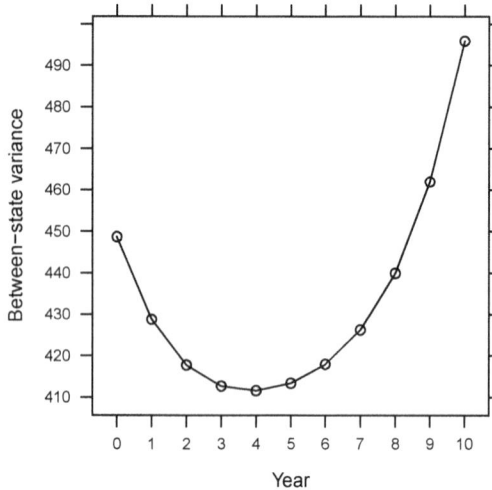

FIGURE 14.7: Estimated between-state variance in number of packs of cigarettes per capita from quadratic random slope model (M4)

As the trajectories are differentiated mainly by their intercepts, the classes can be interpreted as 'low' and 'high' consumption groupings of states, although the 'low' class also experiences a slightly quicker decline in consumption over time. We also note that taking the average of each of the three parameter estimates across the two classes gives values close to the parameter estimates of the average trajectory from the GCM (14.13); this is expected because the two classes are of roughly equal size. In this case, LCGA is of limited value because the small number of states limits the number of classes that can be extracted and we are forced to represent between-state variation by only two classes. In further analysis, you could fit a two-class growth mixture model to investigate whether the between-state variances in the intercept and coefficients of t and t^2 differ between classes.

Health satisfaction in Germany

We illustrate the application of growth curve models with autocorrelated residuals using data from the German Socio-Economic Panel, analysed in Snijders and Bosker (2012, Chapter 5). We analyse annual data on respondents who were age 55 at any time in the period 1984–2006, and follow them up to age 60 (or the year 2006). The dataset contains 5934 observations from 1146 individuals, so the mean number of observations per individual is 5.2.

The response variable is a self-reported measure of an individual's satisfaction with the current state of their health, coded on a 11-point scale from 0 ('completely dissatisfied') to 10 ('completely satisfied') with mean 6.0 and standard deviation 2.3. Linear GCMs are fitted to these data with age (centred at 55) as the time variable. From a random slope model, the estimate of the average slope of age is $\hat{\beta}_1 = -0.026$ (s.e. = 0.014) which is not significant at the 5% level. However, there is significant between-individual variance in the age slope ($\hat{\sigma}_u^2 = 0.071$) with a 95% plausible values range of $(-0.55, 0.50)$. The correlation between the intercept and slope random effects is estimated as -0.38 which, taken with the negative estimate of β_1, implies that individuals with higher satisfaction at age 55 ($u_{0j} > 0$) tend to have larger negative slope effects ($u_{1j} < 0$) and therefore a faster decline in satisfaction with age ($\beta_1 + u_{1j} < 0$).

We next relax the assumption that the time-varying residuals are independent, extending the random slope model by specifying an AR(1) structure for e_{ij}, i.e. $\mathrm{cor}(e_{ij}, e_{i'j}) = \rho^{|i-i'|}$. The autocorrelation parameter is small ($\hat{\rho} = 0.071$) but statistically significant (the likelihood ratio statistic for a test of the null hypothesis that $\rho = 0$ is 7.5 on 1 degree of freedom). In this case, within-individual correlation in health satisfaction is adequately captured by the intercept and slope random effects. However, if we add AR(1) residuals to a random intercept model, the estimate of the residual autocorrelation increases to 0.163 and the likelihood ratio test statistic increases to 67.5. The random intercept model makes the unrealistic assumption that $\mathrm{cor}(y_{ij}, y_{i'j})$ does not depend on the length of time between occasions i and i', which is relaxed in the AR(1) model.

14.10   Software

Growth curve models can be fitted as multilevel models using any of the software options discussed in Chapter 13 or as factor models using software for structural equation modelling (see Chapter 11). Latent class growth analysis (LCGA) is implemented in SAS or Stata using the `traj` packages (Jones and Nagin 2007, Jones and Nagin 2013), and both LCGA and growth mixture models are implemented in various R packages such as `flexmix` (Grün and Leisch 2025). Mplus (Muthén and Muthén 2017) is the most general purpose software which can fit all models discussed in this chapter. Finally, while this chapter has focused on models for continuous responses, all models discussed can be generalised to other types of response (including binary, ordinal and count data) and the software mentioned above can handle these extensions.

14.11   Further reading

We recommend the following textbooks on longitudinal data analysis.

Hedeker, D. and Gibbons, R.D. (2006). *Longitudinal Data Analysis*. London: Wiley.

Rabe-Hesketh, S. and Skrondal, A. (2021). *Multilevel and Longitudinal Modelling using Stata, Volume I* (4th ed). College Station, TX: Stata Press.

Skrondal, A. and Rabe-Hesketh, S. (2004). *Generalized Latent Variable Modelling: Multilevel, Longitudinal, and Structural Equation Models*. Boca Raton, FL: Chapman and Hall/CRC.

Verbeke, G. and Molenberghs, G. (2000). *Linear Mixed Models for Longitudinal Data*. New York: Springer.

References

Agresti, A. (2007). *Introduction to Categorical Data Analysis* (2nd ed.). New York: Wiley.

Agresti, A. and B. Finlay (2008). *Statistical Methods for the Social Sciences* (4th ed.). Englewood Cliffs, NJ: Prentice Hall.

American Psychiatric Association (2013). *Diagnostic and Statistical Manual of Mental Disorders: 5th ed.* Arlington, VA: American Psychiatric Association.

Anderson, S. B. and M. H. Maier (1963). 34000 pupils and how they grew. *Journal of Teacher Education 14*, 212–216.

Arbuckle, J. L. (2006). *Amos 7.0 User's Guide*. Chicago, IL: SPSS Inc.

Atkinson, A. C. (1987). *Plots, Transformations and Regression*. Oxford: Clarendon Press.

Barnett, D., J. Galbraith, C. Roaf, and S. Rutherford (1999). Can secondary school teachers improve literacy and achievement by helping students develop fluent legible handwriting? Technical report, Report to the Teaching Training Agency, UK.

Bartholomew, D. J. (1998). Scaling unobservable constructs in social science. *Applied Statistics 47*, 1–13.

Bartholomew, D. J., M. Knott, and I. Moustaki (2011). *Latent Variable Models and Factor Analysis: A Unified Approach* (3rd ed.). Wiley Series in Probability and Mathematical Statistics. John Wiley & Sons.

Bartholomew, D. J. and S. O. Leung (2002). A goodness-of-fit test for sparse 2^p contingency tables. *British Journal of Mathematical and Statistical Psychology 55*, 1–15.

Bartholomew, D. J. and P. Tzamourani (1999). The goodness-of-fit of latent trait models in attitude measurement. *Sociological Methods and Research 27*, 525–546.

Basilevsky, A. (1994). *Statistical Factor Analysis and Related Methods. Theory and Applications*. Wiley Series in Probability and Mathematical Statistics. New York: John Wiley & Sons.

Bastian, M., S. Heymann, and M. Jacomy (2009). Gephi: an open source software for exploring and manipulating networks. In *Proceedings of the International AAAI Conference on Web and Social Media*, Volume 3, pp. 361–362.

Bates, B., M. Mächler, B. Bolker, and S. Walker (2015). Fitting linear mixed-effects models using lme4. *Journal of Statistical Software 67*(1), 1–48.

Bauer, D. J. and P. J. Curran (2003). Distributional assumptions of growth mixture models: Implications for overextraction of latent trajectory classes. *Psychological Methods 8*, 338–363.

Baum, C. F. (2006). *An Introduction to Modern Econometrics Using Stata*. College Station, TX: Stata Press.

Bell, A. (Ed.) (2021). *Age, Period and Cohort Effects: Statistical Analysis and the Identification Problem*. Abingdon: Routledge.

Bentler, P. M. (2008). *EQS 6 Structural Equations Program Manual.* Encino, CA: Multivariate Software.

Berridge, D. M. and R. Crouchley (2011). *Multivariate Generalized Linear Mixed Models Using R.* Boca Raton, FL: CRC Press.

Bock, R. D. and M. Lieberman (1970). Fitting a response model for n dichotomously scored items. *Psychometrika 35*(2), 179–197.

Bock, R. D. and I. Moustaki (2007). Item response theory in a general framework. In C. Rao and S. Sinharay (Eds.), *Handbook of Statistics, Vol. 26.* Elsevier.

Bollen, K. A. (1989). *Structural Equations with Latent Variables.* New York: Wiley and Sons.

Borg, I. and P. J. F. Groenen (2005). *Modern Multidimensional Scaling.* New York: Springer-Verlag.

Brook, L., B. Taylor, and G. Prior (1991). *British Social Attitudes, 1990, Survey.* London: SCPR.

Browne, M. W. (1984). Asymptotic distribution free methods in analysis of covariance structures. *British Journal of Mathematical and Statistical Psychology 37*, 62–83.

Browne, M. W. (2001). An overview of analytic rotation in exploratory factor analysis. *Multivariate Behavioral Research 36*, 111–150.

Browne, M. W. and R. Cudek (1993). Alternative ways of assessing model fit. In K. A. Bollen and J. S. Long (Eds.), *Testing Structural Equation Models.* Newbury Park, CA: Sage Publications.

Browne, W. J. (2023). *MCMC Estimation in MLwiN, v3.07.* Bristol, UK: Centre for Multilevel Modelling, University of Bristol.

Browne, W. J., H. Goldstein, and J. Rasbash (2001). Multiple membership multiple classification (MMMC) models. *Statistical Modelling 1*, 103–124.

Cai, L., A. Maydeu-Olivares, D. Coffman, and D. Thissen (2006). Limited information goodness-of-fit testing of item response theory models for sparse 2^p tables. *British Journal of Mathematical and Statistical Psychology 59*, 173–194.

Carpenter, J. R. and M. G. Kenward (2013). *Multiple Imputation and its Application.* Chichester: Wiley.

Carton, A., M. Swyngedouw, J. Billiet, and R. Beerten (1993). *Source Book of the Voter's Study in Connection with the 1991 General Election.* Katholieke Universiteit Leuven: Sociologisch Onderzoeksinstituut.

CBSI (1998). Central Bureau of Statistics [Indonesia] and State Ministry of Population/National Family Planning Coordinating Board (NFPCB) and Ministry of Health (MOH) and Macro International Inc. (MI), *Indonesia Demographic and Health Survey 1997.* Calverton, Mayland: CBS and MI.

Chalmers, R. P. (2012). mirt: A Multidimensional Item Response Theory Package for the R Environment. *Journal of Statistical Software 48*(6), 1–29.

Charlton, C., J. Rasbash, W. J. Browne, M. Healy, and B. Cameron (2024). *MLwiN Version 3.12.* Bristol, UK: Centre for Multilevel Modelling, University of Bristol.

Chassagnol, B., A. Bichat, C. Boudjeniba, P.-H. Wuillemin, M. Guedj, D. Gohel, G. Nuel, and E. Becht (2023). Gaussian mixture models in R. *The R Journal 15*, 56–76. https://doi.org/10.32614/RJ-2023-043.

Chen, Y., I. Moustaki, and S. Zhang (2023). On the estimation of structural equation models with latent variables. In R. H. Hoyle (Ed.), *Handbook of Structural Equation Modeling* (2nd ed.). New York: Guilford Press.

Cheng, J., E. Levina, P. Wang, and J. Zhu (2014). A sparse Ising model with covariates. *Biometrics 70*(4), 943–953.

Claeskens, G. (2016). Statistical model choice. *Annual Review of Statistics and Its Application 3*(1), 233–256.

Clausen, S. E. (1998). *Applied Correspondence Analysis*. Thousand Oaks, CA: Sage Publications.

Collins, L. M., B. P. Flaherty, S. L. Hyatt, and J. L. Schafer (1999). *WinLTA 3.1, User's Guide*. The Pennsylvania State University.

Conrad, R. (1964). Acoustic confusions in immediate memory. *British Journal of Psychology 55*, 75–84.

Copas, J. and P. Marshall (1998). The offender group reconviction scale: a statistical reconviction score for use by probation officers. *Journal of the Royal Statistical Society, Series C 47*, 159–171.

Cortez, P. and A. M. G. Silva (2008). Using data mining to predict secondary school student performance. In *Proceedings of 5th Annual Future Business Technology Conference, EUROSIS-ETI, Porto, Portugal, 9–11 April 2008*, 5–12.

Cox, T. F. and M. A. A. Cox (2001). *Multidimensional Scaling* (2nd ed.). London: Chapman and Hall/CRC.

Csárdi, G., T. Nepusz, V. Traag, S. Horvát, F. Zanini, D. Noom, and K. Müller (2025). *igraph: Network Analysis and Visualization in R*. R package version 2.1.4.

Curran, P. J. (2003). Have multilevel models been structural equation models all along? *Multivariate Behavioral Research 38*(4), 529–569.

de Menezes, L. M. and D. J. Bartholomew (1996). New developments in latent structure analysis applied to social attitudes. *Journal of the Royal Statistical Society, A 159*, 213–224.

Diggle, P. J., P. J. Heagerty, K.-Y. Liang, and S. L. Zeger (2002). *Analysis of Longitudinal Data*. New York: Oxford University Press Inc.

Duncan, O. D., A. O. Haller, and A. Portes (1971). Peer influences on aspirations: a reinterpretation. In H. M. Blalock (Ed.), *Causal Models in the Social Sciences*. Chicago: Aldine.

Edwards, D. (2000). *Introduction to Graphical Modelling*. New York: Springer Science & Business Media.

Ehrenberg, A. S. C. (1977). Rudiments of numeracy. *Journal of the Royal Statistical Society, Series A 140*, 277–297.

Ekman, G. (1954). Dimensions of color vision. *Journal of Psychology 38*, 467–474.

Enders, C. K., T. Hayes, and H. Du (2018). A comparison of multilevel imputation schemes for random coefficient models: Fully conditional specification and joint model imputation with random covariance matrices. *Multivariate Behavioral Research 53*, 695–713.

Epskamp, S. (2025a). *IsingSampler: Sampling Methods and Distribution Functions for the Ising Model*. R package version 0.2.4.

Epskamp, S. (2025b). *Psychonetrics: Structural Equation Modeling and Confirmatory Network Analysis*. R package version 0.13.1.

Epskamp, S., A. O. J. Cramer, L. J. Waldorp, V. D. Schmittmann, and D. Borsboom (2012). qgraph: Network visualizations of relationships in psychometric data. *Journal of Statistical Software 48*(4), 1–18.

Epskamp, S. and E. I. Fried (2018). A tutorial on regularized partial correlation networks. *Psychological Methods 23*, 617–634.

Everitt, B. S. and G. Dunn (2001). *Applied Multivariate Data Analysis* (2nd ed.). London: Arnold.

Everitt, B. S., S. Landau, and M. Leese (2001). *Cluster Analysis* (4th ed.). London: Arnold.

Everitt, B. S. and S. Rabe-Hesketh (1997). *The Analysis of Proximity Data.* London: Arnold.

Fischer, G. H. and I. W. Molenaar (Eds.) (1995). *Rasch Models: Foundations, Recent Developments, and Applications.* New York: Springer.

Fleishman, E. A. and W. E. Hempel (1954). Changes in factor structure of a complex psychomotor test as a function of practice. *Psychometrika 19*, 239–252.

Fogle, B. M., J. Tsai, N. Mota, I. Harpaz-Rotem, J. H. Krystal, S. M. Southwick, and R. H. Pietrzak (2020). The national health and resilience in veterans study: A narrative review and future directions. *Frontiers in Psychiatry 11*, 1–10.

Friedman, J., T. Hastie, and R. Tibshirani (2008). Sparse inverse covariance estimation with the graphical lasso. *Biostatistics 9*, 432–441.

Friedman, J., T. Hastie, and R. Tibshirani (2019). *Graphical Lasso: Estimation of Gaussian Graphical Models.* R package version 1.11.

Galton, F. (1886). Regression towards mediocrity in hereditary stature. *Journal of the Anthropological Instistute of Great Britain and Ireland 15*, 246–263.

Galton, F. (1890). Kinship and correlation. *North American Review 150*, 419–431.

Gierl, M. J. and W. T. Rogers (1996). A confirmatory factor analysis of the test anxiety inventory using Canadian high school students. *Educational and Psychological Measurement 56*, 315–324.

Glas, C. A. W. and J. L. Pimentel (2008). Modeling nonignorable missing data in speeded tests. *Educational Psychological Measurement 68*, 907–922.

Goldstein, H. (2010). *Multilevel Statistical Models* (4th ed.). London: Wiley.

Goldstein, H., M. J. Healy, and J. Rasbash (1994). Multilevel time series models with applications to repeated measures data. *Statistics in Medicine 13*(16), 1643–1655.

Goodman, L. A. (1979). Simple models for the analysis of association in cross-classifications having ordered categories. *Journal of the American Statistical Association 74*(367), 537–552.

Goodman, L. A. (1981). Association models and canonical correlation in the analysis of cross-classifications having ordered categories. *Journal of the American Statistical Association 76*(374), 320–334.

Goodman, L. A. (1991). Measures, models, and graphical displays in the analysis of cross-classified data. *Journal of the American Statistical Association 86*(416), 1085–1111.

Gordon, A. D. (1999). *Classification* (2nd ed.). London: Chapman and Hall/CRC.

Greenacre, M. and J. Blasius (Eds.) (1994). *Correspondence Analysis in the Social Sciences.* San Diego: Academic Press.

Greenacre, M. and J. Blasius (Eds.) (2006). *Multiple Correspondence Analysis and Related Methods.* London: Chapman and Hall/CRC.

Grün, B. and F. Leisch (2025). *flexmix: Flexible Mixture Modeling.* R package version 2.3-20.

Haberman, S. J. (1978). *Analysis of Qualitative Data. Vol 1: Introductory Topics.* London: Academic Press.

Hagenaars, J. A. and A. L. McCutcheon (2002). *Applied Latent Class Analysis.* Cambridge: Cambridge University Press.

Hambleton, R. K., H. Swaminathan, and H. J. Rogers (1991). *Fundamentals of Item Response Theory*. Newbury Park, California: Sage Publications.

Hand, D. J., F. Daly, A. D. Lunn, K. J. McConway, and E. Ostrowski (1994). *A Handbook of Small Data Sets*. London: Chapman & Hall.

Haslbeck, J. M. B. and L. J. Waldorp (2020). mgm: Estimating time-varying mixed graphical models in high-dimensional data. *Journal of Statistical Software 93*(8), 1–46.

Heath, A., R. Jowell, J. K. Curtice, J. A. Brand, and J. C. Mitchell (1993). *British General Election Study, 1992*. Colchester, Essex: The Data Archive. [Computer file].

Heckman, J. and B. Singer (1984). A method for minimizing the impact of distributional assumptions in econometric models for duration data. *Econometrica 1*, 271–320.

Hedeker, D. and R. D. Gibbons (2006). *Longitudinal Data Analysis*. London: Wiley.

Heinen, T. (1996). *Latent Class and Discrete Latent Trait Models*. Thousand Oaks: Sage Publications.

Herle, M., N. Micali, M. Abdulkadir, R. Loos, R. Bryant-Waugh, C. Hübel, C. M. Bulik, and B. L. De Stavola (2020). Identifying typical trajectories in longitudinal data: modelling strategies and interpretations. *European Journal of Epidemiology 35*, 205–222.

Hilton, T. L. (1969). Growth study annotated bibliography. Progress Report 69-11, Princeton, NJ: Educational Testing Service.

Ho, Y., Y. Chung, and K.-n. Lau (2010). Unfolding large-scale marketing data. *International Journal of Research in Marketing 27*(2), 119–132.

Hoff, P. D., A. E. Raftery, and M. S. Handcock (2002). Latent space approaches to social network analysis. *Journal of the American Statistical Association 97*(460), 1090–1098.

Hogan, J. W., J. Roy, and C. Korkontzelou (2004). Handling drop-out in longitudinal studies. *Statistics in Medicine 23*, 1455–1497.

Højsgaard, S., D. Edwards, and S. Lauritzen (2012). *Graphical Models with R*. New York: Springer Science & Business Media.

Hosmer, D. W. and S. Lemeshow (2000). *Applied Logistic Regression* (2nd ed.). New York: Wiley.

Howe, L. D., K. Tilling, A. Matijasevich, E. S. Petherick, A. C. Santos, L. Fairley, J. Wright, I. S. Santos, A. J. D. Barros, R. M. Martin, M. S. Kramer, N. Bogdanovich, L. Matush, H. Barros, and D. A. Lawlor (2016). Linear spline multilevel models for summarising childhood growth trajectories: A guide to their application using examples from five birth cohorts. *Statistical Methods in Medical Research 25*(5), 1854–1874.

Hox, J. (2002). *Multilevel Analysis, Techniques and Applications*. New Jersey: Lawrence Erlbaum Associates.

Hoyle, R. H. (Ed.) (2022). *Handbook of Structural Equation Modeling* (2nd ed.). Guilford Press.

Huq, N. M. and J. Cleland (1990). *Bangladesh Fertility Survey, 1989*. Dhaka: National Institute of Population Research and Training (NIPORT).

Ising, E. (1925). A contribution to the theory of ferromagnetism. *Zeitschrift für Physik 31*, 253–258.

Isvoranu, A.-M., S. Epskamp, L. J. Waldorp, and D. Borsboom (2022). *Network Psychometrics with R: A Guide for Behavioral and Social Scientists.* Abingdon, Oxon: Taylor & Francis.

James, G., D. Witten, T. Hastie, and R. Tibshirani (2021). *An Introduction to Statistical Learning: with Applications in R.* New York: Springer.

Jardine, N. and R. Sibson (1971). *Mathematical Taxonomy.* New York: Wiley.

Johnson, J. A. (2014). Measuring thirty facets of the five factor model with a 120-item public domain inventory: Development of the IPIP-NEO-120. *Journal of Research in Personality 51*, 78–89.

Jolliffe, I. T. (1972). Discarding variables in a principal components analysis. I: artificial data. *Applied Statistics 21*, 160–173.

Jolliffe, I. T. (1986). *Principal Component Analysis.* New York: Springer-Verlag.

Jones, B. L. and D. S. Nagin (2007). Advances in group-based trajectory modeling and a SAS procedure for estimating them. *Sociological Methods & Research 35*(4), 542–571.

Jones, B. L. and D. S. Nagin (2013). A note on a Stata plugin for estimating group-based trajectory models. *Sociological Methods & Research 42*(4), 608–613.

Jöreskog, K. G. (1969). A general approach to confirmatory maximum likelihood factor analysis. *Psychometrika 34*, 183–202.

Jöreskog, K. G. (1970). A general method for analysis of covariance structures. *Biometrika 57*, 239–251.

Jöreskog, K. G. (1994). On the estimation of polychoric correlations and their asymptotic covariance matrix. *Psychometrika 59*, 381–389.

Jöreskog, K. G. and I. Moustaki (2001). Factor analysis of ordinal variables: a comparison of three approaches. *Multivariate Behavioral Research 36*, 347–387.

Jöreskog, K. G. and D. Sörbom (1999). *LISREL 8 User's Reference Guide.* Chicago: Scientific Software International.

Jowell, R. (2003). *Human Values: Extract from European Social Survey 2002/2003 (based on ESS1e05.1).* London: Centre for Comparative Social Surveys, City University. [With the Central Co-ordinating Team of the European Social Survey].

Jowell, R., J. Curtice, A. Park, L. Brook, and K. Thomson (1996). *British Social Attitudes: the 13th Report.* Aldershot: Dartmouth.

Jung, T. and K. A. S. Wickrama (2008). An introduction to latent class growth analysis and growth mixture modeling. *Social and Personality Psychology Compass 2*(1), 302–317.

Kaplan, D. (2000). *Structural Equation Modeling.* Thousand Oaks, CA: Sage Publications.

Karlheinz, R. and A. Melich (1992). *Euro-Barometer 38.1: Consumer Protection and Perceptions of Science and Technology.* INRA (Europe), Brussels. [Computer file].

Kateri, M. (2014). *Contingency Table Analysis: Methods and Implementation Using R.* New York: Springer.

Katsikatsou, M. and I. Moustaki (2007). An investigation of the main factors that affect Greek consumers' willingness to pay more for ecological products. Technical Report 229, Department of Statistics, Athens University of Economics and Business.

Katsikatsou, M., I. Moustaki, F. Yang-Wallentin, and K. G. Jöreskog (2012). Pairwise likelihood estimation for factor analysis models with ordinal data. *Computational Statistics and Data Analysis 56*, 4243–4258.

Kish, L. (1965). *Survey Sampling.* New York: Wiley.

Kline, R. B. (2023). *Principles and Practice of Structural Equation Modeling* (5th ed.). New York: The Guilford Press.

Knott, M., M. T. Albanese, and J. Galbraith (1990). Scoring attitudes to abortion. *The Statistician 40*, 217–223.

Kolaczyk, E. D. (2009). *Statistical Analysis of Network Data : Methods and Models*. New York, NY: Springer.

Kolar, M., A. P. Parikh, and E. P. Xing (2010). On sparse nonparametric conditional covariance selection. In *Proceedings of the 27th International Conference on Machine Learning*, pp. 559–566.

Krüger, M. (1994). *Sekseverschillen in Schoolleiderschap [Gender Differences in School Leadership]*. Alphen a/d Rijn: Samson.

Kruskal, J. B. (1964). Multidimensional scaling by optimizing goodness of fit to a nonmetric hypothesis. *Psychometrika 29*, 1–27.

Kruskal, J. B. and M. Wish (1978). *Multidimensional Scaling*. Quantitative Applications in the Social Sciences, Number 11. Sage Publications.

Krzanowski, W. J. and F. H. C. Marriott (1995a). *Multivariate Analysis, Part 1: Distributions, Ordination and Inference*. London: Arnold.

Krzanowski, W. J. and F. H. C. Marriott (1995b). *Multivariate Analysis, Part 2: Classification, Covariance Structures, and Repeated Measurements*. London: Arnold.

Kuha, J., M. Katsikatsou, and I. Moustaki (2018). Latent variable modelling with non-ignorable item non-response: multigroup response propensity models for cross-national analysis. *Journal of the Royal Statistical Society, Series A 181*, 1169–1192.

Lawley, D. N. and A. E. Maxwell (1971). *Factor Analysis as a Statistical Method* (2nd ed.). London: Butterworth.

Linzer, D. A. and J. B. Lewis (2011). poLCA: An R package for polytomous variable latent class analysis. *Journal of Statistical Software 42*, 1–29.

Little, R. J. A. and D. B. Rubin (2019). *Statistical Analysis with Missing Data* (3rd ed.). Hoboken, NJ: Wiley.

Macready, G. B. and C. M. Dayton (1977). The use of probabilistic models in the assessment of mastery. *Journal of Educational Statistics 2*(2), 99–120.

Marmot, M. and E. Brunner (2005). Cohort profile: the Whitehall II study. *International Journal of Epidemiology 34*, 251–256.

Marquart, L. and M. Hayes (2019). Misspecification of multimodal random-effect distributions in logistic mixed models for panel survey data. *Journal of the Royal Statistical Society, Series A 182*(1), 305–321.

Maydeu-Olivares, A. and H. Joe (2005). Limited and full information estimation and testing in 2^n contingency tables: A unified framework. *Journal of the American Statistical Association 100*, 1009–1020.

Maydeu-Olivares, A. and H. Joe (2006). Limited information goodness-of-fit testing in multidimensional contingency tables. *Psychometrika 71*, 713–732.

Maydeu-Olivares, A. and H. Joe (2008). An overview of limited information goodness-of-fit testing in multidimensional contingency tables. In K. Shigemasu, A. Okada, T. Imaizumi, and T. Hoshino (Eds.), *New Trends in Psychometrics*, pp. 253–262. Japan: Universal Academy Press.

Michailidis, G. and J. De Leeuw (1998). The Gifi system of descriptive multivariate analysis. *Statistical Science 13*, 307–336.

Morgan, B. J. T. (1973). Cluster analyses of two accoustic confusion matrices. *Perception and Psychophysics 13*, 13–24.

Morgan, B. J. T. (1981). Three applications of methods of cluster analysis. *The Statistician 30*, 205–223.

Moustaki, I. (2000). A review of exploratory factor analysis for ordinal categorical data. In R. Cudeck, S. Du Toit, and D. Sörbom (Eds.), *Structural Equation Modeling: Present and Future*. Scientific Software International.

Moustaki, I. and M. Knott (2000). Weighting for item non-response in attitude scales by using latent variable models with covariates. *Journal of the Royal Statistical Society, Series A 163*, 445–459.

Mulaik, S. A. (2010). *Foundations of Factor Analysis* (2nd ed.). Statistics in the Social and Behavioral Sciences Series. Boca Raton, FL: Chapman & Hall/CRC.

Muthén, B. and L. Shedden (1999). Finite mixture modeling with mixture outcomes using the EM algorithm. *Biometrics 55*, 463–469.

Muthén, B. O. (1984). A general structural model with dichotomous, ordered categorical and continuous latent variable indicators. *Psychometrika 49*(1), 115–132.

Muthén, B. O. (1994). Multilevel covariance structure analysis. *Sociological Methods and Research 22*, 376–398.

Muthén, L. K. and B. O. Muthén (1998–2017). *Mplus User's Guide* (8th ed.). Los Angeles: Muthén and Muthén.

Nagin, D. S. and K. C. Land (1993). Age, criminal careers, and population heterogeneity—specification and estimation of a nonparametric, mixed poisson model. *Criminology 31*, 327–362.

Nagin, D. S. and C. L. Odgers (2010). Group-based trajectory modeling in clinical research. *Annual Review of Clinical Psychology 6*, 109–138.

Ni, Y., F. C. Stingo, and V. Baladandayuthapani (2022). Bayesian covariate-dependent gaussian graphical models with varying structure. *Journal of Machine Learning Research 23*(242), 1–29.

Nuttall, D. L., H. Goldstein, R. Prosser, and J. Rasbash (1989). Differential school effectiveness. *International Journal of Educational Research 13*, 769–776.

Oh, M.-S. and A. E. Raftery (2001). Bayesian multidimensional scaling and choice of dimension. *Journal of the American Statistical Association 96*(455), 1031–1044.

O'Muircheartaigh, C. and I. Moustaki (1999). Symmetric pattern models: a latent variable approach to item non-response in attitude scales. *Journal of the Royal Statistical Society, Series A 162*, 177–194.

Pan, H. and H. Goldstein (1998). Multilevel repeated measures growth modelling using extended spline functions. *Statistics in Medicine 17*(5), 2755–2770.

Peaker, G. F. (1971). The Plowden children four years later. Technical report, National Foundation for Educational Research in England and Wales.

Quartagno, M. and J. Carpenter (2023). *jomo: A Package for Multilevel Joint Modelling Multiple Imputation*.

Rabe-Hesketh, S. and A. Skrondal (2022). *Multilevel and Longitudinal Modelling using Stata, Volume I* (4th ed.). College Station, TX: Stata Press.

Rasch, G. (1960). *Probabilistic Models for Some Intelligence and Attainment Tests*. Copenhagen: Paedagogiske Institut.

Raudenbush, S. W. and R. T. Congdon (2021). *HLM8: Hierarchical Linear and Nonlinear Modeling*. Chapel Hill, NC: Scientific Software International, Inc.

Raykov, T. and G. A. Marcoulides (2000). *A First Course in Structural Equation Modeling*. Mahwah, NJ: Lawrence Erlbaum Associates.

Reif, K. and E. Marlier (1995). *Eurobarometer 43.1bis: Regional Development, Consumer and Environment Issues.* INRA (Europe), Brussels. [Computer file].

Reiser, M. (1996). Analysis of residuals for the multinomial item response model. *Psychometrika 61*, 509–528.

Reiser, M. (2008). Goodness-of-fit testing using components based on marginal frequencies of multinomial data. *British Journal of Mathematical and Statistical Psychology 61*, 331–360.

Ridge, J. M. (1974). Three generations. In J. M. Ridge (Ed.), *Mobility in Britain Reconsidered.* Oxford University Press. Oxford Studies in Social Mobility, Working paper 2.

Rizopoulos, D. (2006). ltm: An R package for latent variable modelling and item response theory analyses. *Journal of Statistical Software 17*(5), 1–25.

Roaf, M. (1978). A mathematical analysis of the styles of the Persepolis reliefs. In J. Megaw and C. Greenhalgh (Eds.), *Art in Society.*

Roaf, M. (1983). Sculptures and sculptors at Persepolis. *IRAN, Journal of the British Institute of Persian Studies 21*, 1–159.

Rose, N., M. von Davier, and B. Nagengast (2015). Commonalities and differences in IRT-based methods for nonignorable item nonresponses. *Psychological Test and Assessment Modeling 57*, 472–498.

Rosseel, Y. (2012). lavaan: An R package for structural equation modeling. *Journal of Statistical Software 48*(2), 1–36.

Rubin, D. B. (1987). *Multiple Imputation for Nonresponse in Surveys.* Newe York: Wiley.

Satorra, A. and P. M. Bentler (1994). Corrections to test statistics and standard errors on covariance structure analysis. In A. von Eye and C. C. Clogg (Eds.), *Latent Variable Analysis*, pp. 399–419. Thousand Oaks, CA: Sage Publications.

Savalei, V. (2014). Understanding robust corrections in structural equation modeling. *Structural Equation Modeling: A Multidisciplinary Journal 21*, 149–160.

Savalei, V. (2021). Improving fit indices in structural equation modeling with categorical data. *Multivariate Behavioral Research 56*, 390–407.

Schafer, J. L. (1997). *Analysis of Incomplete Multivariate Data.* Boca Raton, FL: Chapman and Hall/CRC.

Schafer, J. L. and M. K. Olsen (1998). Multiple imputation for multivariate missing-data problems: A data analyst's perspective. *Multivariate Bahavioral Research 33*, 545–571.

Sclove, S. (1987). Application of model-selection criteria to some problems of multivariate analysis. *Psychometrika 52*, 333–343.

Scrucca, L., C. Fraley, T. B. Murphy, and A. E. Raftery (2023). *Model-Based Clustering, Classification, and Density Estimation Using mclust in R.* Chapman and Hall/CRC.

Singer, B. and S. Spilerman (1976). Some methodological issues in the analysis of longitudinal surveys. In *Annals of Economic and Social Measurement, Volume 5, number 4*, pp. 447–474. National Bureau of Economic Research, Inc.

Skrondal, A. and S. Rabe-Hesketh (2004). *Generalized Latent Variable Modelling: Multilevel, Longitudinal, and Structural Equation Models.* Boca Raton, FL: Chapman & Hall/ CRC.

Snijders, T. A. B. and R. J. Bosker (2012). *Multilevel Analysis: An Introduction to Basic and Advanced Multilevel Modelling* (2nd ed.). London: Sage.

Spearman, C. (1904). General intelligence, objectively determined and measured. *American Journal of Psychology 15*, 201–293.

StataCorp (2023). *Stata 18 Structural Equation Modeling Reference Manual*. College Station, TX: Stata Press.

Steele, F. (2005). Structural equation modelling: Multilevel. In B. Everitt and D. Howell (Eds.), *Encyclopedia of Statistics in Behavioral Science*. Chichester: Wiley.

Steele, F. (2014). *Multilevel Modelling of Repeated Measures Data: Stata Practical. LEMMA VLE Module 15*. Bristol: Centre for Multilevel Modelling, University of Bristol.

Stroup, W. W., G. A. Milliken, E. A. Claassen, and R. D. Wolfinger (2018). *SAS for Mixed Models: Introduction and Basic Applications*. Cary, NC: SAS Institute Inc.

Tomarken, A. J. and N. G. Waller (2003). Potential problems with "well fitting" models. *Journal of Abnormal Psychology 112*, 578–598.

van Borkulo, C. D., D. Borsboom, S. Epskamp, T. F. Blanken, L. Boschloo, R. A. Schoevers, and L. J. Waldorp (2014). A new method for constructing networks from binary data. *Scientific Reports 4*(1), 1–10.

van Borkulo, C. D. and S. Epskamp (2023). *IsingFit: Fitting Ising Models Using the ELasso Method*. R package version 0.4.

van der Linden, W. J. and R. K. Hambleton (Eds.) (1997). *Handbook of Modern Item Response Theory*. New York: Springer-Verlag.

Varin, C., N. Reid, and D. Firth (2011). An overview of composite likelihood methods. *Statistica Sinica 21*, 5–42.

Verbeke, G. and G. Molenberghs (2000). *Linear Mixed Models for Longitudinal Data*. New York: Springer.

Vermunt, J. K. and J. Magisdon (2005). *Latent GOLD 4.0 User's Guide*. Belmont Massachussetts.

von Stockert, S. H., E. I. Fried, C. Armour, and R. H. Pietrzak (2018). Evaluating the stability of DSM-5 PTSD symptom network structure in a national sample of us military veterans. *Journal of Affective Disorders 229*, 63–68.

Wagstaff, A. and E. van Doorslaer (2000). Income inequality and health: what does the literature tell us? *Annual Review of Public Health 21*, 5432–5467.

Wheaton, B., B. O. Muthén, D. Alwin, and G. Summers (1977). Assessing reliability and stability in panel models. In D. Heise (Ed.), *Sociological Methodology*. San Francisco: Jossey-Bass.

Wilkinson, R. G. (1996). *Unhealthy Societies: The Afflictions of Inequality*. London: Routledge.

Zhang, J. and Y. Li (2023). High-dimensional Gaussian graphical regression models with covariates. *Journal of the American Statistical Association 118*(543), 2088–2100.

Author Index

Subject Index

For Product Safety Concerns and Information please contact our EU
representative GPSR@taylorandfrancis.com
Taylor & Francis Verlag GmbH, Kaufingerstraße 24, 80331 München, Germany